Toxicological Aspects of Energy Production

Charles L. Sanders

BATTELLE PRESS
Columbus • Richland

Distributed by

MACMILLAN PUBLISHING COMPANY
A DIVISION OF MACMILLAN, INC.
NEW YORK

COLLIER MACMILLAN PUBLISHERS
LONDON

Distributed by
MACMILLAN PUBLISHING COMPANY
A DIVISION OF MACMILLAN, INC.
866 THIRD AVENUE, NEW YORK, N.Y. 10022
COLLIER MACMILLAN CANADA, INC.

Printed in United States of America

Library of Congress Cataloging in Publication Data

Sanders, Charles Leonard, 1938-
   Toxicological aspects of energy production.

   Includes index.
   1. Energy industries—Hygienic aspects.   2. Fossil fuels—Toxicology.
3. Radiation—Toxicology.   4. Industrial toxicology.   I. Title.
RC965.E53S26   1984   363.1'19621042   84-21594
ISBN 0-02-948960-1

ISBN 0-02-948960-1 Macmillan Publishing Company, New York

Copyright © 1986, Battelle Memorial Institute

All rights reserved. No part of this book may be reproduced or transmitted in any form or by any means, electronic or mechanical, including photocopying, recording or by any information storage and retrieval system, without permission from the publisher.

Dedicated to the 1982 class, Radiological
Sciences 523, Toxicological Aspects of
Energy Production, Joint Center for Graduate
Study, Richland, Washington, and University
of Washington, Seattle, Washington:
    Cynthia L. Caldwell
    Floyd A. Cheryl
    Sandra J. Gray
    Eric M. Greager
    Bruce W. Killand
    Noelle F. Metting
    Gary C. Schmidtke
    Michael E. Thiede

CONTENTS

	Preface	ix
CHAPTER 1.	Energy Resources and Comparisons	1
CHAPTER 2.	Toxicology and Pathobiology	7
	Basic Toxicology	7
	Metabolism and Carcinogenicity of Chemicals	14
	Absorption, Distribution and Excretion	14
	Toxicokinetics	16
	Biotransformation	19
	Carcinogenicity	20
	Basic Pathobiology	25
	Cell Injury	25
	Inflammation	26
	Host-Parasite Interactions	28
	Immunotoxicology	28
	Biology of Cancer	31
	Short-Term Mutagenicity and Carcinogenicity Tests	34
	Cancer Epidemiology	42
	References	45
CHAPTER 3.	Non-Respiratory Tract Toxicology	49
	Introduction	49
	Kidney	49
	Liver	50
	Skin	50
	Hematopoiesis and Oxygen Utilization	54
	Bone	55
	Neurotoxicology	56
	Eye	58

	Reproductive Toxicology	59
	References	61
CHAPTER 4.	Inhalation Toxicology	63
	History	63
	Pulmonary Structure and Function	63
	Deposition and Fate of Inhaled Particulates ...	64
	Pulmonary Pathology	70
	Infectivity Models	75
	Asbestosis	75
	Silicosis	79
	Welder's Pneumoconiosis	83
	Mixed Silicate Pneumoconiosis	83
	Gases and Vapors	85
	Formaldehyde	86
	Organic Vapors and Solvents	87
	Ethanol and Methanol	89
	Benzene	90
	Phenol	91
	Alkylbenzenes	92
	Inhalation Carcinogenesis	92
	Lung Cancer	92
	Pulmonary Carcinogenesis from Organic Chemicals	94
	Cigarette Smoking	95
	References	99
CHAPTER 5.	Environmental Air Pollution	105
	A Brief History	105
	General Aspects of Air Pollution	107
	Standards for Airborn Contaminants	115
	Polycyclic Aromatic Hydrocarbons	116
	Aromatic Amines	120
	Particles	121
	Nitrogen Dioxide	125
	Ozone	129
	Sulfur Oxides	132
	Hydrogen Sulfide	137
	Carbon Monoxide	139
	Carbon Dioxide	140
	Ammonia	142
	Hydrogen Cyanide	143
	References	144

Contents

CHAPTER 6.	Toxicology of Metals	149
	General Principles	149
	Aluminum	154
	Antimony	154
	Arsenic	155
	Beryllium	157
	Cadmium	158
	Chromium	162
	Cobalt	162
	Copper	163
	Iron	164
	Lead	164
	Manganese	169
	Mercury	169
	Nickel	171
	Palladium and Platinum	172
	Selenium	173
	Uranium	173
	Vanadium	174
	References	175
CHAPTER 7.	Coal and Oil	181
	Introduction	181
	Coal Worker's Pneumoconiosis	181
	Consequences of Carbon Dioxide Production	184
	Acid Precipitation	187
	Coal Fly Ash	189
	Carcinogens from Fossil-Fuel Sources	194
	Aliphatic Hydrocarbons	196
	Crude Oil	196
	Oil Fly Ash	199
	Gasoline	200
	Diesel Fuel	200
	Toxic Effects of Fossil-Fuel Combustion	201
	References	205
CHAPTER 8.	Fossil Fuel Conversion	211
	Coal Liquefaction and Gasification	211
	Shale Oil	219
	Tar Sand	221
	Biological Effects	222
	References	230

CHAPTER 9.	Biomass, Solar and Geothermal	235
	Biomass Conversion	235
	Wood Combustion	237
	Wood and other Biomass Conversion	240
	Biofuel Cells	240
	Solar	241
	Infrared and Ultraviolet Radiations	246
	Geothermal	247
	References	250
CHAPTER 10.	Radiological Health	253
	Health Physics and Ionizing Radiations	253
	Basic Concepts in Radiobiology	260
	Radiation Carcinogenesis	265
	Radionuclide Toxicology	267
	Uranium Mine Exposure	267
	Comparison of Coal and Nuclear Power Plants	270
	Nuclear Fusion	271
	NonIonizing Radiations	274
	Electromagnetic Fields	278
	Magnetohydrodynamics	280
	References	280
	Glossary of Energy-Related Terms	285
	Index	301

Preface

TOXICOLOGICAL ASPECTS OF ENERGY PRODUCTION presents a broad overview of known and potential human health hazards associated with all the major energy technologies. The reader is assumed to have a good background knowledge in biomedical sciences but with limited knowledge in toxicology. Those that will find this book of potential value include students and teachers in high schools, colleges and universities, particularly those interested in the disciplines of public and environmental health, toxicology and pharmacology, industrial and occupational medicine and civil and environmental engineering. In addition, governmental and industrial sections involved with energy production will find this book a valuable tool for the comparative toxicological evaluation of the various energy technologies.

The book is divided into four parts. Part I (Chapters 1 and 2) reviews the principles of toxicology, describes the biological fate of chemicals in the body, discusses basic pathobiology, and reviews short-term toxicity tests. Part II (Chapters 3 and 4) describes the toxicology and pathology of pollutants in several important organ systems. The greatest emphasis is placed on the respiratory tract because of its high probability as a route of exposure to pollutants from energy technologies and its high sensitivity to pollutant related tissue damage. Part III (Chapters 5 and 6) describes the toxicological aspects of specific chemical classes associated with fossil fuels; these include polycyclic hydrocarbons, gases and metals. Part IV (Chapters 7, 8, 9, and 10) describes the biomedical effects associated with each energy technology, including coal and oil, fossil fuel and biomass conversions, solar and geothermal and radiological health aspects associated with uranium mining, nuclear fission and fusion, and with nonionizing radiations and electromagnetic fields. The book concludes with a glossary of some common energy terms.

CHAPTER I

ENERGY RESOURCES AND COMPARISONS

Natural surface sources of oil and gas and coal have been known since the beginning of recorded history. Essentially all fossil fuels originate from the decomposition of organic matter in marine and lacustrine sediments of the earth's crust. The formation of petroleum from plant decomposition occurs in three stages.[6] During the first stage, diagenesis ($<50°C$), methanogenic bacteria form methane from substrates in the sediment. About 15% of the C_{15} to C_{40} hydrocarbons in crude oil are also formed as a result of low-temperature biological and chemical reactions. During catagenesis (50° to 200°) about 85% of the oil and 75% of the gas are formed from the cracking of the organic molecules (kerogen) in the sediment. In the last stage, metagenesis ($>200°C$), only gas is formed in substantial quantities. The overall reaction is:

$$2C_{10}H_{14} \rightarrow C_{10}H_{18} + 2CH_4 + 2C_4H$$
$$\text{kerogen} \quad \text{oil} \quad \text{gas} \quad \text{pyrobitumen}$$

During the last one hundred years the utilization and production of energy have been characterized by profiles of change, from an economy based mostly on wood, up to 1880, to that of today, based upon a myriad of sources – coal, oil, gas, nuclear, solar and other systems. The profile continues to change as rising and falling economies in the world increase and decrease their demands for energy. Countries such as Japan and France, poor in fossil energy sources, opt for a greater emphasis on nuclear power plants. Other countries, such as Brazil, hard hit by foreign exchange money problems and uncertainties of oil supplies abroad, choose biomass conversion to alcohol to fuel their nations motor vehicles. During the late 1970s, the United States initiated a massive program to supply low-sulfur liquid and gaseous fuels from coal; the recent downturn in our economy and the increased emphasis on energy conservation measures have diminished this emphasis.

Coal, petroleum and natural gas provide most of the energy in the U.S. today (Table 1.1). In 1972, 45% of the energy produced was provided by crude oil, 19% by coal and 31% by natural gas. By 1980, the contribution

Table 1.1. Energy sources for the United States.

Natural Gas
Petroleum
    Conventional
    Oil Shale
    Tar Sand
Coal
    Conventional
    Gasification
    Liquifaction
    Magnetohydrodynamics
Biomass
    Direct Burning
    Conversion to Oil
    Conversion to Alcohols
Hydroelectric
Tidal
Nuclear
    Conventional Fission
    Breeder Fission
    Fusion
Geothermal
Solar
Wind
Hydrogen
Fuel Cells

made by oil and natural gas had slightly decreased, while coal's contribution had substantially increased. Nuclear power and hydropower combined account for about 10% of today's energy consumption.

Total energy consumption in 1976 was 73 quads; 1 quad = 1×10^{15} BTU (British Thermal Unit). One quad is equivalent to 180 million barrels of oil, 42 million tons of bituminous coal, 1 trillion cubic feet of natural gas or 293 billion kilowatt-hours of electricity. Cost equivalents for various energy sources are given in Table 1.2. A total of 79 quads were used in the U.S. in 1980; of this, 15.8 quads came from coal; 20.5 quads from natural gas; 20.5 quads from domestic oil; 15.2 quads from imported oil; 2.9 quads from nuclear fission and 3.1 quads from other sources. Energy consumption by the year 1985 is estimated at about 90 quads; in comparison 10 quads were used in 1900 and 30 quads in 1945 at the height of WW II. Projections for the future indicate that energy consumption, by the year 2000, will probably be only slightly changed for natural gas, domestic oil

TABLE 1.2. United States energy production and costs (Columbia Institute for Political Research. 1982. Washington, D.C.); kWh=kilowatt hour.

Energy Source	Energy Production, quads	Current Cost, cents/kWh
Petroleum	20	4.3
Coal	16	2.3
Natural Gas	21	2.5
Nuclear Fission	3	1.9
Hydroelectric	3	0.8
Biomass/Alcohol	2	5-20.
Geothermal	0.1	2.0
Solar Heating	–	6–16.
Solar Electric	–	30-50.*
Wind	–	20.*
Photovoltaics	–	100.*
Ocean Thermal	–	7.0

*Costs are declining rapidly in these energy production systems.

and imported oil but will increase to 34 quads for coal, 11 quads for nuclear fission and 12 quads for other energy sources.

A historical review of energy utilization demonstrates the short time during which fossil fuel, particularly oil, has been used as an energy source (Figure 1.1). In 1800, the U.S. economy overwhelmingly favored the use of wood as an energy source—both in the home and in industry—particularly as charcoal for iron production. Local wood shortages and high costs of transportation, as well as a shift of population into the cities, resulted in a shift from wood to coal as a principal fuel source. Coke from coal replaced charcoal in blast furnaces, and coal replaced wood in urban homes. Coal utilization for transportation remained steady from 1920 to 1945, then decreased rapidly to near zero by 1960. Use of coal and oil for central-station electric power production by utilities rapidly increased following the end of WW II. The current shift from oil to coal occurred, in large measure, because of the rapidly escalating costs of petroleum imports.

Estimates of recoverable energy resources in the U.S. are listed in Table 1.3. The preeminence of coal is due to its abundance and the decreasing supplies and increasing costs of crude oil and natural gas. Coal is ranked according to energy output or BTU: The BTU of anthracite and bituminous

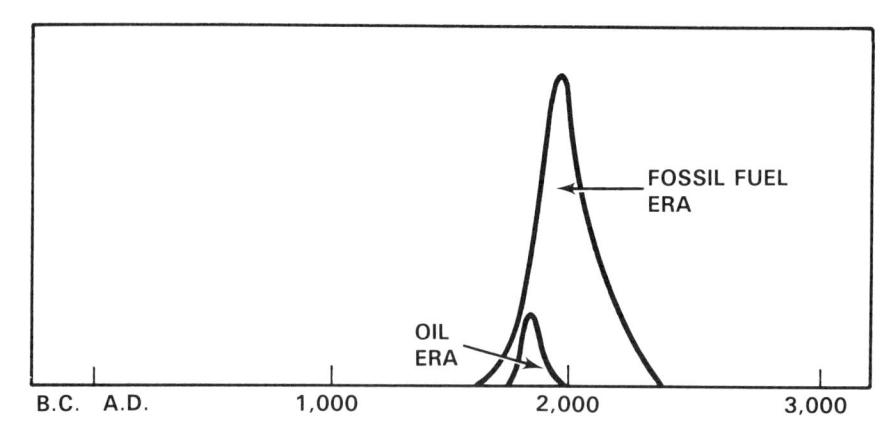

FIGURE 1.1 — Predicted relative production and consumption of fossil fuels. The Y-axis represents a relative quantity, showing the preponderance of coal and oil shale over oil as the main fossil fuel source of the next few centuries.

coals is $12-15 \times 10^3$ BTU/lb; subbituminous coal is $8-11 \times 10^3$ BTU/lb; lignite is $5-7 \times 10^3$ BTU/lb.

Crude oil production is expected to decline in the U.S. by 30% by the year 2000; natural gas is expected to decline by about 75% during this period. Proven reserves of crude oil in the U.S. range from 200–400 billion barrels. Of this amount, 55% is in the lower 48 states, 33% is in off-shore sites, and 12% is in Alaska. Daily consumption of crude oil in the U.S. has declined in the past few years as a result of increased conservation and cost and because of the depressed economy. In order to maintain 1982 production, the oil industry would have to discover 45 billion barrels of oil during the 18 years between 1982 and 2000. In comparison, drillers have found only 44% of the undiscovered oil of the preceding 18 years. To find that much oil in the next 18 years (to maintain 1982 production levels), drillers would have to discover oil fields six times faster than in the past.[2]

In contrast, proven crude oil reserves in the mid-eastern countries are 10 times those of the U.S.; proven petroleum reserves in Africa or the U.S.S.R. are each twice those of the U.S. Renewable energy resources (wood, alcohols from biomass, wind, hydropower, geothermal and solar) could make energy contributions equivalent to 3 million barrels of oil a day in the U.S. by 1985. Nuclear power equivalent energy production could be 4 million barrels of oil a day by 1987;[1] this figure is likely to be decreased because of the high costs and declining construction or terminations of nuclear power plants.

One might ask, "What happened to the energy crisis of the 1970s?" Today the U.S. imports only about 30% of our total daily consumption of 15 to 16 million barrels per day, while countries in the Organization of

TABLE 1.3. Amounts of recoverable energy resources in the United States (A National Energy Plan for Energy Research, Development, and Demonstration: Creating Energy Choices for the Future. 1975. Vol. I: The Plan. U.S. Energy Research and Development Administration, ERDA-48, Washington, D.C.).

Resource	Amount	Degree of Difficulty in Recovery, quad Enhanced Easy	Difficult
Coal	600 billion tons	12,000	1,030
Natural Gas	750 trillion cubic feet	775	1,100
Petroleum	200 billion barrels	1,200	?
Shale Oil	200 billion barrels	1,200	130,000
Uranium	3.6 million tons	1,800	?
Tar Sands*	1 trillion barrels	2,000	?

*Mostly found in Canada

Petroleum Exporting Countries (OPEC) are producing at only 60% of capacity. The reasons are economic recession and improved energy efficiency. In Japan, U.S. and western Europe, energy consumption per unit of gross national product (GNP) has fallen 17% since 1974, while forecasts are for another 22% fall by the year 2000.[3]

Industry accounted for about 40% of U.S. energy consumption in 1984. However in 1982 industry consumed one-third less energy than trends indicated they should have in 1973.[4] Industry's use of fossil fuels have undergone a number of changes since 1970. Coal use for all purposes but coke production has declined throughout most of the 1970s. The continued decline in coal use in industry suggests that federal, state, and local air quality ordinances may be impediments to the greater use of coal as an industrial fuel. Industrial use of wood as a source of energy grew to a level of importance by 1980 surpassing that of all industrial coal used for nonmetallurgical purposes; wood use is concentrated in the lumber and paper industries. Despite a four-fold increase in the price of crude oil, industrial use of this fuel has continued to increase throughout the 1970s. Overall, compared to 1972, relative increases occurred in the use of heating coal, wood energy, and petroleum, while relative decreases occurred in the use of metallurgical coal and natural gas. Except in limited applications, the increased use of coal awaits the formation of cost-competitive, clean-burning, coal combustion technologies.[4]

A large degree of the increase in U.S. energy efficiency is due to our ability to expand the relative use of high-quality fuels such as petroleum, and also to shift fuel useage between sectors of the economy. There has been a marked decline in energy return on investment for all principal fuels in recent decades. Future energy-related economic growth will depend largely on the net energy yield of alternative fuel sources.[5]

Other sources of fossil fuels are available for commercial use. The amount of shale oil present in the Green River formation of Colorado, Utah and Wyoming is estimated at 1,400 billion barrels of oil at grades of 10 to 100 gallons of oil per ton of shale. An additional large oil shale reserve that grades < 25 gallons per ton of shale is also found in the Chattanooga formation of the eastern U.S. At present, the cost of recovering the oil from shale is not competitive with that for crude oil and coal. The bitumen surrounding sand particles of tar sand formations constitutes another large fossil fuel reserve. The major source of tar sands in North America is the Athabasca and Peace River deposits of northern Alberta, Canada, estimated at 1 trillion barrels of oil. U.S. tar sand reserves in the Rocky Mountain states are estimated at 33 billion barrels of oil.

References

1. E. Marshall. 1980. "Planning for an oil cutoff." Science 209:246–247.
2. R. A. Kerr. 1984. How fast is oil running out? Science 226:426.
3. Research & Development, November, 1984, p. 73.
4. R. C. Marlay. 1984. Trends in industrial use of energy. Science 226:1277–1283.
5. C. J. Cleveland, R. Costanza, C. A. S. Hall and R. Kaufmann. 1984. Energy and the U.S. Economy: A biophysical perspective. Science 225:890–897.
6. J. M. Hunt. 1984. Generation and migration of light hydrocarbons. Science 226:1265–1270.

CHAPTER 2

TOXICOLOGY AND PATHOBIOLOGY

Basic Toxicology

Each energy source has its own peculiar set of potential toxicological problems; for example, environmental air pollutants from most energy sources; polycyclic aromatic hydrocarbons (PAH) from fossil fuels; or ionizing radiation from nuclear fission. Such toxicological associations with energy production and utilization can seriously affect how energy sources are used.

Toxicology is the study of poisons or toxic agents, or, more specifically, the investigation of adverse or harmful actions of physical or chemical agents on living organisms. It is the specific dose that determines an agents toxicity; as Paracelsus (1493–1541) noted: "All substances are poisons; there is none which is not a poison. The right dose differentiates a poison and a remedy." The toxicologist examines the harmful effects of toxic agents and ascertains the probability of their occurrence. He may describe what the agent does to living organisms, examine mechanistic or theoretical aspects of the agent or apply data on toxic agents for the purpose of regulating exposure and minimizing harmful effects.[1-9]

There are several important branches in the discipline of toxicology. Environmental toxicology is concerned with the harmful actions of agents on organisms in the ecosphere. Often, environmental toxicology is concerned with the evaluation of toxic agents only with respect to recreational or economic value. The medical and legal aspects of toxic agents, emphasizing detection and quantification by analytical chemistry, is the basis for forensic toxicology. The clinical toxicologist studies diseases associated with exposure to toxic agents and methods to treat or counteract the harmful effects of these agents. The industrial toxicologist examines the adverse effects of agents in the workplace, monitors exposure levels in workers and attempts to anticipate future problems.

Very few industrial chemicals have undergone extensive toxicity testing. A recent National Research Council committee looked at 675 of the approximately 5,000,000 chemicals listed in the chemical literature. Only 18%

of the drugs and 10% of the pesticides of those 675 chemicals had been adequately tested such that sufficient data were available to make a complete health assessment; no toxicity data were found for 38% of the pesticides and 25% of the drugs. Only minimal toxicity testing had been performed for 20% of commercial chemicals, and virtually none for the other 80%.[27] This amounts to a lack of information concerning the toxicities of over 65,000 different industrial chemicals listed as having been in commercial production since 1945.[29]

During the last decade, the field of toxicology has undergone numerous significant changes:

- Many epidemiological studies now examine the long-term effects of exposure to toxic agents.
- A wide variety of short-term toxicity testing methods have been developed to screen chemicals before performing expensive long-term tests in animals.
- A large array of food additives, drugs, pesticides, industrial chemicals, cosmetics, etc. are now routinely tested, requiring the establishment of numerous corporate and other industrial toxicology testing laboratories.
- The formation of the Environmental Protection Agency (EPA) and the passage of the Toxic Substances Control Act (TOSCA), as of January 1, 1977, focused attention on the potential risks of chemical usage in industry, providing a mechanism for testing and regulating such chemicals. Enormous quantities of some chemicals are used by industry in the U.S.; in millions of pounds, these include benzene, 9.5, vinyl chloride, 6.9 and formaldehyde, 5.4, for 1983 alone.[11]
- The Delaney Clause, first published in 1958 as part of the Food Additive Amendment to the Food, Drug and Cosmetic Act, stipulates, in essence, that if an agent is found by testing, to induce cancer in animals or man, this agent may then not be used as an additive in food. This legislation has led to a great deal of effort to determine what constitutes a "safe or acceptable dose," centered around the development of statistical, mathematical models for safety risk evaluation.
- Antivivisectionists have pointed out some abuses in animal experimentation and have been successful in some places in obtaining legislation to reform and regulate the use of animals in research with toxic agents.
- Analytical chemists have developed a series of sophisticated and highly sensitive measuring tools that can detect chemicals in the parts-per-trillion range. When dealing with known or suspected human carcinogens, the detection of only a few parts per trillion can have very significant political and economic influences on society.
- Growing emphasis has been placed on investigating the metabolic

transformation and fate of chemicals in tissues and how these influence toxic effects.

- Labor unions, state legislatures and the general public are more concerned about the toxic actions of environmental and industrial pollutants; these and other citizen groups are a potent political force for minimizing exposure.
- The National Toxicology Program (NTP) was established by The Department of Health, Education and Welfare in 1978 for the purpose of broadening the toxicological testing of chemicals and developing and validating new testing methods. The three specific goals of NTP are: (1) to broaden the toxicological characterization of the chemicals being tested; (2) to increase the rate of chemical testing; and (3) to develop and begin to validate a series of protocols more appropriate to regulatory needs. These research activities of NTP cover genetic toxicology, carcinogenesis, chemical deposition, general toxicology, immunotoxicology, neurotoxicology, inhalation toxicology and reproductive and developmental toxicology. Chemicals or chemical types are nominated by governmental agencies for testing by NTP. An executive committee then determines which chemicals will undergo testing. Each selected chemical undergoes a possible four-phase study (Table 2.1). In Phase 1, the toxic potential of the chemical is determined and then confirmed in Phase II. Tests aimed

TABLE 2.1 Interrelationships of major testing activities of the National Toxicology Program (NTP) (D. A. Carter. 1981. Role of the regulatory agencies in the activities of the National Toxicology Program. Regulatory Toxicology and Pharmacology 1:8–18).

Phase	Genetic Toxicology	Carcinogenesis	Toxicology
I	Mutagenesis, in vitro	Cell transformation	Rodent screening tests
II	Mutagenesis in Drosophila	Mouse lung adenoma test	Organ system toxicities
III	Mutagenesis in mice	Lifetime rodent bioassay	Characterization of specific toxicities
IV		Evaluation of all data	

at achieving a more definitive result are carried out in Phase III, which is the basic carcinogenesis program area. In Phase IV, the data are analyzed for scientific and public health implications.[10]

Heart disease is the leading cause of death in the U.S., accounting for 38% of all mortality, followed by cancer (21%), stroke (9.0%) and lung infections (3.0%). However, the focus of toxicology has emphasized the roles of cancer and pulmonary disease rather than those of the cardiovascular diseases. Old age increases sensitivity to toxic agents since normal physiological functions decrease with advancing age, providing less "tolerance" or repair capacity for the damaging actions of toxic agents. Very young children also exhibit a greater sensitivity to toxic agents because of the greater cellular sensitivity of rapidly proliferating and differentiation tissues seen in growing children. Other high risk groups include pregnant women and their children, exposed in utero, persons with pre-existing disease, the obese, cigarette smokers and those with inherited genetic disease which are sensitive to specific toxic agents (Table 2.2).

The biological response of a population to a toxic chemical is dependent on dose. With increasing dose, the following categories of response occur:[1]

- The toxic chemical is measurable in large numbers of the population but does not produce any physiological change.
- The toxic chemical is associated with physiological change of uncertain significance in a small portion of the population.
- Sentinel physiological changes are seen in a small group, associated with specific disease states but without clinical symptoms.
- A small group exhibits significant morbidity and/or clinical symptoms.
- A few individuals may die from exposure to the toxic chemical at the highest dose levels.

On the other hand, some chemicals, such as selenium, are essential and beneficial to life at low doses; chronic dietary deficiencies in these elements may lead to morbidity or even mortality. Normal doses promote optimum health; higher doses cause toxicity. The nutritionist is concerned with deficiency diseases; the toxicologist is concerned with disease caused by excess exposure. The deleterious effects of both deficiency and excess states are usually higher in fetal or early postnatal development than in adults, with a smaller optimal dose range.

Serial sacrifices of experimental animals after administration of several dose levels of a toxic agent are required to assess the biological responses with respect to incidence, dose and time. Statistical procedures are then applied to the data for risk assessment. A probit plot is often used in acute lethality tests to fit the dose-mortality curve to a line so that the LC_{50} (lethal concentration of a gas or vapor or other aerosol that causes early death

TABLE 2.2. Genetic subpopulations that may exhibit increased sensitivity to various chemical and physical toxic agents.

High-Risk Group	Toxic Agent(s)
Glucose-6-phosphate dehydrogenase deficiency	Carbon monoxide, lead, nitrite, ozone, ionizing radiation
Glutathione deficiency	Lead, ozone
Gilbert's syndrome	Polychlorinated biphenyls
Albinism	Ultraviolet radiation
Cystinosis, cystinuria	Cadmiun, lead, mercury, uranium
Catalase deficiency	Ozone, ionizing radiation
Porphyria	Chloroquine, hexachlorobenzene, lead
Wilson's disease	Lead, vanadium
Thalassemia (Cooley's anemia)	Benzene and benzene derivatives, ozone
Sickle cell trait	Aromatic amino and nitro compounds, carbon monoxide, cyanide
Cystic fibrosis	Respiratory irritants
Phenylketonuria	Ultraviolet radiation
Immune hypersensitivity	Isocyanates
Methemoglobin reductase deficiency	Nitrates, nitrites, ozone, vanadium
Depressed inducibilty of aryl hydrocarbon hydroxylase	Polycyclic aromatic hydrocarbons

in 50% of the population) or LD_{50} (dose of toxic agent that causes early death in 50% of the population) may be calculated for a specific species, sex, age group and route of exposure.

Exposures to toxic agents may be at a high dose level that maximizes and produces immediate biological responses or at lower levels, over a continuous, prolonged or life-span period. Potential routes of exposure include oral, percutaneous, intramuscular, subcutaneous, intravenous, inhalation, nasal, intrathecal, intraperitoneal, ocular, intraocular, intrapleural, intradermal, rectal, intrauterine, sublingual, vaginal or intracranial.

The three basic dose categories for toxic agents are: effective dose (ED), toxic dose (TD) and lethal dose (LD). The concept of ED was developed by pharmacologists; it is the amount required to produce a physiological response, regardless of intensity or significance. TD is the quantity required to produce clinically important effects. LD is the amount required to kill in a short period of time. As generally defined, portions of the dose-response curves for ED, TD and LD may overlap. The therapeutic index used by pharmacologists is equivalent to the LD_{50}/ED_{50} and is useful in estimating a median dose of physiological effectiveness, while also indicating the margin of safety for the test compound.

The amount of a chemical required to produce toxic effects varies enormously among the chemical species (Table 2.3). For example, about 10 gram of saccharin is required to produce cancer in rats, as compared to 50 milligram of carbon tetrachloride, 1 milligram of urethane or 2-acetylaminofluorene, microgram quantities of dimethylnitrosamine and nanogram

TABLE 2.3. **Ranking of chemicals according to class of toxicity** (C. D. Klaassen and J. Doull. 1980. Evaluations of safety: Toxicological evaluation. In, "Casarett and Doull's Toxicology" Second Edition, MacMillan Publishing Co., NY, pp. 11–27).

Chemical	LD_{50}(mg/kg)	Toxicity Class and Dose	Toxicity Rating
Glucose	>10,000	Almost non-toxic	1
Ethyl alcohol	10,000	Slightly toxic	2
Sodium chloride	4,000	Moderately toxic	3
Phenobarbital	150	Very toxic	4
Parathion	7	Extremely toxic	5
Strychnine	2	Super toxic	6
Nicotine	1		
Dioxin	0.001		
Botulinus toxin	0.00001		

quantities of stericmatocystin or aflatoxin. Various methods have been proposed for comparing toxic injury from various agents. One method measures the ranking hazard as the sum of transport factors and the hazard index measured in log units as an estimate of toxicity.[2] Another approach measures acute lethality, pT, according to the equation:

$$pT = -\log(T),$$

where $T = LD_{50}$. The pT_{50} value is expressed in order of magnitude units and is corrected for the molecular weight of the chemical to give moles/kg body weight. For intraperitoneally injected chemicals in mice, typical pT values range from 16 for botulinus toxin to 4.7 for mercuric chloride and 1.4 for sodium chloride.[3] Both methods allow for comparison of lethality data of chemicals over a broad dose range.

A great variety of studies may be carried out to evaluate the toxicity of a chemical. These might include human epidemiological and clinical studies, or infectivity, mortality, mutagenicity, teratogenicity, immunological, neuro-behavorial, pathophysiological and reproductive or other organ specific studies in animals.

Toxic effects may be immediate or delayed, reversible or irreversible, local or systemic, idiosyncratic or with a genetic predisposition and allergic or hypersensitization reactions. Study intervals may be short- or long-term, acute or chronic. Intermediate-term studies in animals may include subchronic feeding and inhalation studies.

Toxic agents are classified according to target organ(s) or tissue(s) in which they exert the most pronounced effects, source of the toxic agent and most likely route of exposure, types of effects produced, physical and chemical state of the toxic agent, and mechanisms of action, including the agents biotransformation in tissues and overall toxicity rating.

Factors that influence toxic potential include the ability of the chemical to penetrate or pass through cell membranes; metabolic activation or inactivation; method of transport and concentration of the chemical in the target tissues; interaction of the chemical or its metabolites with sensitive macromolecules, such as DNA; and the degree of cellular repair possible after exposure.

Exposure to two or more toxic agents may result in an additive toxic response, in which the agents act independently of one another; the total response is the sum of the effects of each of the agents. Action of two agents may be synergistic, in which each effect is exaggerated or enhanced, so that total response is greater than the additive effect expected from either agent acting independently. Another type of toxic interaction is potentiation, where one agent produces the response and another agent enhances it. The opposite reaction is antagonism, in which there is a negative interaction of agents, producing a protective effect, so that the total response is less than that of either agent individually.

Metabolism and Carcinogenicity of Chemicals

Absorption, Distribution and Excretion

To produce harmful effects in the body, toxic agents must first penetrate the normal barriers of uptake; these include the intact epithelial layers of the skin, gastrointestinal tract and the respiratory tract. The more important modes of chemical deposition, along with their routes of translocation and excretion are shown in Figure 2.1. Having gained access to the body tissues, chemical agents may be detoxified, or their toxicity may be enhanced by enzymatic biotransformation. Biotransformation may also enhance the renal and biliary excretion of chemicals. To penetrate tissues and achieve sufficient concentration to cause toxicity, a chemical must pass through the cell membranes of several tissues into the cell cytoplasm or surface, interstitium or plasma (Figure 2.2). This is accomplished by:

- Simple diffusion
- Filtration
- Facilitated diffusion
- Active transport
- Pinocytosis
- Phagocytosis

The dominant form of transport for most chemicals is by simple diffusion across cell membranes, with the rate of chemical uptake dependent on lipid solubility as measured by the lipid-water partition coefficient. Small hydrophilic molecules can pass through aqueous channels in the membranes, directly following the path of water. Larger lipophilic organic molecules may diffuse directly through the cell membrane, with the rate of transport being a function of lipid solubility. Highly ionized molecules are poorly diffused through cell membranes. A change in pH in a direction that shifts the equilibrium reaction towards a non-ionized form from an ionized form often enhances lipid solubility and uptake of the chemical by simple diffusion.

Filtration occurs when large amounts of water are transferred across porous membranes, as is seen in the kidney for the rapid and efficient transport of low molecular-weight chemicals. Facilitated diffusion is a method of transport in which a lipophobic chemical may be bound to a carrier in the cell membrane, thereby facilitating the diffusion of the chemical across the lipid layer of the cell membrane. In both simple and facilitated diffusion, cell uptake of the chemical occurs with increasing concentration gradient; these forms of cell transport do not require an endogenous source of energy, such as ATP.

2 – Toxicology and Pathobiology

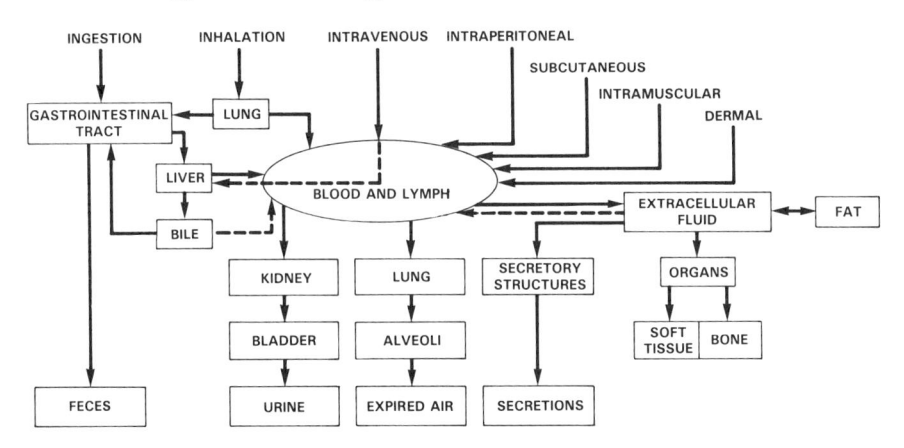

FIGURE 2.1—Mode of deposition and common routes of translocation and excretion in the body following deposition of chemicals (L. J. Casarett and J. Doull. 1975. Toxicology. The Basic Science of Poisons. Macmillan Publishing Co, Inc, NY, p. 27).

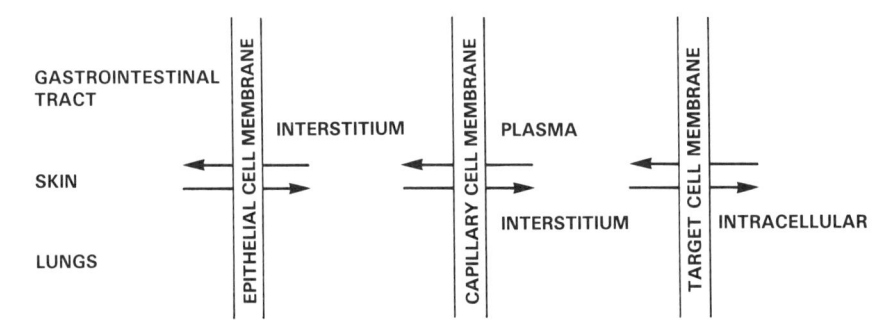

FIGURE 2.2—Scheme for uptake and distribution of chemicals in the body. Methods of cell transport, distribution of chemical binding receptor sites, the affinity of the chemical for cell receptors and the type and degree of biotransformation reactions at each site will influence the concentration and retention of the chemical at that site.

In active transport, the toxic chemical is selectively introduced into the cell against an electrochemical concentration gradient. Active transport thus requires an endogenous source of energy and a membrane carrier that binds to the chemical. Few toxic chemicals are taken into cells by active transport.

Pinocytosis is a form of cell transport whereby the cell "drinks" extracellular fluids containing solutes and colloids by creating small surface membrane invaginations that result in intracellular vesicle formation.

In phagocytosis, the cell actively moves around particulate material and pulls it into the cell, forming structures called phagosomes. The intracellular phagosomes unite with lysosomes to form phagolysosomes, and lysosomal enzymes digest the phagocytized material. Phagocytosis requires a source of endogenous energy such as ATP, divalent cations such as calcium and, in most cases, protein coating of the particulate material.

Nonionized lipophilic chemicals are taken up by the gastrointestinal epithelium by simple diffusion; amino acids, sugars, electrolytes and other basic building blocks are taken up mostly by active transport. Intestinal bacteria can transform some chemicals; for example, nitrites can be transformed to more toxic nitrosamines; the pesticide DDT, to DDE.

Various tissues act as storage depots for these absorbed chemicals. Plasma proteins may bind chemicals and inactivate them. In many cases, the fraction of non-bound, ionized chemical in the plasma is what determines the toxicity of a chemical. Fat is a depot for many organic chemicals; bone stores many metals.

The tightly joined capillary endothelial cells that surround astrocytes in the brain and spinal cord constitute the blood-brain barrier, which effectively limits the transport into the brain of hydrophilic but not lipophilic chemicals. On the other hand, the placenta is a poor barrier to transport of many chemicals into the fetus, such as chemicals bound to the surface of red blood cells.

Chemicals may be excreted from the body through body fluids and excreta: e.g., urine, feces, bile, mucus, sweat, tears, milk, menstrual bleeding and exhaled air. However, the most significant elimination of chemicals from the body occurs by the urine and feces. The glomeruli of the kidneys filter most toxic chemicals with molecular weights of < 60,000 from the blood, unless they are bound to plasma proteins. Significant passive resorption occurs in kidney tubules by diffusion if the chemical is lipophilic. Many chemicals are transported into the intestines via bile drainage from the liver, with or without prior metabolic biotransformation. Fecal excretion includes depositions in the gastrointestinal tract from bile, saliva, pancreatic and gastric secretions, material cleared from the lung and swallowed as well as undigested food constituents.

Toxicokinetics

Toxicokinetics is the discipline that attempts to describe and predict the concentration of a toxic chemical in a dynamic system. This determination is useful in defining aspects of chemical absorption, translocation, biotransformation, binding to macromolecules and excretion (Figure 2.3).

2 – Toxicology and Pathobiology

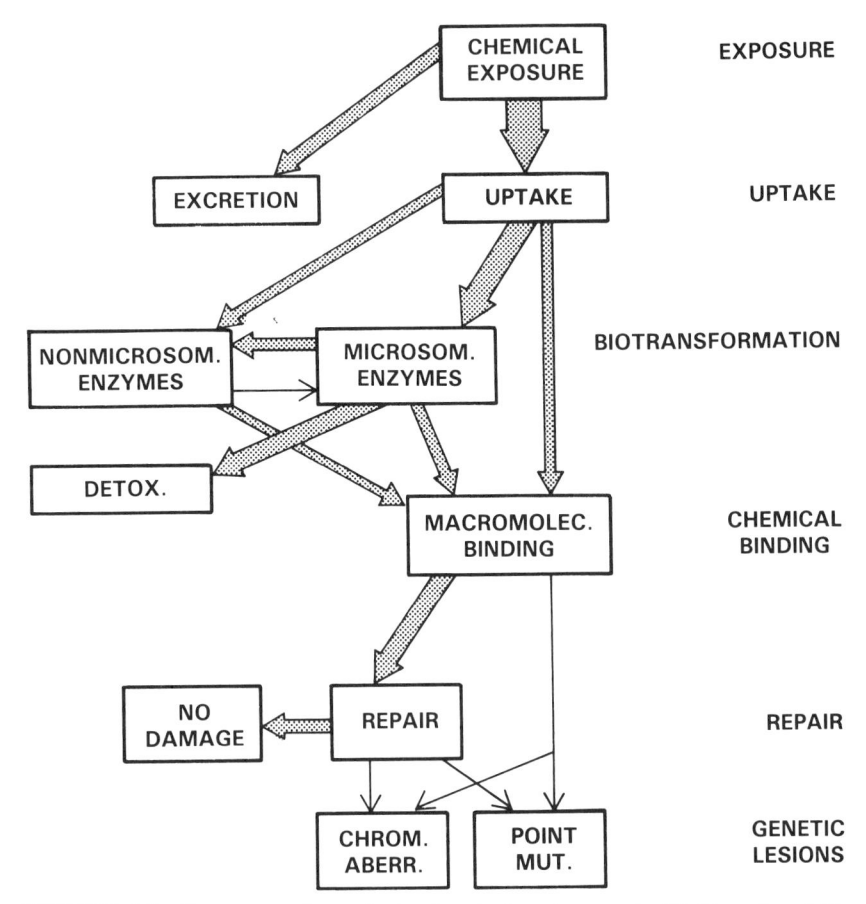

[Flow diagram showing sequence of events in chemical mutagenesis with the following boxes and stages:
- CHEMICAL EXPOSURE → EXPOSURE
- EXCRETION, UPTAKE → UPTAKE
- NONMICROSOM. ENZYMES, MICROSOM. ENZYMES → BIOTRANSFORMATION
- DETOX., MACROMOLEC. BINDING → CHEMICAL BINDING
- NO DAMAGE, REPAIR → REPAIR
- CHROM. ABERR., POINT MUT. → GENETIC LESIONS]

FIGURE 2.3—Sequence of events in chemical mutagenesis (C. Mamel and U. Rannug. 1980. Short-term mutagenicity tests. J. Toxicology Environmental Health 6:1065–1075).

Many toxic chemicals act on specific cellular receptors. The chemical binds to the receptor prior to producing its biological action. Where binding occurs is determined by the receptor site's affinity for the chemical, cellular permeability to the chemical, the number and concentration of receptor sites and the degree of enzymatic biotransformation of the chemical. Membrane receptors may exhibit a high degree of selectivity and specificity for certain chemicals, such as that seen with stereoisomers. The chemical may bind to the receptor covalently, by hydrogen bonding or by van der Waals bonding.

Reaction rates may follow either zero-order or first-order kinetics. In zero-order kinetics, the rate at which the reaction occurs is independent of the concentration of the chemical. For example, the enzymatic degradation of ethanol does not increase as the blood alcohol increases; this presents a metabolic bottleneck in the body's ability to handle further alcohol intake. In first-order kinetic reactions, the rate of the reaction is proportional to the amount of chemical present at any time. Zero- and first-order kinetic reactions apply to toxic chemicals, with first-order reactions being the most prevalent.

Little is known about the toxicokinetics of many chemicals. In practice, a series of reactions is simultaneously occurring in tissues as the chemical is being transported, metabolized, translocated and eliminated from the body. Metabolism plays a critical role in determining the toxicity of many chemicals. For some chemicals, metabolic alterations produce highly toxic metabolites from innocuous or weakly toxic parent compounds (Table 2.4). For some highly toxic, parent compounds, biotransformation produces much less toxic compounds or metabolites, which are more rapidly

TABLE 2.4. Increased toxicity of chemicals resulting from biotransformation reactions in tissues of humans.

Originally Deposited Chemical Specie	Transformed Chemical Specie of Increased Toxicity
Sulfanilamide	Acetylsulfanilamide
Ethylene glycol	Oxalic acid
Methanol	Formaldehyde
Fluoroacetate	Fluorocitrate
Parathion	Paraoxon
Tremorine	Oxytremorine
Tri-O-cresyl phosphate	Cyclic phosphate
Dimethylnitrosamine	Diazomethane
Heptachlor	Heptachlor epoxide
Pyridine	n-Methyl pyridinum chloride
Chloral hydrate	Trichloroethanol chloride
Nitrobenzene	Nitrosobenzene
Acetanilid	Aniline
Pentavalent arsenicals	Trivalent arsenicals
Selenate	Selenite
2-Naphthylamine	2-Amino-1-naphthol
Codeine	Morphine
Phenylthiourea	Hydrogen sulfide
Tetraethyl lead	Triethyl lead
Benzo(a)pyrene	Benzo(a)pyrene diol epoxide

eliminated from the body. These metabolic and excretory pathways are usually capacity-limited, being saturated at high substrate levels, so that biological damage is not related in a linear manner to the delivered dose.[12] The development of comprehensive toxicokinetic descriptions of these metabolic processes with respect to toxicity at the receptor sites will provide a better understanding of different responses among different species and extrapolation of toxicity data obtained in animal experimentation to humans.

Biotransformation

To accurately predict toxicity, we must understand the various biotransformation processes to which a chemical may be subjected.[14] We must also know the tissues with the greatest sensitivity to the chemical and those that will contain the greatest concentration of the chemical. Homeostasis involves regulatory and adaptive mechanisms in the body the attempt to contain, prevent or inhibit the harmful actions of toxic chemicals.[13] Toxicity consists of both a breakdown in the homeostatic mechanisms which control regulation within the cell and an attack on the cell through a series of feedback mechanisms.

The most important aspect of endogenous chemical surveillance or detoxification of toxic chemicals involves an enzyme system called microsomal mono-oxygenase. The cytochrome P-450 mono-oxygenase system, located in the smooth endoplasmic reticulum of hepatocytes, is the most important chemical detoxification and biotransformation center in the body. Biotransformation reactions, which often act to make lipid-soluble chemicals more water-soluble, have two phases. In the first phase, the chemical is subjected to enzymatic oxidation, reduction or hydrolysis; in the second phase, the metabolites are subjected to conjugation reactions. The majority of biotransformation reactions occur in the liver; however, most other tissues retain a limited ability to carry out these reactions, which include: Aromatic hydroxylation, aliphatic hydroxylation, N,O,S-dealkylation, epoxidation, desulfuration, sulfoxidation, N-hydroxylation, azo reduction and aromatic nitro-reduction. In addition to the mono-oxygenases, amine oxidase, epoxide hydratase, amidases, esterases and many other enzymes participate in these reactions.

Conjugation reactions occur most frequently after a chemical has undergone oxidation, reduction or hydrolysis. Compounds that most frequently participate in the conjugation of chemicals are glucuronic acid, acetate, sulfate, phosphorylated ribose, glycine, cysteine, glutathione, glutamine and S-adenosyl methionine. Phenoxic or alcoholic hydroxyl, aliphatic or aromatic carboxylic acid, aromatic amine and sulfhydryl groups are the most likely functional groupings to undergo conjugation with glucuronic acid. Chemicals that induce mono-oxygenases also will likely induce glucuronyl transferase activity, thus promoting glucuronide conjugation. The liver is also the most active site for conjugation reactions, with hepatocytes having the highest concentration of glutathione of any cell in the body. Glutathione rapidly reacts with electrophilic centers of some chemicals, forming a conjugate, or it directly destroys peroxides through redox reactions.[16]

Levels of biotransforming enzymes are often low in the newborn, making them more susceptible to toxic effects. Low levels of glucuronyl transferase in the newborn human liver results in the failure to transform many chemicals, such as chloramphenicol, to nontoxic glucuronides. In addition to age differences, there are also differences between species. For example, 2-amino-naphthol is metabolized in humans and dogs into a carcinogen, 2-amino-1-naphthol; in the rabbit, it is metabolized into noncarcinogenic chemicals, 2-amino-6-naphthol and 2-acetamido-naphthalene.

Carcinogenicity

Chemical carcinogens (procarcinogen, proximate carcinogen or ultimate carcinogen), or their electrophilic metabolites, induce genetic changes which directly or indirectly result in malignant transformation of normal cells to tumor cells.[22] Several classes of organic and inorganic chemical carcinogens are found in the general and occupational environment. Polycyclic aromatic hydrocarbons (PAH) are largely the result of fossil fuel utilization and combustion and are of particular importance in potential cancer induction.

Paracelsus described a fatal disease in miners of Schneeberg which, 300 years later, was diagnosed as lung cancer resulting from inhalation of high levels of natural radioactivity within the mines. Sir Percival Pott, in 1775, found scrotal cancer in chimney sweeps and related this tumor to exposure to soot. A century later, scrotal cancer was also found in workers employed in distillate production from coal and shale oil. In 1895, Rehn described the carcinogenic action of aniline derivatives in coal tar workers, mentioning naphthylamine as the most probable cause of the tumors. In 1928, leukemia was associated with exposure to benzene.[17] Today there is strong evidence that human exposure to at least 25 different agents will cause cancer (Table 2.5). By the end of WW II, nearly one million chemical compounds had been recorded; today that number is nearly five million. Not only have the numbers of new chemicals increased enormously, but the amounts used in industry have grown tremendously; production of synthetic fibers has increased over 6,000%, plastics over 2,000%, oil consumption over 2,000% and synthetic organic chemical and solvent production over 1,000% since the end of WW II. The great majority of these compounds have molecular weights < 1,000; about 60,000 are in common use today.[18]

TABLE 2.5. Physical and chemical agents strongly indicated as causes of cancer in humans (R. Doll and R. Peto. 1981. The causes of cancer: Quantitative estimates of avoidable risks of cancer in the United States today. J. National Cancer Institute 66:1193–**1309**).

Agents	Organs in which Carcinogenic
Aromatic hydrocarbons	Lung, skin, bladder, scrotum
Benzene	Bone marrow
Aromatic amines, amides and nitro compounds	Bladder, liver
Mustard gas	Lung
Antitumor Chemotherapy agents	Bone marrow, lymph nodes, etc.
Isopropyl alcohol	Nasal cavity
Vinyl chloride	Liver, brain
Bis(chloromethyl) ether	Lung
Aflatoxins	Liver
Oils, tars, pitches, soots	Skin, lung
Cigarette smoke	Lung, colorectum, oral cavity, etc.
Ethyl alcohol	Head and Neck, esophagus
Chloramphenicol	Bone marrow
Diethylstilbesterol	Vagina, breast
Steroid contraceptives	Liver
Arsenic compounds	Skin, liver, lung
Chromium compounds	Nasal cavity, lung
Cadmium compounds	Prostate
Nickel compounds	Nasal cavity, lung
Uranium ore	Lung
Thorotrast	Bone marrow, liver, connective tissues
Asbestos	Lung, pleura, peritoneal mesentery
Wood dust	Nasal cavity
Ultraviolet light	Skin
Ionizing radiations	All organs

The relative carcinogenicity of various chemicals was published by the EPA Office of Toxic Substances in the mid-1970s.[19] EPA's ordering plan resulted in the assignment of several-digit numbers (the so-called "adjusted ordering number") for all chemicals entered in the NIOSH Suspected Carcinogen List (Table 2.6). The numbers assigned by the EPA plan are an

TABLE 2.6. Adjusted ordering numbers for several inorganics and organics (Hittman Associates, Inc. 1979. Environmental Assessment Report: Solvent Refined Coal (SRC) Systems. U.S. Department of Commerce, NTIS, PB–300 383).

Chemical	Adjusted Ordering Number
Beryllium	16,000,000
Benzo(a)pyrene	3,314,500
Dibenz(a,h)anthracene	754,833
N-Nitrosodimethylamine	59,053
3-Methylcholanthrene	18,683
Cadmium	7,329
Chromium	7,327
Selenium	6,426
N,N'Dimethylhydrazine	2,208
Cobalt	1,682
Dibenz(a,i)pyrene	1,612
Benz(a)anthracene	1,562
Dibenz(c,g)carbazole	679
Aminotolune	638
N-Nitrosodiethylamine	577
Nickel	477
a-Aminophthalene	423
Dibenz(a,h)acridine	312
Dibenz(a,j)acridine	284
Ethylenamine	211
Lead	136
Diazomethane	78
Benzo(b)fluoranthene	78
Dibenzo(a,l)pyrene	65
4-Aminobiphenyl	54
4-Nitrobiphenyl	54
Phenanthrene	44
Indol(1,2,3-cd)pyrene	43
Formaldehyde	43
Methyl chrysene	39
Tetraethyl lead	36
p-Dimethylaminoazobenzene	35
Chrysene	32
Nickel carbonyl	26
Benzo(e)pyrene	23
Hydrazine	11
Mercury	11
Indole	6.5
Benzidine	3.5
Silver	1.7
Anthracene	1.3
Naphthalene	1.2
Pyrene	0.3

indication of the relative degree of concern that might be warranted for a particular substance regarding its possible carcinogenic potential. The numbers have an enormous range, covering at least nine orders of magnitude. Very large ordering numbers indicate that a small amount of the chemical is carcinogenic; small numbers indicate a high dose is required. Chemicals with adjusted ordering numbers of <1.0 are not treated as suspected carcinogens.

There is no apparent common structural feature among the dozen or so classes of known carcinogenic chemicals. Most chemical carcinogens have a molecular weight of <500 and are generally lipophilic, although inorganic carcinogens are exceptions. The predictability of carcinogenicity in humans is poor when carcinogenicity has been demonstrated in only one species, but much better when it has been shown in several species. Attempts have been made to predict carcinogenicity for organic compounds using structural-activity relationships.[20] Of the criteria for such chemical compounds are:

- Polymers form in the absence of stabilizers
- Cause cell transformation in vitro
- Alkylate DNA
- Purine or pyrimidine analogs that substitute for bases in DNA
- Trigger DNA repair
- Cause a loss of ribosomes from endoplasmic reticulum
- Induce microsomal sulfhydryl group formation
- Stimulate cellular hyperplasia
- Uncouple oxidative phosphorylation
- Denature nucleic acids
- Cause teratogenic effects
- Cause mutagenic effects
- Inhibit spermatogenesis
- Display electrophilicity
- Bind covalently to DNA

Chemical carcinogenesis results from the delivery of the ultimate carcinogen to the sensitive cellular target at a dose and dose-rate that cause an irreversible alteration in cell proliferation, survival and/or differentiation. Though some carcinogens do not require metabolic biotransformation to form a ultimate carcinogen, many metabolic reactions play a role in carcinogenesis (Table 2.7). One of the most intensively studied classes of carcinogens are PAH; of these compounds, the one most studied is benzo(a)pyrene (BaP). Among the metabolic products of BaP metabolism in tissues are epoxides and other metabolites that bind covalently to DNA and other macromolecules. Some strongly electrophilic chemicals, such as alkylating or acylating agents, do not require metabolic activation to induce tumors.

TABLE 2.7. Biotransformation reactions that play a role in chemical carcinogenesis (Mid-America Toxicology Course notes, April 20–25, 1980, Kansas City, MO, pp. 29–32).

I. Phase I Metabolism

Oxidations
  Aliphatic/aromatic C-oxygenation
    (hydroxy-, epoxidation)
  N-, O-, or S-deakylation
  N-oxidation or N-hydroxylation
  S-oxidation
  Oxidative deamination
  Dehalogenation
  Metallo-alkane deakylation
  Desulfuration
  Alcohol or aldehyde
    dehydrogenation
  Purine oxidation
  Tyrosine hydroxylation
  Monoamine oxidation
  Diamine oxidation
  Aromatization

Reductions
  Azo reduction
  Nitro reduction
  Arene oxide reduction
  N-hydroxyl reduction
  Quinone reduction

Hydrolyses
  Ester hydrolysis
  Amide hydrolysis
  Peptide hydrolysis
  Epoxide hydrolysis

Phase II Metabolism

Conjugations
  Glucuronidation
  Sulfate conjugation
  Glutathione conjugation
  Acetylation
  Glycine conjugation
  Serine conjugation
  N-, O-, or S-methylation
  Ribonucleoside or
    ribonucleotide
    conjugations
  Glycoside conjugations
Post Conjugation Reactions
  C-oxygenation
  Loss of glucuronide by
    beta-glucuronidase
  Loss of glycoside
  Deacetylation

Berenblum[19] applied to the skin of mice a variety of irritants which exhibited null or weak carcinogenic activity. The skin tumor response increased when a single, low dose of BaP was given at the same time. These irritants, such as croton oil, he termed promoters of carcinogenesis; BaP was termed the initiator of carcinogenesis. More than two centuries earlier, John Hill anticipated Berenblum's two-stage model of chemical carcinogenesis in a paper entitled, *Cautions Against the Moderate Use of Snuff*:[21]

"Whether or not the tumors ... which occur in snuff-takers are absolutely caused by that custom, or whether the principles of the disorder were there before, and snuff only irritated the parts, and hastened the mischief I shall not pretend to determine. Even supposing the latter only to be the case, the damage is certainly more than the indulgence is worth. No man should venture upon snuff who is not sure that he is not liable to cancer, and no man can be sure of that."

Berenblum postulated that the initiator of tumor formation causes rapid, irreversible genetic damage to cells. Tumors occur if exposure to a promoter stimulates the progression and expression of the initiating event. Complete carcinogens act as both initiators and promoters. Possible promoters are cigarette smoke in lung cancer, phorbol esters in skin cancer, phenobarbital in liver cancer and saccharin in bladder cancer.

"Chemical carcinogens are metabolized by numerous pathways catalyzed by enzymes in the endoplasmic reticulum and other parts of the cell. Reactions in which functional groups are created (epoxidation and epoxide hydration, catalyzed by cytochrome P-450-linked mono-oxygenase and epoxide hydratase, respectively) are especially important in the activation of polycyclic hydrocarbon carcinogens to ultimate carcinogenic forms, and may also participate in the activation of other chemical carcinogens. Numerous factors, genetic as well as environmental, affect the activities and the balance of different enzymes that participate in carcinogen activation and detoxification. The reasons why carcinogens act on specific target tissues are incompletely understood, although differences in enzyme profiles between tissues certainly contribute to the target tissue variability. Also, the location of where activation takes place is not known. It has been demonstrated that conjugated metabolites of carcinogens may be activated by spontaneous or enzymatic hydrolysis; this raises the possibility of transport of metabolites to distant target tissues. The concept of metabolic activation of carcinogens by the body's own enzymes has led to the development of short-term assay systems, which essentially measure the production of biologically active metabolites from potential carcinogens."[23]

Basic Pathobiology

Cell Injury

The action of toxic agents on mammalian tissues that causes subcellular and cellular injury constitutes the basis of pathological processes. It is at the macromolecular and biochemical levels that disease is first detected, followed closely by alterations and disruption of subcellular organelles. The sensitivity of different cell types to toxic agents is wide ranging. Deprivation of oxygen is lethal to neurons in 3–4 minutes, compared with 15–20 minutes in heart muscle cells and 1–2 hours in kidney tubule epithelium; the result is necrosis, with denaturation and hydrolysis of cell components.

Many necrotic processes result from the intracellular release of lysosomal enzymes. Lethal damage may be expressed as nuclear depolymerization and condensation (pyknosis), fragmentation of pyknotic nuclei (karyyorrhexis) or swelling and lysis of the nucleus (karyolysis). The later stages of cell necrosis leave only membranous, myelin bodies where once there was a viable cell (Figure 2.4).

Sublethal injury to a cell which may not result in cell necrosis, is characterized by vacuole and inclusion formations in the cytoplasm, altered fluid and energy metabolism, mitochondrial swelling and disruption, changes in phagolysosomes, dilation of endoplasmic reticulum and altered membrane permeability.

Inflammation

Inflammation is the response of body tissues to injury or irritation. Inflammation is beneficial in fighting infections and in wound healing but may be harmful if it becomes chronic. The classic signs of inflammation (heat, redness, swelling and pain) are the result of fine vascular changes occurring in the region of injury. The heat is due to the increased blood flow in the injured area; the redness to the dilation of the fine vasculature; the swelling to extravascular edema; and the pain to pressure on nerves or the formation of kinins. These processes are usually followed by healing and restoration of normal tissue integrity.

There are three basic parts to the inflammatory process. The first part is characterized by hemorrhage and necrosis, with release of hydrolytic enzymes from necrotic cells and of histamine from degranulating basophils and mast cells. The second part is characterized by vasodilation and migration of leukocytes into the injured area within minutes to hours following the injury. There is active congestion of the edges of the injured site, engorgement of capillaries with blood, accumulated serous fluid in extracellular spaces, lysis of necrotic cells and accumulation of polymorphonuclear leukocytes and macrophages.

The inflammatory reaction may then proceed to a chronic phase, with formation of fibrous tissue encapsulating the initial area of injury, and the infiltration of chronic inflammatory cells. This organized inflammatory reaction is a sign of failure of the acute inflammatory response to resolve the irritant stimulus.

The third part of the inflammatory response is the repair/proliferative phase. In most cases, granulation tissue fills the wound or injured site and normal parenchymal tissues regenerate and renew the site. Fibrosis or scar tissue formation occurs with incomplete renewal, long-term irritation or following damage to tissues with limited renewal capability, such as that of cardiac muscle following myocardial infarct.

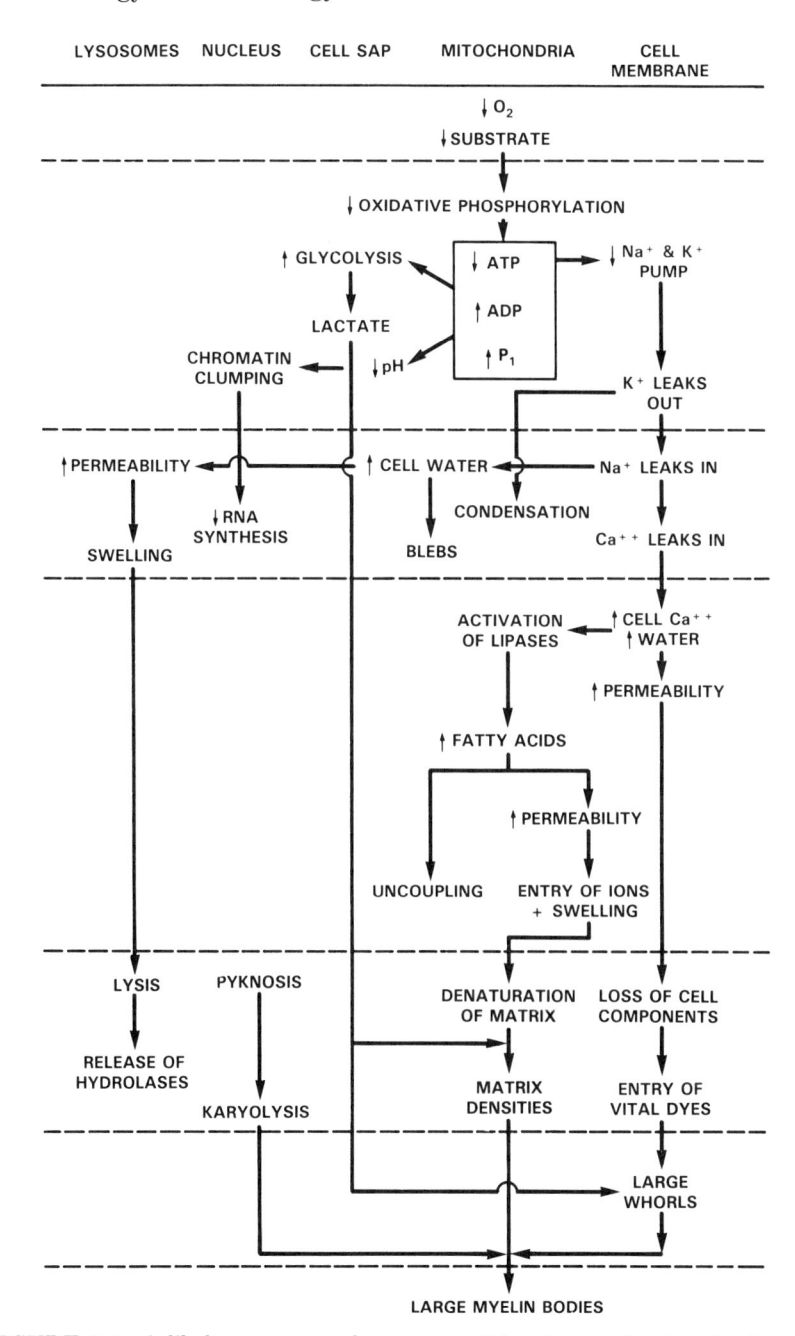

FIGURE 2.4 — A likely sequence of events resulting from cell injury leading to cell death (B. F. Trump and A. U. Arstilia. 1975. Cellular reaction to injury. In, "Principles of Pathobiology", Oxford University Press, NY, pp. 9–96).

Host-Parasite Interactions

An inflammatory response follows infection with pathogenic microorganisms, resulting in the mobilization of phagocytic cells and the erection of defense barriers at the site of invasion. Extracellular bacteria evoke a rapid polymorphonuclear leukocyte response; intracellular bacteria produce a more delayed inflammation that is characterized by mononuclear cell proliferation. Peripheral blood neutrophilia often follows infection with extracellular bacteria, whereas an increase mononuclear cell count in the blood is associated with intracellular infection.

Granulomatous tissue forms if the phagocytic cells are unable to kill the bacteria. In tuberculosis, the host defense process becomes part of the disease, with secondary tissue injury occurring from necrosis within granulomas. Delayed hypersensitive and allergic reactions in response to infection may result in immunologic injury; local lymphoid hyperplasia in areas near sites of infection is a common indicator of an immunogenic reaction. Systemic metabolic, hormonal and cardiovascular reactions may follow parasitism for a variety of microorganisms. Shock may occur from the release of vasoactive substances, bacterial toxins or following intravascular coagulation. Many fatalities from infection are the result of cardiovascular collapse.

Communicability is a measure of the transfer of a parasite from one individual to another, but is not a measure of pathologic response. The ability of the parasite to induce disease in the host is a measure of its pathogenicity. The severity of the parasite's pathogenicity is a measure of its virulence. Three factors are associated with pathogenicity: (a) invasiveness, (b) toxigenicity and (c) immunologic hypersensitivity. Encapsulation of pathogenic bacteria and the presence of certain extracellular enzymes on bacterial surfaces may enhance invasiveness. Toxigenicity is due to the elaboration of exotoxins or endotoxins that are toxic to the host. Endotoxins are found in some gram-negative bacteria, usually consisting of antigenic and pyrogenic bacterial cell wall complexes. Exotoxins are simple proteins associated with growth, death and lysis of proliferating bacteria. Most parasites display not only host specificity but also a predilection for certain tissues or organs within the host. Exposure to toxic agents increases cell injury and the sensitivity of tissues to infection with pathogenic bacteria.

Immunotoxicology

Immunotoxicology is concerned with toxic agents that require a component of the immune system for manifestation of clinical symptoms or those that impair normal immune function. The scope and content of immunotoxicology interactions are not well known, although understanding of this area is rapidly increasing. Because of the polymorphism and complexity of the immune system, the question of what agents are immunotoxic is difficult to determine, requiring a battery of bioassays to adequately evaluate the toxicity to the immune system. In general, chemical and physical agents can produce abnormal stimulation or suppression of immune system components. The immune system functions in resistance to infection. Antibodies produced in response to bacterial antigens facilitate phagocytosis and destruction of bacteria, neutralize bacterial toxins and prevent the entry of viruses into the cell. Immune responses are host defense mechanisms in which toxic agents are recognized as foreign and are neutralized, eliminated or metabolized with minimal injury to the host tissues. Immunologic injury may impair the ability of the host to combat microorganism disease infection, or the host may become sensitized to its own tissues, or cross-reacting antigens may result in autoimmune disease, with widespread pathologic manifestations.[24]

T-lymphocytes, which function in cell-mediated immunity, are involved in intracellular infections, delayed hypersensitivity, cytotoxic activities and modulation of B-lymphocytes. B-lymphocytes differentiate into plasma cells, which secret large amounts of immunoglobulins that are responsible for humoral immunity and are useful in combating extracellular infection. Antibody synthesis does not occur during fetal life until the differentiation of T- and B-lymphocytes from thymus and bone marrow, respectively. This cellular differentiation provides the basis for the later development of immunocompetency.

Antibodies, or immunoglobulins, have been well-characterized; they are divided into five classes: IgM or Macroglobulin, IgG, IgA, IgD and IgE. Human immunoglobulin levels reach adult levels in early childhood, usually by the age 1 to 4. Immunoglobulins are characterized by:

- A very high degree of specificity for antigens
- Memory capability, causing rapid antibody synthesis when the host is exposed to the antigen at a prolonged period of time following initial antigen exposure.
- Self-recognition capabilities, whereby they can discriminate between the host's own and very similar foreign macromolecules.

There are basic mechanisms or types of immune system malfunction (Table 2.8). Direct cellular injury by circulating antibodies may result in cell lysis. This response is helpful if the lysed cells are tumor-producing but harmful if they are normal host cells, such as destruction of red blood cells during hemolytic anemia. In indirect injury due to circulating antibodies, there may be an immediate hypersensitivity because initial antigen exposure has sensitized the host. Succeeding antigen exposures may then produce an immediate asthmatic or anaphylactoid response, mediated through the release of histamine from disrupted basophils and mast cells. Immune complex formation (Arthus's reaction) results from the precipita-

TABLE 2.8. **Immunopathologic responses to tissue injury** (R. P. Sharma. 1983. Immunologic Considerations in Toxicology, Vol. I, CRC Press, Boca Raton, FL, pp. 2–43).

Immunologic Nature	Type of Response	General Mechanisms	Examples of Associated Problems
Humoral	I	Anaphylactic reactions; release of active substances by antigen reaction with IgE fixed on target cells	Hay fever, skin or systemic anaphylaxis, penicillin allergy
Humoral	II	Fixation of antigen on a target cell, reaction with antibody, resulting in complement-induced cytotoxicity	Hemolytic anemia, transfusion reactions, autoallergic disease
Humoral	III	Antigen-antibody complex precipitating on cells; reaction is complement-dependent	Arthus's reaction, serum sickness, glomerulonephritis, amyloidosis
Cellular	IV	Delayed hypersensitivity; antigen-sensitized cells causing cytotoxic effect	Tuberculin reaction, graft rejection, tumor-associated conditions
Humoral	V	Antibody production to biologically important molecules (e.g., insulin or clotting factors) or tissues	Allergic diabetes, hemophilia, myasthenia gravis, pernicious anemia, allergic encephalomyelitis

tion of antigen-antibody complexes (particularly IgG) on the vascular endothelium of skin, joints or kidney tubule epithelium, causing an acute inflammatory response. Cell-mediated immunity or delayed hypersensitivity reactions result from a series of cell reactions that cause the production of specifically reactive T-lymphocytes. The prototype reaction is the granulomatous response to Mycobacterium tuberculosis. Many examples of contact dermatitis and allergies are a result of delayed hypersensitivity reactions.

A variety of biological tests are available to detect immune injury.[26] Immunocellular stem cell numbers can be determined by using colony-forming cell assays. Cell-mediated immunity (CMI) can be measured by in vitro and in vivo assay systems. In in vitro systems, the functional capacities of lymphocytes to stimulant mitogens or antigens are examined. For in vivo systems, delayed and contact hypersensitivity or rejection of allographs are tested following toxic agent exposure. Humoral-mediated immunity (HMI) is tested by examining the levels of circulating immunoglobulins and antibody production in response to antigen exposure.

In 1979, NIEHS came to a conference concensus with 35 participating immunologists and recommended a list of suitable bioassays for evaluating chemically-induced immunotoxicity.[30] Among these were assays that tested:

- Pathotoxicology-hematology, liver, chemistry, serum proteins, lymphoid organ weights and histology
- Host resistance-tumor challenge and infectious agent challenge
- Radiometric delayed hypersensitivity
- Lymphoproliferative responses-PHA, Con A, LPS and MLC
- Humoral immunity-Ig levels, specific antibody titer, plaque forming cell assay
- Macrophage function assays
- Bone marrow progenitor cell assays.

Agents such as ionizing radiation or p-dioxin are general immunosupressants. Other agents may exhibit more limited and specific immunologic injuries (Table 2.9). For example, photoallergens such as salicylanilides, sulfonamides and ragweed plants, may result in contact dermatitis only following contact with subsequent exposure to sunlight.[25]

Biology of Cancer

A fundamental problem in cancer biology is defining the nature of the initiating event leading to the formation of tumors. Far from being simply the result of rapid, disorderly cell proliferation in inappropriate locations, cancer can also be a highly logical, coordinated process in which certain cells acquire a unique combination of specialized abilities.[32] Most scientific observations of the nature of cancer are consistent with a mutational mechanism for transforming cells into tumors. A mutation does not automatically mean cancer since the carcinogen must cause the mutation in genes that are related to cancer formation. Only if the mutation occurs in the right place does the cell complete the first step in the carcinogenesis process. A single mutation is not sufficient to initiate cancer formation; subsequent mutation(s) must occur in the proper genes to initiate the pro-

TABLE 2.9. A partial list of chemical agents that have been reported to be immunotoxic (M. I. Luster, J. H. Dean and J. A. Moore. 1982. Evaluation of immune functions in toxicology. In, "Principles and Methods of Toxicology," Raven Press, NY, pp. 561–586).

Chemical	HMI	CMI	Monocyte	Host Resistance
Polyhalogenated aromatics				
TCDD	D	D	–	D
PCB	D	NC	D	D
TCDF	D	D	NC	NC
PBB	D	D	–	D
HCB	NC	D	D	D
Metals				
Lead	D	D	NC	D
Cadmium	D	–	NC	D
Arsenic	D	NC	NC	D
Methylmercury	D	NC	NC	D
Dialkytins	D	D	–	D
Inhaled Pollutants				
Nitrogen dioxide	NC	D	NC	D
Sulfur dioxide	NC	D	D	NC
BaP	D	NC	I	NC
Tobacco smoke	NC	NC	NC	D
Ozone	NC	NC	D	D

D = decrease
I = increase
NC = no change or some studies increase and some studies decrease

cess. Recent research indicates that there are a series of genes within the cell, termed oncogenes, each with specified functions, such as coding the synthesis of one enzyme. Oncogenes are thought to control cell proliferation at various points throughout the cell. Mutations in several oncogenes within the cell results in the transformation of the normal cell into a tumor cell. Included in the supporting data for mutagenesis as initiating event in carcinogenesis is evidence that most carcinogens are also mutagens, that carcinogens or their metabolites often bind to DNA and that those humans that are deficient in DNA repair enzymes (xeroderma pigmentosum) or with chromosome abnormalities (Bloom's syndrome), have high risks of developing cancer.

Most human tumors are clonal in origin, resulting in formation of a defect or block in the ability of cancerous cells (derived from normal stem cells) to differentiate in a formal manner. The most common cancers are found in tissues with high rates of cell proliferation (e.g., skin, gastrointestinal tract, uterine mucosa). Tumor development is often seen as a problem in cell differentiation and loss of cell proliferation control, even though the tumor may initially have resulted from an epigenetic event.

Berenblum postulated a two-stage mechanism for the origin of tumors that requires an external initiator and promoter of tumor progression before full expression of the tumor is evident (Figure 2.5). The idea was that latent tumor cells are continuously being formed, but that few are expressed as tumors for lack of promotion, or because of control by the immune surveillance system in tissues. Tumor progression has been refined as the gradual evolution of a tumor toward increased autonomy by a series of stepwise changes along a variety of alternative pathways. Characteristics of tumor progression include increased proliferation and growth, invasiveness, undifferentiated tumor cell morphology and metastasis. In humans, the association of immune surveillance failure and cancer induction has been indicated from several situations following exposure to physical or chemical immunosuppressive agents. Most spontaneous tumors have little antigenicity; rapidly arising tumors caused by laboratory oncologic viruses or chemical carcinogens are often strongly antigenic.

Tumors may be benign, growing slowly by expansion, or malignant, growing slowly or rapidly by expansion, invasion and/or metastasis. Benign tumors are life-threatening only if they compress a vital structure; malignant tumors are life-threatening because they destroy or replace vital tissues as well as compressing vital structures. Tumors are assisted in their

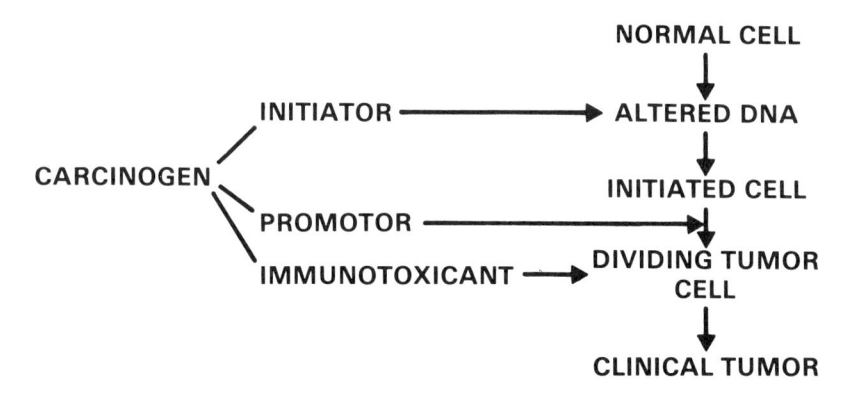

FIGURE 2.5 – A model of tumor formation.

invasiveness by their general lack of adhesiveness, their ability, in many cases, to move by ameboid motion and the presence of surface enzymes such as hyaluronidase, which digests connective tissues that are in the way of the advancing tumor. Malignant tumors may exhibit a normal cell proliferation rate but abnormally prolonged cell survival, or they may have a high proliferation rate. Growing tumors require a continuing development of blood vessels to provide for increasing needs of nutrients and oxygen. This neovascularization of the tumor, is mediated in the host, by the production of potent angiogenesis factors within the tumor that promote the proliferation of capillary endothelium. Since tumors are normally angiogenic and normal tissues are not, the ability to induce capillary proliferation is probably acquired during the early stages of tumor development.

Well-differentiated tumors retain some of the structures and functions of the normal cells from which they evolved. For example, a well-differentiated thyroid carcinoma retains the ability to take up iodine. Poorly differentiated tumors or anaplastic tumors have a more embryonic appearance and have lost many of the functions of the normal differentiated cell. Usually, the more undifferentiated the tumor cell, the greater the degree of malignancy (that is, increased proliferation, invasiveness and metastasis) and the worse the prognosis.

Among the host factors in carcinogenesis that can be either inherited or acquired or that can operate at the systemic or cell level are:

- Age
- Sex
- Hormones, growth factors and receptors
- Immunologic factors
- Nutritional status
- Acquired diseases
- Prior exposure to environmental promoters or initiators
- Carcinogen metabolism
- DNA repair
- Chromosomal defects
- Cellular proliferation
- State of differentiation and gene expression
- Oncogenes

Short Term Mutagenicity and Carcinogenicity Tests

Ever since the discovery in 1942 that mustard gas is mutagenic, scientists have been attracted to the idea that carcinogenesis of a chemical might involve mutagenic events.[45] Induced or "spontaneous" mutations occur in both somatic and germ-line cells. The field of environmental mutagenesis arose in the late 1960s with the objective of identifying and exploring mechanisms of chemicals that alter genetic material. It is estimated that germinal mutations are reflected as hereditary disease in about 4% of newborns.[45] Somatic mutations are thought to be the initiating events in cancer formation and have been postulated to result in atherosclerosis, aging, teratogenicity and spontaneous abortions. The measurement of mutations in short-term tests provides relatively inexpensive data to estimate life-span carcinogenic potential from pollutant exposure. Ionizing radiations were the first man-made agent shown to be both mutagenic and carcinogenic. The chromosome damaging effects of X-rays was easily demonstrated in peripheral blood lymphocytes of patients undergoing radiotherapy for cancer; chromosomal damage has also been seen in nuclear dockyard workers, plutonium workers and in survivors of Hiroshima and Nagasaki. The incidence of structural chromosomal aberrations increases lineally with an increase in radiation dose. However, ionizing radiations do not increase sister-chromatid exchanges (SCEs), nor do chemical mutagens that produce genetic damage like ionizing radiation (e.g., radiomimetics such as bleomycin and other chemotherapeutic alkylating agents). Besides radiation and alkylating agents, several other chemical agents have been shown to produce chromosome aberrations in humans (Table 2.10).

TABLE 2.10. Occupationally exposed populations that show a clear relationship between exposure to an agent and an increased frequency of chromosomal aberrations in culture peripheral blood lymphocytes (H. J. Evans. 1982. Cytogenetic studies on industrial populations exposed to mutagens. In, "Indicators of Genotoxic Exposure," Cold Springs Harbor Laboratory, Banbury Report 13:325–335).

Agent	Population Size
Ionizing Radiations	>1,000
Arsenic	. 33
Benzene	>190
Chloromethylether	12
Chloroprene	>50
Epichlorhydrin	>100
Organophosphates	>180
Styrene	>50
Vinyl chloride	>500

In 1978, the value of chemicals produced in the U.S. was 113 billion dollars. The toxicity of the vast majority of these chemicals has not been studied for long-term, chronic health effects as biomedical endpoints. Most chemicals are not tested because government facilities and trained toxicologists and pathologists are limited.; expense and time required to perform the tests are also factors. Carcinogenicity tests in rodents, on the average, cost >$300,000 for each chemical and take at least three years to complete.

As an alternative to long-term tests, a variety of short-term, rapid and inexpensive assays can be used to screen potentially toxic chemicals. To improve risk assessments, an index of potential carcinogenicity is required, to help in setting future research priorities. Short-term bioassays, such as the Ames test for mutagenicity, provide an index over at least a million-fold chemical concentration range. Short-term tests which use microorganisms, fungi, insects, amphibians or mammals may include:[34]

- Tests using prokaryotic microorganisms
- Tests using eukaryotic microorganisms
- Mutagenesis tests in mammalian cell cultures
- Tests that measure DNA repair and replication inhibition
- In vitro transformation tests
- In vivo tests in mammals
- In vitro and in vivo mammalian cytogenetics tests

Some chemicals are potent in one test system but weak in another. For example, 1-aminoacetylfluorene is potent in the Ames test but weak in the hamster embryo cell transformation test. Therefore, it is necessary to perform a battery of tests to evaluate toxicity potential in humans.

Carcinogens may either be genotoxic (capable of reacting with and damaging DNA) or epigenetic (unable to damage DNA to any detectable extent).[46] The emphasis for so-called genotoxic agents is on the ability to detect damage to DNA. Genotoxic agents may cause:

- Point mutation
- Gene rearrangements
- Altered DNA methylation
- Altered chromatin structure

A given mutagen, when interacting with cells may cause different kinds of genetic lesions, ranging from a single base change to a chromosome break or even loss of a chromosome. The interactions which are genetically important are those which cause a change in base-pair sequences in DNA. Thus, cytogenetic lesions may be visible changes in chromosomes or invisible changes in base-pairs, as in point mutations (Figure 2.6).

Numerous protocols have been proposed for multi-level testing schemes that may be used for both short-term and subsequent long-term testing

2 – Toxicology and Pathobiology

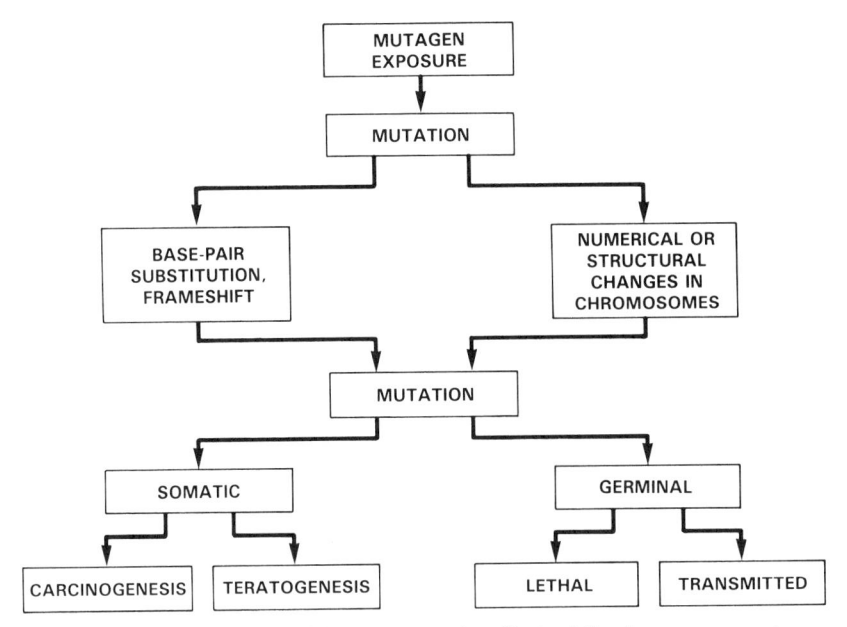

FIGURE 2.6—Categories of human genetic effects following exposure to mutagenic agents (D. J. Brusick. 1978. The role of short-term testing in carcinogen detection. Chemosphere 5:403-417).

of chemicals following positive results with short-term tests (Figure 2.7; Table 2.11). A complex series of biochemical activation or deactivation events may occur following exposure of living cells to a chemical that may alter the chemical's carcinogenic potential.[35]

The Salmonella/Ames test is the most widely used short-term test.[42] It takes about three days and uses strains of Salmonella typhimurium for detection of frameshift or base-pair substitution using reverse mutation as an end point. Tissue homogenates from tissues such as rat liver are often used to metabolize chemicals to active mutagens. Mitotic crossover (reciprocal exchange between homologous chromosomes), gene conversion (restoration of wild-type genotype) and mitotic nondisjunction (resulting in aneuploidy, typically with trisomic and monosomic daughter cells) have been measured in a variety of microorganisms.

The sex-linked recessive lethal test typically uses the fruit fly, Drosophila melanogaster. In this test, lethal mutagenic changes in the X-chromosome are detected by treating adult, male flies with the test chemical and mating them with untreated female flies. First-generation females are mated to untreated, normal male flies. If one of the X-chromosomes in the daughter flies carries the lethal mutation inherited from the treated male parent, then the eggs of this mating will not hatch.

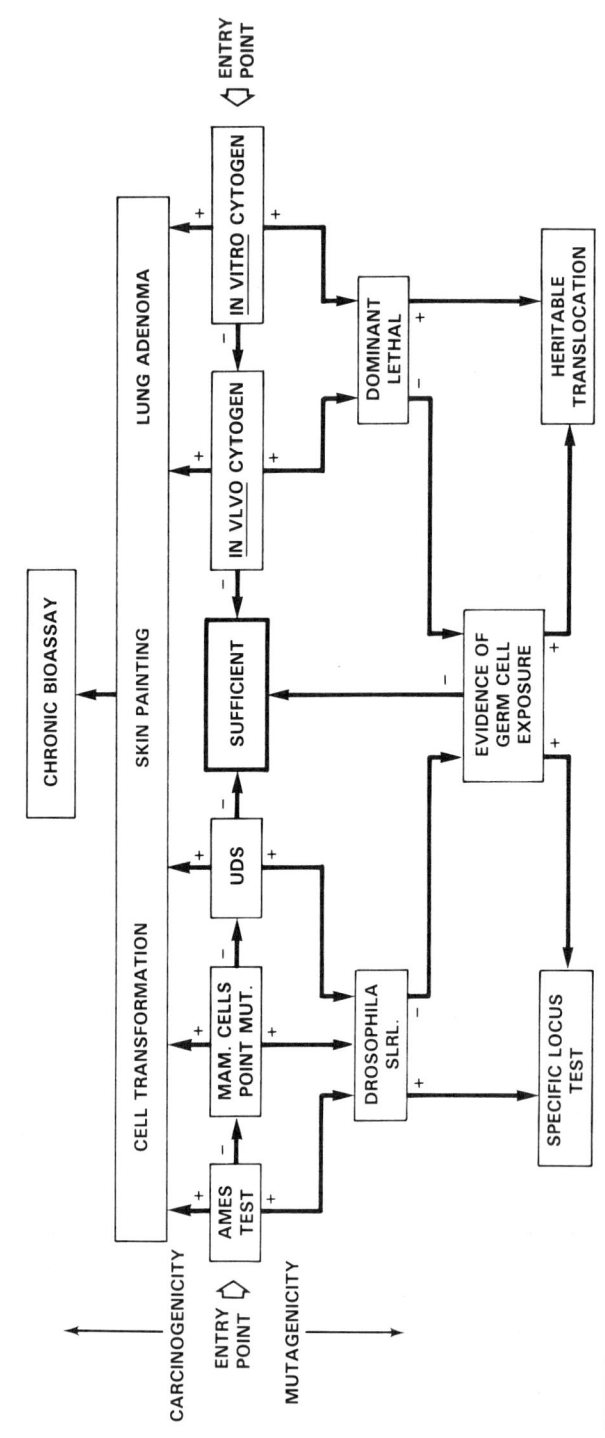

FIGURE 2.7 – A proposed three-level testing scheme for the EPA to use in the systematic evaluation of chemicals for mutagenicity and carcinogenicity (D. Brusick. 1982. Genetic toxicology. In, "Principles and Methods of Toxicology", A. W. Hayes, Ed., Raven Press, NY, NY, pp. 223–272).

TABLE 2.11. Tiered biological characterization procedures Used for the toxicological evaluation of coal gasification products and pollutants. (C. A. Reilly, M. J. Peak, T. Matsushita, F. R. Kirchner and D. A. Haugen. 1981. Chemical and biological characterization of high-Btu coal gasification (The HYGAS process). IV. Biological Activity. In, "Coal Conversion and the environment," CONF-801039, NTIS, Springfield, VA, pp. 310-324).

Tier	Measurement	System
Level 1	Genetic toxicity	
	Mutation	Ames Salmonella assay
	Sister chromatid exchange	Mouse myeloma cells
	Cytotoxicity	
	Growth inhibition	Mouse myeloma cells
	Lethality and phagocytosis	Rabbit alveolar macrophages
Level 2	Genetic toxicity	
	Mutation	V79 Chinese hamster cells
	Cytotoxicity	
	Growth inhibition and lethality	Mouse myeloma cells
	Lethality	V79 Chinese hamster cells
Level 3	In vitro transformation	C3H10T1/2 cells
	Whole animal toxicity	
	Acute	
	Oral, LD50	Mouse
	Intraperitoneal, LD50	Mouse
	Ocular	Rabbit
	Dermal, LD50	Rabbit
	Chronic	
	Skin	Rabbit
	Delayed-type hypersensitization	Albino guinea pig
	Carcinogenesis	Hairless mouse

The missing male offspring are evidence of lethal chromosome damage in the treated males. Other tests used to detect dominant lethal mutations may utilize the Habrobracon wasp.

The majority of mutagenesis tests in cultured mammalian cells use established cell lines with stable karyotypes having loci such as hypoxanthine-guanine phosphoribosyl transferase, thymidine kinase or membrane Na$^+$/K$^+$ ATPase (ouabain-resistant). Cell types as Syrian hamster embryo (SHE), Chinese hamster V79 or Chinese ovary cells (CHO), mouse lymphoma or adult rat liver are used in these assays. Human cells, also used in these tests, are normal diploid fibroblasts and lymphoblasts and xeroderma pigmentosum diploid fibroblasts deficient in DNA repair enzymes. In addition to mutagenic changes, tests may measure DNA repair, unscheduled DNA synthesis, DNA fragmentation or DNA replication inhibition. The effectiveness of in vivo, covalent binding of chemicals to DNA may be related to the mutagenicity and carcinogenicity of the chemical (Table 2.12).

The mouse specific-locus test, originally designed by Russell,[36] is used for the in vivo testing for heritable germ-line mutations. In this test, phenotypic changes such as coat or eye color and ear length that result from mutations at any of seven loci are detected in the progeny of mutagen-treated males mated to females that are homozygous-recessive at these loci.

In the sister-chromatid exchange test (SCE) the symmetrical exchange

TABLE 2.12. Covalent binding of various chemicals to DNA in rat liver. (Data taken from: G. Zbinden. 1979. Application of basic concepts to research in toxicology. Pharmacol. Rev. 30:605–616).

Compound	Route of Administration	Hours After Administration	CBI*
Aflatoxin B1	Oral	6	10,000
Aflatoxin M1	Oral	6	14,000
Benzo(a)pyrene	Intraperitoneal	50	10
Benzene	Inhalation	10	1.7
Estrone	Oral	8	1.1
Toluene	Inhalation	5	0.1
Saccharin	Oral	50	<0.1

*Covalent Binding Index = umole bound chemical/mole DNA-P/-mmole chemical/kg body weight

of DNA between sister chromatids in metaphase chromosomes is measured. The SCE test often gives positive results along with the Ames, micronucleus and chromosome aberration tests (Table 2.13). In the unscheduled DNA synthesis test (UDS), a test chemical is injected intraperitoneally in rodents with an intratesticular injection of 3-thymidine. DNA synthesis, as well as sperm counts and morphology, may be examined in this test.[37] The degree of positive response among the various tests varies according to the class of chemical.[44]

TABLE 2.13. Results from various test systems with commonly studied carcinogens (M. Kirsch-Volders, M. Radman, P. Jeggo and L. Verschaeue, 1983. Molecular mechanisms of mutagenesis and carcinogenesis. In, "Mutagenicity, Carcinogenicity, and Teratogenicity of Industrial Pollutants," Plenum Press, NY, NY, pp. 5-61).

Chemical	Ames	SCE	Chromosome Aberrations	Micronucleus
2-Aminoacetylfluorene	+	+	−	+
N-Hydroxy-2-AAF	+	+	−	
N-acetoxy-2-AAF	+	+	+	
Benzidine	+			+
Cyclophosphamide	+	+	+	+
Nitrogen mustard	+	+	+	
Captan	+		+	
Benzo(a)pyrene	+	+	+	−
Benz(a)anthracene	+	+	−	
3-Methylcholanthrene	+	+	+	+
Myleran	+	+	+	−
Ethyl methane sulfonate	+	+	+	+
Dimethylsulfate	+	+	+	
1,3-Propane sultone	+	+	+	
1,2,3,4-Diepoxybutane	+	+	+	
Dimethylnitrosamine	+	+	+	+
Diethylnitrosamine	+	+	+	−
N-methyl-N-nitrosourea	+	+	+	
Aflatoxin B1	+	+	+	+
Mitomycin C	+	+	+	+
Hycanthone	+	+	+	+
Triazoquone	+	+	+	+
Ethidium bromide	+	−	+	

+ positive test result
− negative test result

Chromosome abnormalities account for one-third of all spontaneous abortions, while one in two hundred live births carries a chromosome aberration. Nearly all agents that produce specific-locus mutations also produce chromosome abnormalities. In vivo tests, using chromosome abnormalities in thymus, bone marrow, spleen or spermatogonia as test endpoints, are useful methods for evaluating mutagenicity and carcinogenicity. One such test uses in vivo exposure to a test chemical, followed by a series of in vitro cultures of peripheral blood lymphocytes to examine metaphase-arrest mitotic figures for chromosome abnormalities. The micronucleus test is based on the observation of extranuclear bodies in the cell cytoplasm due to chromosomal fragments that remain after cell division. Bone marrow cells or polychromatic erythrocytes may be examined in this test after in vivo exposure to a chemical agent.

In the granuloma pouch assay, the target cell population is croton oil-induced granulation tissue growing on the inside of an air pocket under the skin on the backs of rats. Test chemicals are injected into the pouch, and fibroblasts are isolated and analyzed for DNA breaks and point mutations, or the granuloma is examined later for tumor development.[43]

In vitro transformation tests are related to in vivo cancer formation since transformed cells often will cause cancer or develop into tumors when re-injected into rodents. Transformation in embryonic cell lines is a more efficient measure of toxicity since embryonic cells can metabolize to their active forms a greater number of procarcinogens than adult cells, while spontaneous transformation is low in embryonic cells. A high correlation between cell transformation and in vivo carcinogenesis has been obtained for a wide variety of known carcinogens and noncarcinogens. Clonal growth assays that use relative plating efficiencies to compare control and chemically treated cell cultures are also useful in toxiclogical testing.[5] Other tests examine damage to cell organelles.[39-41]

Routine genetic toxicology tests most frequently performed by governmental, contract, and industrial laboratories are the Ames test, in vitro cytogenetics, in vitro SCE, UDS mutation assays in CHO cells at the HGPRT locus and in vivo bone marrow assays.[43] The International Commission for Protection against Environmental Mutagens and Carcinogens supports the 1980 report on the Biological Effects of Ionizing Radiation (BEIR III, National Research Council) in the use of linear extrapolation between the lowest reliable dose data and the spontaneous rate. A wide variety of genetic, cytological, biochemical and morphological tests may be designed to evaluate the potential long-term toxicity of chemicals. The challenge is to use these tests that most accurately predict carcinogenicity in the least time and money.

Cancer Epidemiology

In 1976 there were 675,000 new cases of cancer diagnosed in the United

States, equally distributed among males and females, and 370,000 deaths due to cancer. These data excludes most skin tumors, other than melanoma. The most frequent tumor is basal cell carcinoma of the skin, which is almost never invasive and nearly 100% curable by simple surgical excision. About one-third of the U.S. population will develope cancer during their lifetime. In 1981, the actuarial five-year survival for all cancer patients, excluding basal cell carcinoma of the skin, was 41%; survival was better in females than in males. The leading site of lethal cancer in males was the lung, followed by colorectal and prostate cancer; in females, breast cancer resulted in the most deaths, followed by cancer of the lung and colorectum (Table 2.14). Cancer mortality is markedly age-dependent, increasing from a rate of 10 cases/100,000 at ages <20 to over 1,000 cases/100,000 annually by age 60 (Figure 2.8). In 1900, the average life expectancy at birth was 47 years, compared to 71 years in 1970. Since the death rate for cancer at 71 years of age is 10 fold the rate at age 47, it follows that the cancer mortality rate today would be greater than in 1900.

The majority of cancer in humans is attributable to lifestyle factors and is therefore subject to preventive actions. Lung cancer in men, a comparatively rare disease in 1930, had increased 20 fold by 1970. Cigarette smoking alone is responsible for 90% of lung cancer, 40% of all cancers in males, and 30% of all cancers in males and females.[28] Alcohol consump-

TABLE 2.14. 1978 Estimates of cancer deaths in the United States by site and sex (American Cancer Society).

Site	Percent of All Cancer Deaths Male	Percent of All Cancer Deaths Female	Annual Death Rate for Both Sexes
Lung	34	12	95,000
Colorectum	12	15	53,000
Prostate	10	–	22,000
Breast	*	19	35,000
Leukemia & Lymphoma	9	9	19,000
Pancreas	5	5	21,000
Urinary Tract	5	3	17,000
Head and Neck	3	1	8,000
Skin	2	1	2,000
Ovary	–	3	6,000
Uterus	–	6	11,000
All Other	20	26	

*About 100 deaths per year from male breast cancer

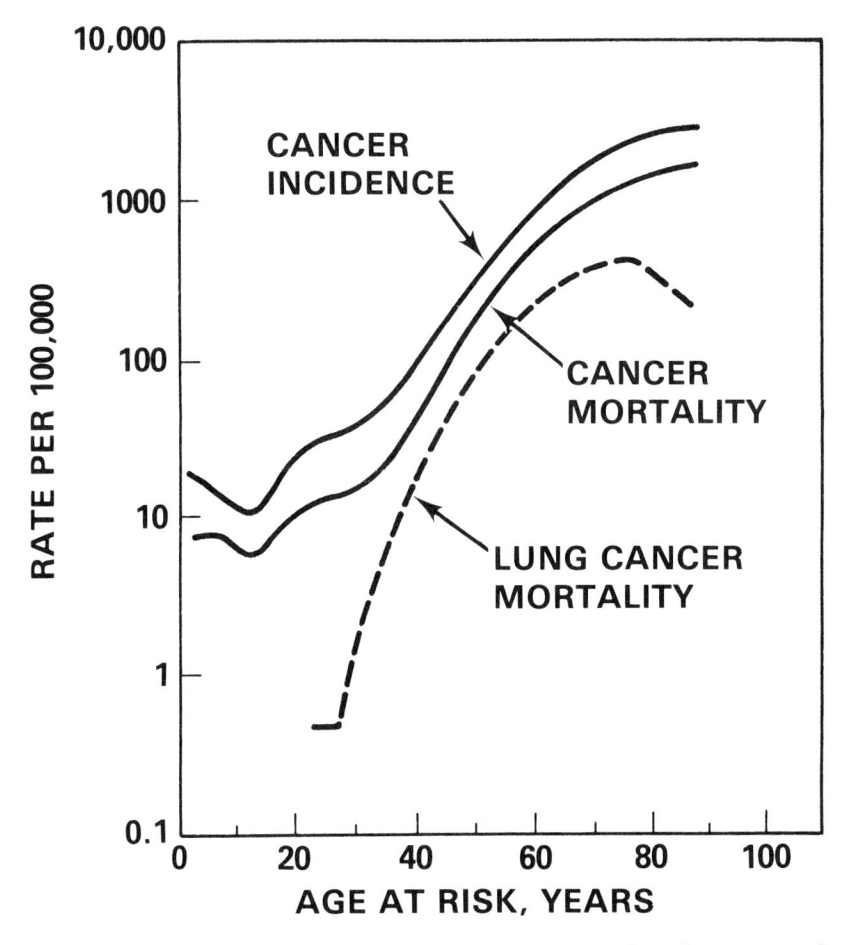

FIGURE 2.8—Age-specific mortality and incidence rates for all cancers and lung cancer in U.S. males (C. L. Sanders and R. L. Kathren. 1983. Ionizing Radiation. Tumorigenic and Tumoricidal Effects. Battelle Press, Columbus, OH, p. 69).

tion is responsible for about 5% of all cancers in males. About 50% of all cancers in both sexes are due to diet.[28] The age-adjusted incidence rates for cancer vary by a factor of three around the world; for cancers in certain anatomical locations, such as esophagus and liver, variability is 100-fold. Occupational exposures to toxic agents account for <5% of all cancers in humans, but those that occur tend to be in specific organs, such as lung, liver, urinary bladder and nasal cavity.[28] Asbestos, PAH, heavy metals and benzene are some of the chemicals implicated as carcinogens in the workplace. The number of agents that are known to cause human cancers in an occupational setting are small (about 20 counting all carcinogenic polycyclic hydrocarbons as one).[33]

Drugs account for 2–3% of all human cancers. Synthetic estrogens taken during pregnancy can subsequently produce vaginal and cervical carcinomas in female offspring; this is an example of transplacental carcinogenesis. Alkylating agents and other drugs used to treat cancer can greatly increase the risk of secondary cancer formation. For example, chemotherapy for ovarian carcinoma increases the risk of acute nonlymphocytic leukemia by about 170-fold. Immunosuppressive drugs increase the risk of cancer. Ultraviolet light increases the risk of skin cancer. Overall, ionizing radiations, mostly used for therapy of cancer, have been implicated in about 2% of all human cancers.

References

1. J. F. Finklea, D. J. Hammer, T. A. Himners and C. Pinkerton. 1971. Human Pollutant Burdens. Proc. A.C.S. Symposium on Determination of Air Quality, Los Angeles, CA, pp. 49–55.
2. J. Doull, C. D. Klaassen and M. O. Amdur. 1980. Casarett and Doull's Toxicology, Second Edition, Macmillan Publishing Co., NY, NY.
3. T. D. Luckey and B. Venugpal. 1977. Metal Toxicity in Mammals, Vol. 1., Plenum Press, NY, NY.
4. B. Ballantyne. 1977. Current Approaches in Toxicology, John Bright & Sons Ltd., Bristol, U.K.
5. A. J. Finkel, Editor. 1983. Hamilton and Hardy's Industrial Toxicology, Fourth Edition, John Wright PSG, Inc., Boston, MA.
6. A. Goldstein, L. Aronow and S. M. Kalman. 1974. Principles of Drug Action, John Wiley, NY, NY.
7. T. A. Loomis. 1974. Essentials of Toxicology. Second Edition, Lea & Febiger, Philadelphia, PA.
8. K. P. DuBois and E. M. K. Geiling. 1959. Textbook of Toxicology, Oxford University Press, NY, NY.
9. E. J. Ariens, A. M. Simonia and J. Offermeier. 1976. Introduction to General Toxicology. Academic Press, NY, NY.
10. D. A. Canter. 1981. Role of the regulatory agencies in the activities of the National Toxicology Program. Regulatory Toxicology and Pharmacology 1:8–18.
11. Chemical and Engineering News, May 7, 1984; p. 9.
12. M. E. Anderson. 1981. Saturable metabolism and its relationship to toxicity. CRC Critical Reviews in Toxicology 9:105–150.
13. L. Golberg. 1979. Toxicology: Has a new era dawned? Pharmacol. Rev. 30:351–370.
14. The Scientific Committee, Food Safety Council. 1980. Proposed System

for food safety assessment. Food Cosmet. Toxicol. 16, Suppl. 2:1–136.
15. E. D. Wills. 1981. The role of glutathione in drug metabolism and the protection of the liver against toxic metabolites. In, "Testing for Toxicity," Taylor and Francis, Ltd, London, U.K.
16. H. G. S. VanRaalte. 1979. Suspicion and confidence in toxicology and occupational medicine. In, "Toxicology and Occupational Medicine," Elsevier/North Holland, Inc., Amsterdam, The Netherlands, pp. 81–91.
17. J. A. Miller and E. C. Miller. 1979. Perspectives on the metabolism of chemical carcinogens. In, "Environmental Carcinogenesis," Elsevier/North Holland Inc., Amsterdam, The Netherlands, pp. 25–42.
18. Office of Toxic Substances. 1976. An ordering of the NIOSH suspected carcinogens. List PFS 251-851, U.S. Environmental Protection Agency, Washington, D.C.
19. I. Berenblum. 1979. Theoretical and practical aspects of the two-stage mechanisms of carcinogenesis. In, "Carcinogens: dentification and Mechanisms of Action," Raven Press, NY, NY, pp.
20. F. F. Becker. 1979. Keynote Address: Evolution, chemical carcinogenesis and mortality: The Cycle of Life. In, "Carcinogens: Identification and Mechanisms of action," Raven Press, NY, NY, pp. 5–24.
21. D. J. Brusick. 1977. In vitro mutagenesis assays as predictors of chemical carcinogenesis in mammals. Clin. Toxicol. 10:79-109.
22. O. Pelkonen and K. Vahakangas. 1980. Metabolic activation and inactivation of chemical carcinogens. J. Toxicol. Environ. Health :989-
23. M. F. LaVia. 1975. Immunologic injury. In, "Principles of Pathobiology," Oxford University Press, Oxford, UK, pp. 175–202.
24. T. L. Pazdernik. 1980. Notes on immunotoxicology. Mid-America Toxicology Course Notes, April 20–25, 1980, Kansas City, MO.
25. W. L. Morison, J. A. Parrish and J. H. Epstein. 1979. Photoimmunity. Arch. Dermatol. 115:350-
26. R. P. Sharma. 1983. Immunologic Considerations in Toxicology, Vol. l., CRC Press, Boca Raton, FL, pp. 2–43.
27. Research and Developement. 1984. May, p. 50.
28. R. Doll and R. Peto. 1981. The causes of cancer: Quantitative Estimates of avoidable risks of cancer in the United States Today. J. Natl. Cancer Inst. 66:1193–1309.
29. P. H. Abelson. 1984. Environmental risk management. Science 226:1023.
30. M. I. Luster, J. H. Dean and J. A. Moore. 1982. Evaluation of immune functions in toxicology. In, "Principles and Methods of Toxicology," Raven Press, NY, pp. 561–586.
31. I. B. Weinstein and F. P. Perera. 1982. Molecular cancer epidemiology. In, "Indicators of Genotoxic Exposure," Cold Spring Harbor Laboratory, Banbury Report 13:3–15.
32. B. Rensberger. 1984. Cancer: The new synthesis. Science (September), pp. 28–33.

33. R. Doll. 1984. Occupational cancer: Problems in interpreting Human evidence. Ann. Occup. Hyg. 28:291-305.
34. M. Hollstein and J. McCann. 1979. Short-term tests for carcinogens and mutagens. Mutat. Res. 65:133-226.
35. G. Zbinden. 1979. Application of basic concepts to research in toxicology. Pharmacol. Rev. 30:605-616.
36. W. L. Russell. 1951. X-ray induced mutations in mice. Cold Spring Harbor Symp. Quant. Biol. 16:327-336.
37. G. A. Saga, J. G. Owens and R. B. Cumming. 1976. Studies of DNA repair in early spermatid stages of male mice after in vivo treatment with methyl-, ethyl-, propyl-, and isopropyl methane sulfonate. Mutat. Res. 36:193-212.
38. M. E. Frazier and T. K. Andrews. 1979. In vitro clonal growth assay for evaluating toxicity of metal salts. In, "Trace Metals in Health and Disease," Raven Press, NY, NY, pp. 71-81.
39. B. A. Fowler, J. S. Woods and C. M. Schiller. 1977. Ultrastructural and biochemical effects of prolonged oral arsenic exposure on liver mitochondria in rats. Environ. Health Perspect. 19:197-204.
40. J. D. Lockhart. 1972. How toxic is hexachlorophene? Pediatrics 50:229-235.
41. M. E. Billingham, J. W. Mason, M. R. Bristow and J. R. Daniels. 1978. Anthracycline cardiomyopathy monitored by morphologic changes. Cancer Treat. Rep. 62:865-872.
42. B. N. Ames. 1984. The detection of environmental mutagens and potential carcinogens. Cancer 53:2034-2040.
43. N. E. McCarroll, M. G. Farrow, K. L. McCarthy and H. E. Scribner. 1984. A survey of genetic toxicology testing in industry, contract laboratories and government. J. Appl. Toxicol. 4:66-72.
44. S. Parodi et al. 1983. Quantitative predictivity of the transforming in vitro assay compared with the Ames test. J. Toxicol. Environ. Health 12:483-510.
45. C. Auerbach and J. Robson. 1946. Nature 157:302.
46. B. L. Harper and D. L. Morris. 1984. Implications of multiple mechanisms of carcinogenesis for short-term testing. Teratogenesis, Carcinogenesis, and Mutagenesis 4:483-503.

CHAPTER 3

NON-RESPIRATORY TRACT TOXICOLOGY

Introduction

The lung, skin and gastrointestinal tract are the major routes of exposure for toxic agents. The lung is the most important route of exposure and the most often damaged organ by toxic agents associated with energy production. However, non-respiratory organs and tissues are also often criticial in the toxicological evaluation of chemicals both as primary sites of deposition and sites of chemical translocation, concentration, biotransformation and excretion. The organ of greatest initial exposure may or may not be the organ that exhibits the greatest damage from a toxic agent.

Kidney

The functional unit of the kidney is the nephron; the adult human kidney contains about one million nephrons. Collectively, they filter about 180 liters of blood per day, reabsorbing 179 liters of fluid and producing about 1 liter of urine. Renal function is clinically assessed by examining renal concentrating ability (urine osmolarity, amino acid and glucose levels in urine) or filtration ability (paraaminohippurate, phenol red or phenosulfonaphthalein clearance). Rough indicators of glomerular damage can be made from comparisons of plasma and urine creatine levels. The epithelial cells lining the kidney tubules contain a high level of catalase and alkaline phosphatase; urinary increases in catalase or phosphatase are indicative of tubular kidney damage.

Renal damage from exposure to toxic agents is due to abnormal hemodynamic regulation, damage to the fine vasculature of the kidney, or damage to glomeruli or specific portions of nephron tubules.[1] Biopsy of kidney tissues may detect some of these irregularities but may miss the cause of more subtle renal dysfunctions. Specific nephrotoxins are the heavy metals, especially mercury, cadmium and uranium, and certain chlorinated hydrocarbons, such as carbon tetrachloride and chloroform.

Liver

The liver parenchymal mass is defined in terms of acinar functional units.[2] The acinus is an irregular mass of hepatocytes arranged around a portal venule, a hepatic arteriole, a bile ductule and lymph vessels and nerves. Phagocytic Kupffer cells line the vascular sinusoids.

Hepatotoxins are a diversified group of toxic agents that produce a variety of hepatic lesions that are collectively termed toxic hepatitis. Type I hepatotoxins produce distinctive lesions in which severity is related to dose; they usually appear after a brief but predictable latent period. Hepatotoxins may themselves be biotransformed in the liver or they may influence the biotransformation of other chemicals in the liver; for example, phenobarbitol alters the biotransformation of many chemicals by inducing biotransforming enzymes. Type I hepatotoxins produce liver damage by causing an accumulation of fatty acids in cells (e.g., ethionine), cell necrosis (e.g., tannic acid), both fatty acid accumulation and cell necrosis (e.g., carbon tetrachloride), cholestasis (e.g., chlorpromazine) or liver cancer (e.g., aflatoxin). Type II hepatotoxins act on the liver in an unpredictable manner, influencing only a small portion of the exposed population. Type II hepatotoxins induce viral-like hepatitis reactions, resulting in later cell necrosis or cholestasis; examples of type II hepatotoxins are iproniasid, once used for the treatment of tuberculosis; halothane anesthesia; and indomethacin, used to treat gout.

Liver damage may be evaluated by measuring hepatic dye clearance, plasma enzyme levels (serum glutamic oxaloacetic transaminase or SGOT or serum glutamic pyruvic transaminase or SMGPT), bile flow and histopathology of biopsied tissues.[3]

An increasing number of chemicals have been found to induce liver tumors in rodents.[4] Such tumors have been referred to as type A or B nodules, hyperplastic nodules, hepatocellular adenoma, hepatoma or hepatocellular carcinoma. Carcinomas in the liver may have two cellular components: A biliary epithelial component which, on carcinogenic transformation, becomes a cholangiocarcinoma; and a hepatocyte component, which is transformed into a hepatoma. Presumed human hepatocarcinogens are benzidine, auramine, 2-naphthylamine, aflatoxin and vinyl chloride. A list of chemicals that show hepatocarcinogenicity in rodents are given in Table 3.1.

Skin

The skin constitutes a primary natural defense for man against hazardous agents in the environment, protecting against parasite infection, chemical and physical agent damage, trauma, UV irradiation, water and electrolyte

TABLE 3.1. The carcinogenicity of chemicals in the liver and other tissues of mice and rats and their mutagenicity in salmonella typhimurium (J. M. Ward, R. A. Griesemer and E. K. Weisourger. 1979. Toxicol. Appl. Pharmacol. 51:389-397).

Chemical	Tumors in Mice Liver	Tumors in Mice Other Tissues	Tumors in Rats Liver	Tumors in Rats Other Tissues	Mutagen (Ames Test)
2-Acetylaminofluorene	+	+	+	+	+
N-Hydroxy-2-acetylaminofluorene	+	+	+	+	+
Benzidine	+	–	+	+	+
6-Aminochrysene	+	+	–	–	+
4-Aminobiphenyl	+	+	+	+	+
4, 4$^{5/8}$-(Imidocarbonyl) bis	+	+	+	+	+
2, 7-Bisacetylaminofluorene	+	+	+	+	+
3$^{5/8}$-Methyl-4-dimethylaminobenzene	+	+	+	–	+
4-Dimethylaminobenzoate	+	+	+	+	–
2-Nitrosonaphthalene	+	+	–	–	+
o-Aminoazotoluene	+	+	+	–	+
2-Naphthylamine	+	+	–	–	+
N-Hydroxy-2-napthalamine	+	+	–	–	+
N-Hydroxy-1-naphalamine	+	+	–	–	+
7,12-Dimethylbenz(a)anthracene	+	+	–	+	+
Benzo(a)pyrene	+	+	–	+	+
3-Methylcholanthrene	+	+	–	+	+
Dimethylnitrosamine	+	+	+	+	+
Diethylnitrosamine	+	+	+	+	+
Dibutylnitrosamine	+	+	+	+	+
N-Nitrosopiperidine	+	+	+	+	+
Nitrosomorpholine	+	+	+	+	
Nitrosomethylurea	+	+	–	+	+

loss and temperature extremes, in addition to providing sensory information required for response to changes in the environment. The skin's physiological protection mechanisms include the epidermal barrier of the stratum corneum, the bacteriostatic and fungistatic properties of long-chain fatty acids produced by sebaceous glands, the shielding of melanocytes from UV radiation, the thermal regulatory processes of the skin blood vasculature and sweat glands, the dilution of toxic chemicals on the skin by sweat, the mechanical resiliency against physical penetration provided by plastic fibers and collagen in the dermis and immunological reactions to percutaneous exposure from toxic agents. Overall, occupationally induced diseases of the skin account for 40–50% of all toxicity problems in the workplace.[5]

A chemical deposited on the surface of the skin may pass through the horny layer, through the epidermis, into the corium. At that point, percutaneous absorption has been accomplished. The chemical passes into the systemic circulation by two routes: Transfollicular penetration, along the route of the hair follicles; or through the epidermis. Transfollicular absorption is rapid but accounts for only a very small part of total skin absorption. Most chemicals penetrate the skin by massive transepidermal transport according to Fick's law. This principle can be paraphrased, in this connection, to say that absorption is a function of the partition coefficient of the chemical, the diffusion coefficient through the epidermis, the thickness of the skin and the difference in concentration across the stratum corneum. Factors that influence percutaneous absorption are disease, trauma, ambient temperature, hydration, area of skin exposed and pH and concentration of the chemical. Application of large amounts of some chemicals to the skin of humans has caused death; these include the use of pure boric acid powder applied to diaper dermatitis; salicyclic acid applied to psoriatic skin and undiluted hexachlorophene baths for infants. Relatively small amounts of pesticides, such as parathion, have also caused death in humans.

Pesticides and PAH compounds have received considerable attention with respect to percutaneous absorption.[6] Many PAH compounds, when topically applied to skin, are inducers of skin carcinoma in animals. Once absorbed from the skin, they are metabolized to ultimate carcinogens or to water-soluble metabolites that are largely conjugated in the liver and excreted in the bile. Inducible enzymes in the skin are responsible for the rapid breakdown and excretion of most deposited PAH compounds. Benzo(a)pyrene and dimethylbenz(o)anthracene rapidly penetrate the skin of mice and their metabolites are excreted in the bile. The amount of PAH absorbed from the skin decreases with increasing applied dose due to saturation of the transport and biotransformation processes in epidermal cells (Figure 3.1), thus enhancing the tumorigenicity of the PAH in the skin.

Pathological reactions of the skin to toxic agents may be classified as early, short-term, inflammatory, or long-term carcinogenic responses. Early

3 – Non-Respiratory Tract Toxicology

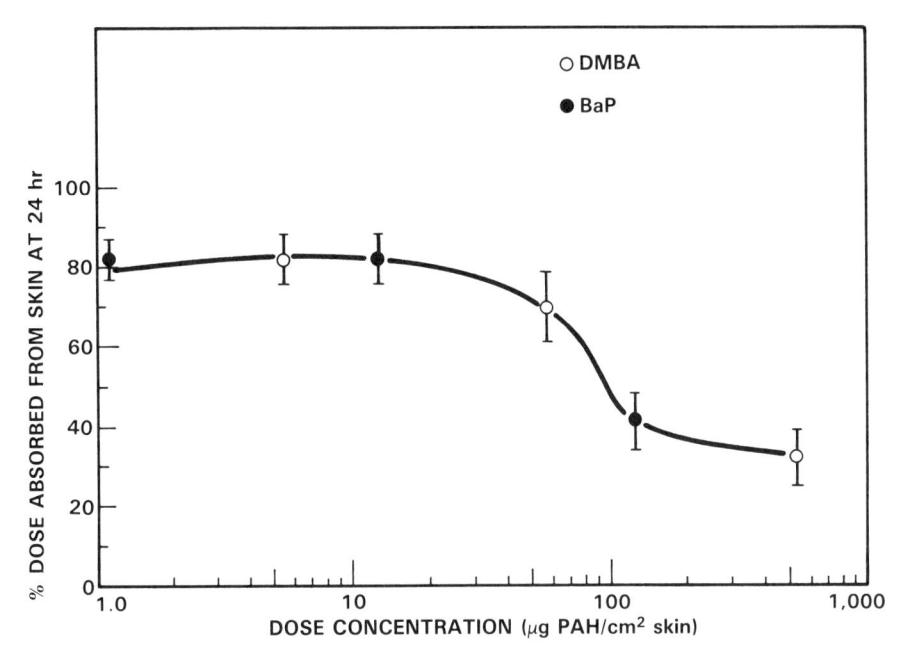

FIGURE 3.1—Influence of amount of percutaneous 14-C-PAH on their absorption from the skin of mice during the first day after application (C. L. Sanders, C. Skinner and R. A. Gelman. 1984. Percutaneous absorption of 7, 10–14C benzo(a)pyrene and 7, 12 –14 dimethylbenz(a)anthracene in mice. Environmental Research 33:353–360.).

changes include eczematous inflammation with epidermal swelling, resulting from direct action of a chemical, to sensitization to an allergen or to phototoxicity. The signs of primary dermatitis are erythema, edema, exfoliation, desquamation and necrosis of skin. These effects are usually produced at the site of chemical contact with skin and are not mediated through systemic allergic reactions. Watch tests are useful in detecting sensitivity to many chemicals through the formation of contact dermatitis.

Acne and changes in blood vessels, connective tissues, skin pigmentation and sweat glands are seen after exposure to certain toxic environments. A major cause of morbidity in American soldiers serving in the south Pacific in WW II was "prickly heat" caused by plugged and ruptured sweat glands. Proliferative lesions in the epidermis associated with long-term irritation of the skin may result in skin tumors. Many PAH compounds, lactones, epoxides and peroxyl compounds yield a high incidence of skin tumors when applied to shaved skin of mice. However, care must be taken to distinguish between irritants and carcinogens in such tests, since both types of agents will cause epidermal hyperplasia when applied repeatedly to the skin of mice. Only carcinogens increase the thickness (epidermal cell numbers) of the epidermis as well as increasing epidermal cell proliferation and altering cell orientation and morphology. Hyperplastic papillomas appear early after application of many test chemicals which may later regress or may progress to malignant tumors.

Biotransformation of applied chemicals alters their carcinogenicity. BaP is a potent carcinogen whose metabolites may show strong hyperplastic activity but little carcinogenicity when applied to the skin of mice, or may result in stronger carcinogenic activity than the parent compound (Table 3.2). Tumor location and age are also important in PAH carcinogenesis: injected diol-epoxide, a BaP metabolite, is 50 times more carcinogenic in the lung of newborn mice than is BaP, but it is 50 times less carcinogenic on the skin than is BaP.

Hematopoiesis and Oxygen Utilization

Stem cells in the bone marrow undergo cell division and maturation, leading to the production of erythrocytes (erythropoiesis), platelets (thrombopoiesis), neutrophils, eosinophils and basophils (granulopoiesis), lymphocytes (lymphopoiesis) and monocytes and macrophages (monopoiesis). The sum of all these marrow proliferation groups is called hematopoiesis. Hematopoiesis is carried out in the fetal spleen, yolk sac, liver and bone marrow; only bone marrow persists as a site of hematopoiesis shortly after birth. In humans, hematopoiesis ceases in the femur and tibia after age 20–25, continuing only in the sternum, vertebrae and ribs and pelvis. Mature erythrocytes contain hemoglobin, which functions in oxygen transport. Lymphocytes function in humoral and cell-mediated immunity. Platelets derived from megakaryocytes in the bone marrow function in blood clot formation. Granulocytes and monocytes (macrophages) are phagocytic cells that assist in combating infection and wound healing.

Chemicals can cause both a generalized depression in hematopoiesis and, often, a more profound depression in granulopoiesis. The responses are varied; aspirin inhibits platelet aggregation; excess cobalt causes polycythemia; carbon monoxide interferes with oxygen transport.

Impaired ability to transport oxygen from the lung leads to hypoxia and damage to hypoxia-sensitive tissues, such as neurons in the brain. Changes in the heme molecule alters its affinity for oxygen and, thus, oxygen transport. Hydrogen ions, carbon dioxide or 2,3-diphosphoglycerate all decrease the affinity of heme for oxygen.[7] Shifts in the oxygen dissociation curve, towards either a greater or a lesser affinity for oxygen are generally considered undesirable. Carbon monoxide binds more tightly to heme than does oxygen, lowering the tissue level of oxygen even at normal PO_2 levels; a similar effect is seen with anemia or with nitrite exposure. Cyanide poisons cells by reversibly complexing with cytochrome c oxidase, blocking oxidative metabolism and phosphorylation.

TABLE 3.2. Carcinogenicity of benzo(a)pyrene and its metabolites on mouse skin. Mice were treated with BaP or its derivative once every two weeks for 60 weeks (W. Levin et al. 1977. Carcinogenicity of benzo-ring derivatives of benzo(a)pyrene on mouse skin. Cancer Research 37:3356-3361).

Compound	Dose (μμmoles)	Percent of Mice with Skin Tumors
Benzo(a)pyrene	0.020	2
	0.025	7
	0.050	59
	0.100	60
	0.400	100
Diol-epoxide 1	0.020	0
	0.100	4
	0.400	7
BaP 7, 8-dihydrodiol	0.025	22
	0.050	76
	0.100	92
H4-7, 8-diol	0.100	0
H4-9, 10-epoxide	0.050	0
	0.400	4

Bone

The supporting framework of humans is comprised of bone and cartilage. Zones of cartilage in growing bone act as centers for new bone growth, with osteoblasts forming new bone and osteoclasts resorbing old bone, in a continual pattern of bone remodelling. Osteocytes maintain the bone as living tissue. The majority of the body's collagen and >90% of the body's citrate are found in bone. The mineral portion of bone, hydroxyapatite $[Ca_{10}(PO_4)_6(OH)_2]$, also contains smaller amounts, of sodium, magnesium, citrate, and fluoride.[8] Bone metabolism is under the control of the parathyroid gland through the hormone calcitonin. Hormones from the thyroid, adrenal, ovary or testis and pancreas and vitamins A, C and D play important roles in maintaining normal bone metabolism.

Some metal ions, such as Ca, Sr, Ra, and Ba, may be substituted by ion exchange in bone crystal without disturbing crystalline symmetry. Other elements, such as Pu, Am, Y, Th and La, bind to biochemical ligands on bone surfaces and are subsequently buried and redistributed by bone remodelling processes. The bone also acts as a storage site for toxic metals such as lead and beryllium.

Damage to bone may occur in utero from teratogenic actions of toxic chemicals. In utero or post-natal exposures to toxic chemicals may cause growth-stunting due to damage to endochondral regions. Bone injury may result from damage to the vascular system supplying nutrients and oxygen to bone cells, or to loss in calcium causing osteoporosis and bone fracturing, or to the induction of malignant tumors such as osteosarcomas from osteoblasts or chondrosarcoma from chondroblasts.

Neurotoxicology

The influence of toxic agents on the nervous system, whether behavioral, biochemical or histopathological, has been the subject of considerable interest during the last decade. Neurotoxicity is manifested following damage to the cellular elements of the brain, spinal cord or nerves, or to damage to the blood-brain barrier resulting in increased blood vessel permeability. The brain is particularly sensitive to hypoxia.

Behavior is considered an indicator of the net sensory, motor and integrative processes occurring in the central and peripheral nervous system and a sensitive indicator of toxicity.[9] Neurotoxins are classified according to the type of neurological function affected or the site of primary damage (Tables 3.3 and 3.4). Chemicals like lead, mercury and arsenic cause diffuse damage to the CNS; manganese causes localized damage. Myelin is the target for carbon monoxide toxicity; the neuromuscular junction for botulinum toxin; peripheral neuropathies result from excessive exposure to ethyl alcohol or carbon disulfide. Lead toxicity influences learning; mercury, visual discrimination; and carbon monoxide, locomotor activity. In methylmercury poisoning, increased vascular permeability is noted in cranial nerves and in sensory dorsal root ganglia. Sensory neuropathy due to increased vascular permeability has been seen with adriamycin, chloramphenicol, dinitrobenzene, thallium and disulfiram.[12]

Methods used to assess sensory damage in animals include maze discrimination tests, operant conditioned and unconditioned behavior and externally triggered behavior, such as circadian activity. Screening tests for motor dysfunction include assessments of body posture, muscle tone, gait, fighting reflexes and the ability to swim or maintain balance.

Behavioral modification may result from chemical exposure. Behavior is classified as unconditioned (unlearned) or as conditioned (learned). Respondent unconditioned behavior is elicited by known, observable stimuli; operant unconditioned behavior is caused by unknown but observable

TABLE 3.3. Symptoms commonly reported by humans following exposure to neurotoxins (H. A. Tilson and C. L. Mitchell. 1982. Models for Neurotoxicity. Rev. Biochem. Toxicol. 2:265–294).

Function Affected	Symptomatology
1. Sensory	Anosmia; paresthesias in feet, toes, fingers; visual deficits, photophobia, nystagmus; auditory deficits, tinnitus; perceptual dysfunctions, pseudohallucinations
2. Motor	Weakness in hands, arms, legs; paralysis; incoordination, dizziness; fatigue; tremor, convulsions; hyperactivity; slurred speech
3. Nervous System Excitability	Nervousness, irritability, agitation; euphoria; psychosis; apathy, lethargy; depression, compulsive behavior
4. Associative	Impaired short-term or long-term memory; confusion, disorientation
5. Physiologic	Disrupted sleep-awake cycles; hypothermia; sweating; loss or stimulated appetite

stimuli. Toxicant-induced changes in behavior might be related to changes in sensory processing, such as with visual or auditory impairments.[13]

Nervousness and irritability are common indicators in humans of exposure to toxic chemicals. Benzene and kepone produce hyperirritability in humans, while polybrominated biphenyls produce lethargy.[9] The induction threshold for audiogenic or electroshock grand-mal type seizures in animals is often used to test neurological excitability. Low-level exposures to environmental agents for long periods of time can produce a gradual alteration in nervous system function, followed by an overt, and frequently irreversible, neurological effect. Behavioral changes suggestive of nervous system malfunction may be among the first signs of toxicity.[9]

TABLE 3.4. Summary of the principal types of chemical-induced structural damage to the nervous system (A. J. Dewar. 1981. Neurotoxicity testing—with particular reference to biochemical methods. In, "Testing for Toxicity," Taylor & Francis, Ltd., London, UK, p. 202).

Type	Pathological Reaction	Examples
Neuronopathy	Nerve cell degeneration	Methylmercury Adriamycin Aluminum * Vincristine
Axonopathy	Distal axon degeneration	n-Hexane Acrylamide Carbon disulfide Organophosphates Isoniazide Thallium Arsenic Ethanol
Myelinopathy	Segmental demyelination	Buckthorn toxin** Diphtheria toxin Hexachlorophene Cuprizone* Lead Triethyl tin Acetyl ethyl

*Toxic to only the central nervous system
**Toxic to only the peripheral nervous system

Eye

A number of chemicals at low concentrations are capable of causing reflex tearing; these chemicals are called lacrimators. At threshold dilutions, lacrimators cause instant tearing without tissue damage. Alpha-chloroacetophenone (MACE) is a powerful lacrimator used in law enforcement and personal protection. In photochemical smog, nonirritating air pollutants are converted by irradiation to peroxyacyl nitrates, with accompanying lacrimator activity, particularly in the smogs of Los Angeles and other cities of the southwestern states.[10] Eye irritation, manifested by itching, burning, swelling and lacrimation, occurs commonly among nonsmokers passively exposed to cigarette smoke; acrolein and other agents in the smoke are the offending chemicals.[11]

Thallium salts induce cataracts in rats by damaging the lens epithelium. Optic neuritis results from human exposure to thallium, as was seen in the 1930s with the use of thallium acetate in cosmetic creams. The basis of thallium toxicity is its active transport into the optic nerve and lens and substitution for potassium. Pentavalent arsenic, which has been used in the treatment of trypanosomiasis, results in contraction of visual fields in about 1% of treated patients. Constriction of visual fields has also been observed in methylmercury poisoning.

Reproductive Toxicology

The influence of toxic agents on the reproductive system and on in utero development is divided for convenience into three areas of study: (1) teratology, (2) placental function and (3) reproduction. Table 3.5 lists data

TABLE 3.5. **Intrauterine development for several mammalian species** (B. A. Schwetz. 1980. Reproductive Toxicology. In, "Mid-America Toxicology Course Notes," Kansas City, MO, pp. 173–203).

Species	Implantation Begins	Primitive Streak	Organogenesis Completed	Birth
Hamster	4.5–5	6	14	16–17
Mouse	4.5–5	7	16	20–21
Rat	5.5–6	8.5	17	21–22
Rabbit	7	6.5	20	30–32
Guinea pig	6	10	25	65–68
Sheep	10	13	35	145–152
Swine	10–12	11	34	110–116
Ferret	10–12	13	–	40–43
Cat	12–13	–	30	60–65
Dog	13–14	13	30	60–65
Rhesus monkey	9	18	45	164–170
Baboon	9	19	47	172–178
Human	6.5–7	18	55	260–280

Time After Conception, Days

TABLE 3.6. Environmental pollutants that exert adverse effects in reproduction.

Agent	Known Effects on Conceptus or Reproductive Function
Lead	Abortion, mental deficiencies
Mercury	Abortion, menstrual disorders, birth defects
Cadmium	Retarded fetal growth
Selenium	Abortion
Dioxins	Abortion, birth defects, stillbirth
Polychlorinated biphenyls	Retarded growth, neural depression
2, 4-D and 2, 4, 5-T	Abortion, stillbirth, birth defects
Benzene, toluene	Menstrual dysfunction, anemia
Carbon monoxide	Fetal death, brain damage
Ozone	Abortion, birth defects
Ionizing radiations	Genetic mutations, birth defects, embryonic, fetal, neo-natal death, growth stunting, mental retardation
Thalidomide	Birth defects
Diethylstilbesterol	Vaginal adenocarcinoma in offspring
Alcohol	Growth retardation, mental retardation

on reproduction for several mammalian species. The period of greatest sensitivity to teratogens occurs during organogenesis, a time of rapid cell proliferation and differentiation. Organ-specific malformations are produced mostly by a teratogen given during the embryonic development of that organ. Sterility results from damage to gametes, malformations from damage during in utero development and functional and growth abnormalities when exposure to toxic agents occurs during later fetal development. The spontaneous incidence of malformations in humans is 2-3%, most of which are minor, non-life-threatening types. In contrast, other species, such as the rabbit, have a spontaneous malformation incidence of 10%, with about 1% being major abnormalities.

The first attention given to teratogenic agents occurred in 1941 when the association of death, blindness and deafness was seen in the offspring of women exposed to rubella (German measles) during pregnancy. About 200,000 birth defects are recorded in the U.S. each year, while about 550,000 spontaneous abortions, stillbirths, miscarriages and infant deaths are caused annually from defective fetal development. About 200 chemicals are known or suspect teratogens or cause fetal toxicity in humans (Table 3.6).

The response of mammals to teratogens is highly variable. The teratogenic dose for thalidomide in the monkey and rat is 10 and 50 times greater, respectively, than in humans. Thus, the teratogenicity of thalidomide in humans was woefully underestimated even when based only on studies in animals. About 20% of all malformations in humans are due to known genetic disease, 3-5% from chromosomal aberrations, 2-3% from infections such as rubella, cytomegalovirus, herpes, and syphilis, 1-2% from endocrine imbalance, and 2-3% from drugs; the remaining 60-70% of malformations are due to unknown causes.[14]

Toxicological evaluation of reproduction often considers the affects of chemicals on gametogenesis, libido, fertilization, implantation, embryonic growth and survival, fetal growth and survival, neonatal growth and survival, lactation and post-weaning growth and maturity.

References

1. E. C. Foukes and P. B. Hammond. 1975. Toxicology of the kidney. In, "Toxicology. The Basic Science of Poisons," Macmillan Publishing Co., Inc., NY, NY, pp. 190-200.
2. A. M. Rappaport. 1969. Anatomic considerations. In, "Disease of the Liver," J. B. Lippincott Co., Inc., Philadelphia, PA, pp. 1-49.
3. G. L. Plaa. 1975. Toxicology of the liver. In, "Toxicology: The Basic Science of Poisons," Macmillan Publishing Co., Inc., NY, NY, pp. 170-189.
4. L. Tomatis, C. Partensky and R. Montesano. 1973. The predictive value of mouse liver tumor induction in carcinogenicity testing-A literature survey. Int. J. Cancer 12:1-20.
5. R. R. Suskind. 1977. Environment and the skin. Environ. Health Perspectives 20:27-37.
6. H. I. Maibach, R. J. Feldmann, T. H. Milby and W. F. Serai. 1971. Regional variation in percutaneous penetration in man. Environ. Health 23:208-211.
7. R. P. Smith. 1975. Toxicology of the formed elements of the blood. In, "Toxicology. The Basic Science of Poisons," Macmillan Publishing Co., Inc., NY, NY, pp. 225-243.
8. A. Budy. 1975. Toxicology of the Skeletal System, In, "Toxicology. The Basic Science of Poisons," Macmillan Publishing Co., Inc., NY, NY, pp. 244-260.
9. H. A. Tilson and C. L. Mitchell. 1982. Models of neurotoxicity. Review Biochem. Toxicol. 2:265-294.

10. A. M. Potts and L. M. Gonasun. 1975. Toxicology of the eye. In, "Toxicology. The Basic Science of Poisons," Macmillan Publishing Co., Inc., NY, NY, pp. 275–309.
11. F. Speer. 1968. Tobacco and the nonsmoker. A study of subjective symptoms. Arch. Environ. Health 16:443–446.
12. J. M. Jacobs. 1982. Vascular permeability and neurotoxicity. In, "Nervous System Toxicology," Raven Press, NY, NY, pp. 285-298.
13. H. A. Tilson and G. J. Harry. 1982. Behavioral principles for use in behavioral toxicology and pharmacology. In, "Nervous System Toxicology," Raven Press, NY, NY, pp. 1–27.

CHAPTER 4

INHALATION TOXICOLOGY

History

In the late 1800s, scientists realized that disease could be transmitted by bacteria, through the air, and started to investigate the particulate matter suspended in air. J. Tyndall published "Essays on the Floating Matter of the Air" in 1882, and M. Arnold published "Untersuchungen uber Staubinhalation und Staubmetastase" in 1885. These were only a few of the pioneers in inhalation research up to the early 1900s, when it was realized that the lung disease of workers in dusty environments, such as mines and foundries, was related to particulate inhalation exposures.

During the 1920s, many studies on the inhalation, retention, and toxicology of dusts were conducted. One of the first was published by J. P. Baumberger in 1923: "The Amount of Smoke Produced by Tobacco and Its Absorption in Smoking as Determined by Electric Precipitation." It was not until the early 1950s that inhalation toxicology became a well-defined research discipline.

Pulmonary Structure and Function

The lung is comprised of a varied number of cell types; these include cells of the conducting airways (mucus-secreting goblet cells, ciliated cells, Clara cells, and endocrine cells of the bronchial-bronchiolar epithelia) and cells of the alveoli (attenuated type I and cuboidal type II epithelial cells, and alveolar macrophages).

The respiratory tract is divided into the nasopharyngeal, tracheobronchial and alveolar regions, each lined by specialized epithelial cells. Ciliated cells predominate over goblet cells in the nasopharyngeal region and the trachea; ciliated cells become less numerous towards the terminal bronchioles, while the alveoli are devoid of both ciliated and goblet cells. The human lung contains 23 branching generations from the trachea to the alveoli.[1] The thin cytoplasm of type I alveolar cells is only about 0.2 microns in thickness as it lines the alveolar walls, providing a short diffusion distance for oxygen to adjuncting capillaries filled with red blood cells. The much less frequent, cuboidal, type II alveolar cells produce a surface-active lipoprotein, called surfactant, that assists in keeping alveoli inflated. The alveolar macrophages, originating from blood monocytes, remove inhaled particulate material by phagocytosis and migrate to the terminal bronchioles where they are carried by mucociliary flow up the conducting airways to the pharyngeal region, where swallowing occurs.

The principal physiological function of the lung is gas exchange, which takes place in the alveoli. The blood-air barrier (attenuated type I cell cytoplasm lining + attenuated capillary lining with separating interstitial space) is normally only about 0.5 micron thick.

The lung volume is divided into fractions for physiological study. The tidal volume is the volume of air inspired or expired during each normal breathing cycle. The inspiration capacity is the maximum volume of air that can be taken into the lung. The vital capacity is the maximum volume of air that can be expelled from the lung by forceful effort following a maximal inspiration. The residual volume is the volume of air remaining in the lung following maximum expiration. Total lung capacity is equal to vital capacity + residual volume. The lung volumes of a healthy, resting male at sea level are 3600 ml inspiration capacity, 1200 ml expiration reserve volume, 4800 ml vital capacity, 1200 ml residual volume and 6000 ml total lung capacity. Normal resting tidal volume is 500 ml; with a breathing frequency of 12 breaths per minute, this gives a minute volume of 6000 ml.

The partial pressure of oxygen is about 100 mm Hg in alveolar air at sea level, decreasing to <20 mm Hg at the level of the cell. The partial pressure of carbon dioxide in expired air is about 40 mm Hg. Arterial blood is 97% saturated with oxygen at an alveolar, partial oxygen pressure of 100 mm Hg.

Deposition and Fate of Inhaled Particulates

The respiratory tract is the most common pathway by which toxic substances cross a body surface and penetrate living cells. Each day the average person breathes 20,000 liters of air laden with particles of various sizes and types, many of which are deposited on an alveolar surface of about 70 m². The deposition and distribution of inhaled particles depends on the physical-chemical characteristics of the aerosol, including particle size, density, shape and solubility. Aerosols of widely distributed particle sizes are continually encountered in the environment (Figure 4.1). For determining deposition in the lung, the most useful and predictive

4 – Inhalation Toxicology

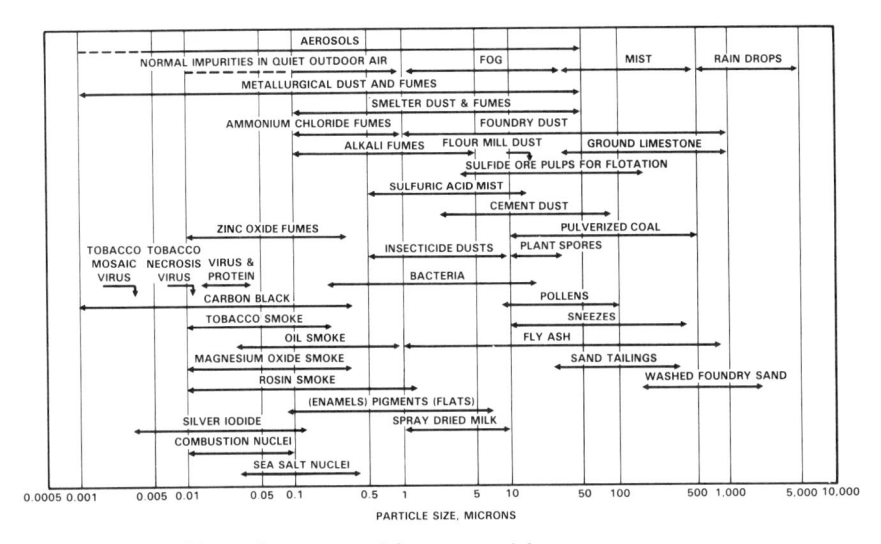

FIGURE 4.1—Sizes of common airborne particles.

characteristic of an aerosol is its frequency distribution of aerodynamic diameters. Aerosols are statistically described in terms of distribution of particle numbers and surface area (mass and shape), as a function of aerodynamic diameter (Figure 4.2). Many heterodisperse aerosols fit a lognormal size distribution, in which a plot of the number of particles in a given size interval versus the log of particle size produces a distribution curve that may be described by the Count Median Diameter (CMD) and the Geometric Standard Deviation (GSD, ratio of the diameter of the particles that are larger than 84.2% of all others to the CMD).[2]

Aerosols are characterized as to total mass concentration, mass concentration in the respirable size range, and particle size distribution. These are determined by the use of cascade impactors, elutriator aerosol spectrometers or by direct measurement on photographs, as well as by solubility, particle shape and surface morphology, and chemical and mineral composition. The distribution of the toxic components within the total mass distribution of the aerosol must also be considered. For example, Natusch and Wallace[34] described the concentration of trace elements in the small-particle fraction of fly ash; this concentration of toxic trace elements arises through their adsorption onto the surface of particles in high-temperature processes like combustion of fossil fuels.

The Task Group on Lung Dynamics[69] divides the human respiratory tract into three regions: the nasopharynx, the tracheobronchial tree and

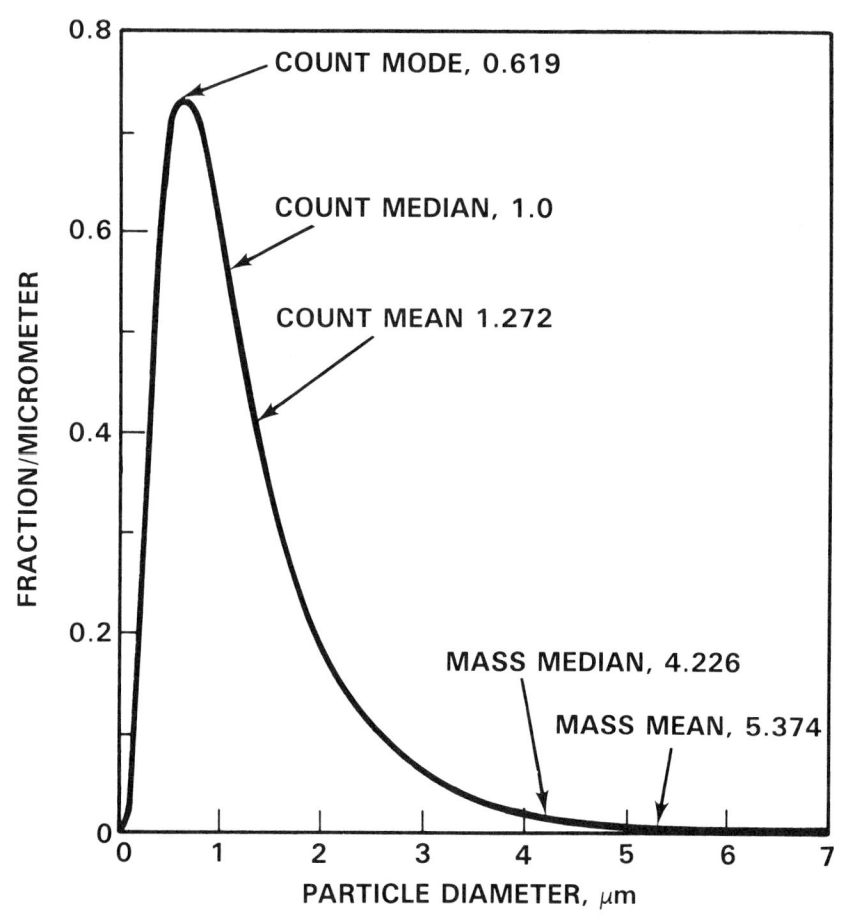

FIGURE 4.2 — Linear form of a log-normal distribution of aerosol particle sizes for count median diameter equal to 1.0 and geometric standard deviation equal to 2.0 (O. G. Raabe. 1971. Particle size analysis utilizing grouped data and the log-normal distribution. Aerosol Science 2:289).

the pulmonary or deep lung region. A model of aerosol deposition for each region is then developed as a function of aerosol aerodynamic diameter for several tidal volumes. The Task Group also noted that mouth breathing substantially increases pulmonary deposition compared to nose-only breathing.

Particles are removed from inhaled air to the respiratory tract by sedimentation as a result of gravity; by inertial impaction and intervention as a result of particle velocity, separately from gravity; by electrostatic

4 – Inhalation Toxicology

attraction between particles and conducting airway surfaces; and by Brownian motion, resulting from collision of very small particles or molecules with gas molecules. Hygroscopicity of particles, exercise, mouth or nose-breathing, and the presence of pulmonary disease influence the deposition of particles in the lung.

Only particles <10 µm in diameter are considered in the respirable range; exceptions are long fibers, such as those of asbestos, whose long axis may be >10 µm. Particles in the range of 0.1 µm to 2.0 µm are often of greatest interest in air pollution studies since this particle size range has the greatest deep lung penetration and retention (Figure 4.3).

Particles deposited on the mucus lining the nasal or the tracheobronchiolar regions are rapidly cleared from the lung by mucociliary flow. Retention half-times for inhaled polystyrene, iron oxide or manganese oxide, in humans, range from 2–3 minutes for particles deposited on the trachea, 20–30 minutes for upper bronchial deposition, 1–6 hours for lower bronchiolar deposition and 1–2 months for alveolar deposition.[4] Particles deposited in the alveoli are rapidly phagocytized by macrophages and, to a lesser degree, by type I alveolar epithelium (Figures 4.4 and 4.5). The phagocytosis of particles by macrophages decreases the probability of phagocytosis by the alveolar epithelium, thereby increasing the clearance rate of deposited particles. Irritation or high dust loads in the alveoli stimulate the migration and accumulation of macrophages in the alveoli

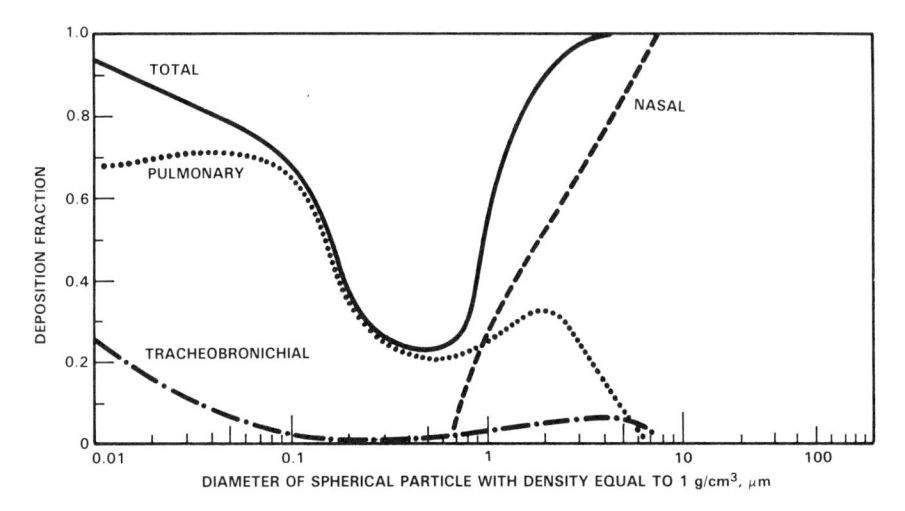

FIGURE 4.3 — Deposition of inhaled particles in the total lung and in regional areas of the human lung (O. G. Raabe. 1979. Deposition and clearance of inhaled aerosols. DOE Report UCD–472–503, NTIS, Springfield, VA).

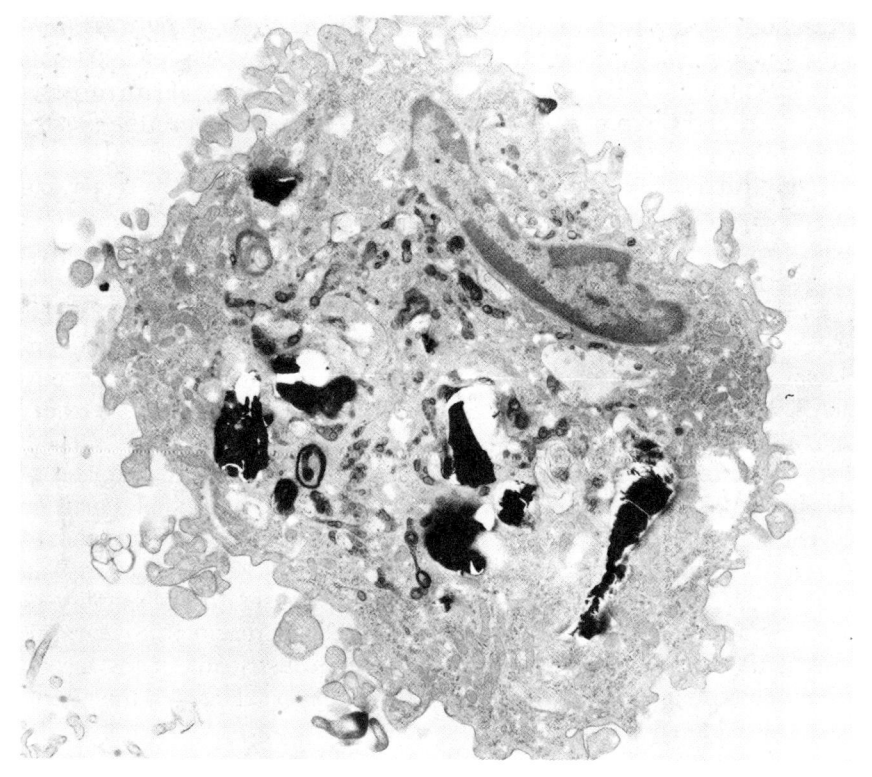

FIGURE 4.4 — Phagocytosis of Mount St. Helens volcanic ash by a rabbit pulmonary macrophage (photograph provided by Al Robinson, Battelle, Richland, WA).

and also increase penetration of particles into alveolar septae and pulmonary lymphatics.

Inhaled particles may exhibit significant solubilization after deposition in the alveoli, with passage of solutes from the air space, across the blood-air barrier into the capillary blood, and then into the tissues of the body. The International Commission on Radiological Protection (ICRP) has developed a lung model[3] which takes into consideration the site of particle deposition and rates of particle solubilization. The model classifies inhaled inorganic compounds according to residence time in the lung: Class D (days), Class W (weeks) and Class Y (years); clearly, Class Y compounds are the least soluble.

Potentially, inhalation provides the most damaging route of entry for toxic chemicals into the body. As percent of administered dose, metals are absorbed about 10 times more effectively from the lungs than from the

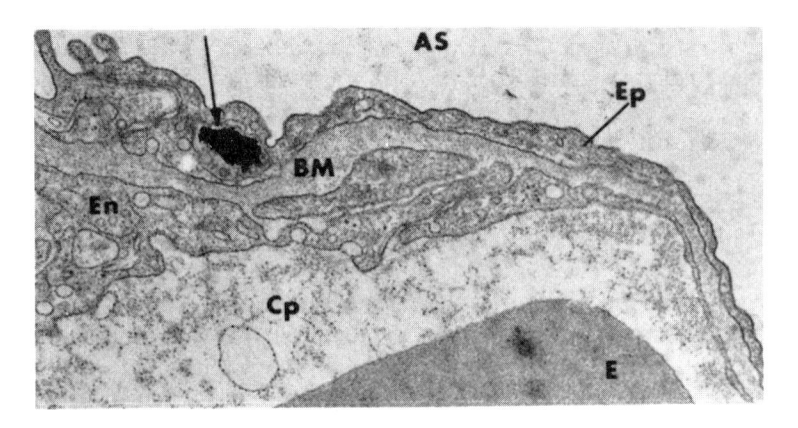

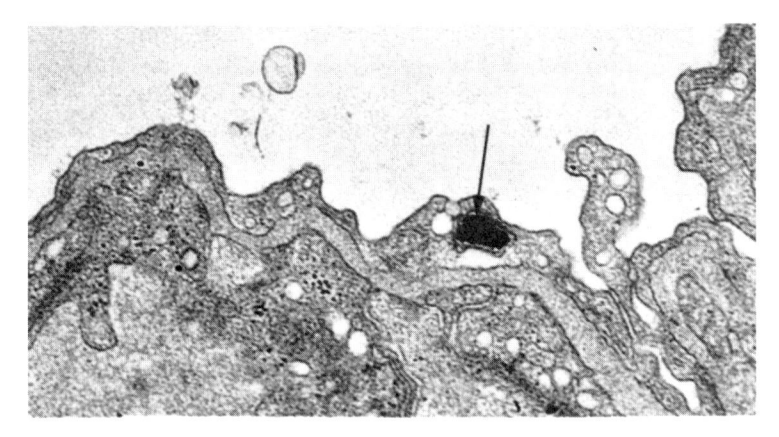

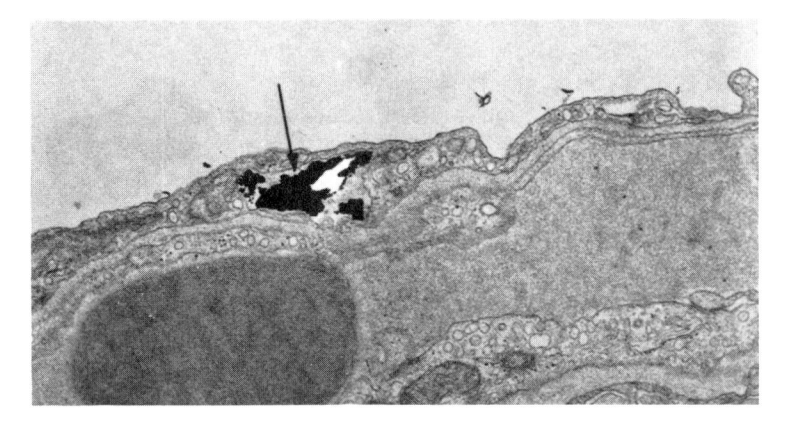

FIGURE 4.5—Engulfment of inhaled ^{239}PuO$_2$ particles by type I alveolar epithelium at 60 minutes, 90 minutes, and 7 days post-exposure (C. L. Sanders and R. R. Adee. 1970. Ultrastructural localization of inhaled ^{239}PuO$_2$ particles in alveolar epithelium and macrophages. Health Physics 18:293–295).

gastrointestinal tract. In humans, the surface area of the lung is about 35 times greater than the surface area of the skin, making pulmonary absorption much more likely than percutaneous absorption, irrespective of anatomical differences that make skin absorption less likely.

Pulmonary Pathology

The respiratory tract is constantly exposed to a wide variety of toxic agents. The lung is the portal of entry for toxic chemicals that damage other organs more significantly than they damage the lung (e.g., uranium or cadmium in kidney). Conversely, some chemicals (e.g., paraquat), when given systemically, cause the greatest damage in the lung. A wide variety of possible chemical, biological, and physical factors are associated with the development of noncancerous respiratory disease. Many chemicals/pharmacological agents exhibit significant side effects in the lung, particularly with respect to the induction of latent interstitial lung disease (Tables 4.1 and 4.2).

A list of some important pulmonary pathological terms, and their definitions, is as follows:

- Acute upper respiratory disease – "common cold", probably due to virus infection.
- Bronchopneumonia – inflammatory exudate extending primarily along bronchi and adjacent alveoli.
- Lobar pneumonia – inflammatory exudate filling the alveoli of a lobe(s).
- Bronchial constriction – narrowing of the airways of the upper respiratory tract.
- Bronchitis – inflammation of the bronchi.
- Bronchiolitis – inflammation of the bronchioles.
- Bronchiolitis obliterans – blockage of inflamed bronchioles by inflammatory, fibrosing, and proliferative reactions, often leading to emphysema.
- Atelectasis – collapse of the alveoli.
- Bronchiectasis – dilation of bronchi or bronchioles due to infection or other conditions (e.g., mucoviscidosis of cystic fibrosis).
- Pulmonary edema – fluid accumulation in intercellular tissue spaces due to increased capillary pressure, lowered osmotic pressure or damage to capillaries.
- Pulmonary alveolar lipoproteinosis – accumulation of lipoprotein in the alveoli due to hyperactive and hyperplastic type II alveolar epithelial cells.
- Pulmonary thrombosis – obstruction of pulmonary blood vessels due to blood clots or other materials in blood.

TABLE 4.1. Factors associated with the development of non-cancerous respiratory disease.

CHEMICAL FACTORS

Air pollutants
  Sulfur, nitrogen and carbon oxides
  Ozone
  Metals
  Quartz
  Fly ashes
  Hydrocarbons
  Asbestos
  Tobacco Smoke
  Other particulates
  Pesticides
Pollutants of water, food and drugs

BIOLOGICAL FACTORS

Viruses
Bacteria
Fungi
Allergens
Genetic predisposition

PHYSICAL FACTORS

Sunlight
Temperature
Humidity
Ionizing Radiation
Electrostatic charge
Oxygen pressure

- Emphysema – dilation of alveoli and destruction of alveolar septa leading to large air sac formation in pulmonary tissues and decreased pulmonary surface area.
- Pulmonary fibrosis – increase in interstitial connective tissues, such as collagen, and replacement of pulmonary endothelium and epithelium with scar tissues.
- Asthma – bronchoconstriction due to release of vasoactive amines, such as histamine, following inhalation of an allergen or irritant.
- Granuloma – consolidated inflammatory response composed of

TABLE 4.2. **Drugs known to produce disease of the human respiratory tract** (E. C. Rosenow. 1972. The spectrum of drug induced pulmonary disease. Ann. Internal Medicine 77:977–991).

ANTIBIOTICS

Colistin
Gentamicin
Kanamycin
Neomycin
Nitrofurantoin
Para-aminosalicyclic acid
Penicillin
Polymyxin B
Streptomycin
Sulfonamides

CHEMOTHERAPEUTIC AGENTS

Azothiopurine
BCNG
BCNU
Bleomycin
Busulfan
Chlorambucil
5-Fluorouracil
Melphalan
Methotrexate
Methyl CCNU
Mitomycin
Neocarzinostatin
Procarbazine
Chlordiazepoxide

ANALGESICS

Aspirin
Heroin
Indomethacin
Methadone
Propoxyphene

ENDOCRINE AGENTS

Corticosteroids
Oral contraceptives

INHALANTS

Acetylcysteine
Cromolyn sodium
Isoproterenol
Mineral Oil
Oxygen

NEUROACTIVE AND VASOACTIVE AGENTS

Diphenylhydantoin
Hexamethonium
Mecamylamine
Methysergide
Pentolinium
Propranolol

INTRAVENOUS

Blood
Lymphangiograms-contrast media
Drug contaminants

MISCELLANEOUS

Anticoagulants
Apresoline
Carbomazepine
Chlorpromazine
D-penicillamine
Furodantin
Gold
Mephenesin
Paraquat
Procainamide
Proctolal

cellular nodules (mixed lymphocytes, macrophages, plasma cells and epithelioid cells).
- Alveolitis – inflammation of the alveoli.
- Pneumonitis – inflammation of the pulmonary region of the lung.

In addition to bacterial and viral infections in the lung, several pre-existing mycotic infections may be exacerbated following exposure to toxic agents. These include coccidioidomycosis, a usually mild and undetected infection; aspergillosis, usually seen following immunosuppression; histoplasmosis, a transitory infection of reticuloendothelial cells; actinomycosis, a common "pathologic" infection in the lung; and yeast infections with cryptococcosis and moniliasis, often found as complicating infections following immunosuppression with cancer therapies.

Chronic bronchitis is characterized by excessive mucus production in the upper airways. Cigarette smoke is much more effective in causing chronic bronchitis than is air pollution; the latter has a more significant impact on nonsmokers. In chronic obstructive disease that interferes with ventilation in the lung, the motion of air in and out of the lung is obstructed by partial closure of the airways. Even children living in highly polluted urban areas may experience increased airway resistance due to pulmonary obstructive disease.[7] Emphysema is also much more prevalent among smokers, possibly due to the release of elastase or inactivation of alpha-antitrypsin.[7]

Upper respiratory infections, influenza, inflammation of the oral cavity and sinuses and hay fever are the most frequent reasons for visiting physicians in the U.S. More Americans suffer from hay fever (9% of the population) than from any other medical condition requiring treatment by a physician. Respiratory diseases, often serious enough to cause hospital admission, are: pneumonia, chronic bronchitis, emphysema, asthma, and a series of other pulmonary diseases (Table 4.3). Chronic obstructive pulmonary disease (bronchitis, emphysema and asthma) are the fifth leading cause of death in the U.S., which may partially reflect an increasing exposure to environmental pollutants. The rate of appearance of these diseases has steadily increased during the past 25 years; deaths due to chronic obstructive lung disease have doubled about every 10 years, from 1935 to 1975.

The term "interstitial lung disease" refers to a large, diversified group of chronic respiratory diseases characterized by alveolitis, accumulation of inflammatory and immunogenic cells in the alveoli and, eventually, by interstitial fibrosis. These disorders are chronic, progressive, debilitating and often fatal. Most of the 100 or more interstitial lung diseases are produced by the inhalation of inorganic and organic aerosols and vapors. Etiology is known in only 70% of all cases of interstitial diseases.[6] Pulmonary fibrosis in interstitial diseases affects primarily the alveolar

TABLE 4.3. Measures of respiratory disease frequency (in thousands) in the United States. The number of physician visits was determined for 1975–1976 using principal diagnoses only (NIAID Task Force Report. 1979. NIH Publ. No. 79-387, Washington, D.C.). The remaining data were obtained from unpublished sources; World Health Organization Annual Epidemiological and Vital Statistics for 1979, Hospital Discharge Survey, NCHS for 1977, and National Center for Health Statistics, the Health and Nutrition Examination Survey for 1971–1973.

Disease	Number of Visits to Physician	Prevalence in Health Interviews	Hospital Discharges	Percent Adults with Disease
Acute upper respiratory infections	93,000	–	–	–
Influenza	10,000	–	–	–
Pneumonia	5,200	–	300	–
Chronic bronchitis	1,600	7,100	320	4.1
Emphysema	5,200	2,100	–	–
Asthma	11,000	6,000	200	3.5
Pharyngitis and nasopharyngitis	2,500	–	–	–
Sinusitis	8,300	23,000	–	–
Hay fever	17,000	16,000	–	8.7
Pneumoconioses	–	320	7	–
Other interstitial disease	–	130	24	–
Cystic fibrosis	–	–	11	–
Pulmonary infarction	–	–	92	–
Pulmonary heart disease	–	–	10	–
Sarcoidosis	–	–	12	–
Acute pulmonary edema	–	–	28	–

region of the lung. Fibrosis is characterized by deranged interstitial collagen and profound changes in parenchymal cell populations. Pneumoconiosis designates a fibrotic lung disease; the character of the fibrosis is dependent on the composition and the amount of inhaled dust.

There are also several specific, occupational allergic or hypersensitive pulmonary diseases that have an immunological basis. Allergic pneumonitis may occur almost immediately following exposure or several days after exposure to the offending agent.

Infectivity Models

The resistance of the respiratory tract to infection is a result of a variety of factors, including pathogenicity, stress, pre-existing disease, diet and immunocompetency. Air pollutants impair the ability of the host to resist infection. Animal model systems, termed infectivity models, have been developed to test the influence of air pollutants on microbe-host interactions in the lung. The models reflect the varied responses of the respiratory tract to air pollution, which may increase the infectivity of a microbe.

In evaluating infectivity, animals are exposed either to filtered room air or to a test air pollutant, such as ozone or cadmium. After exposure to the air pollutant, the animals are exposed to an aerosol of a pathogenic microorganism, such as streptococcus pyogenes, a species of Diplococcus or Klebsiella or to an influenza virus. The test endpoint is usually death within a period of 2–4 weeks. To be most sensitive as a test system, the microorganisms must cause low mortality in control animals and the pollutant must cause a dose-related increased mortality in test animals.[8] Infectivity models have been developed for mice, rats, hamsters and squirrel monkeys; mice are the most sensitive and monkeys the least sensitive to this type of test.

One model has measured increased mortality in mice exposed to oxidant gases or metal compounds over a 15-day period following inhalation of S.pyogenes.[8] Strains of this bacterium are the causative agents of tonsillitis, scarlet fever, pneumonia, erysipelas, and septicemia in humans; they were also often implicated in deaths from pneumonia prior to the use of antibiotics. In mice, death due to S.pyogenes involves necrosis of lung tissue, hemorrhaging and thrombi formation in lung, dehydration due to esophageal infection and, occasionally, to myocardial lesions. S.pyogenes is an intracellular pathogen, that causes marked cytotoxicity to alveolar cells (Figure 4.6). The alveolar macrophage, because of its ability to phagocytize bacteria, is particularly sensitive to damage from air pollutants with respect to subsequent infectivity from inhaled bacteria.[9]

Increased infectivity has been observed in experimental animals following exposure to ozone; nitrogen dioxide; sulfur dioxide; nickel, platinum and cadmium salts; cigarette smoke and automobile exhaust. The use of these animal models that mimic human disease conditions should significantly increase the ability to assess human risk from airborne contaminants.

Asbestosis

Asbestos is a generic name for a group of hydrated mineral silicates that occur naturally in a fibrous form. The subdivision of asbestos into fine fibers by longitudinal fracturing, produces particulate material that is readily

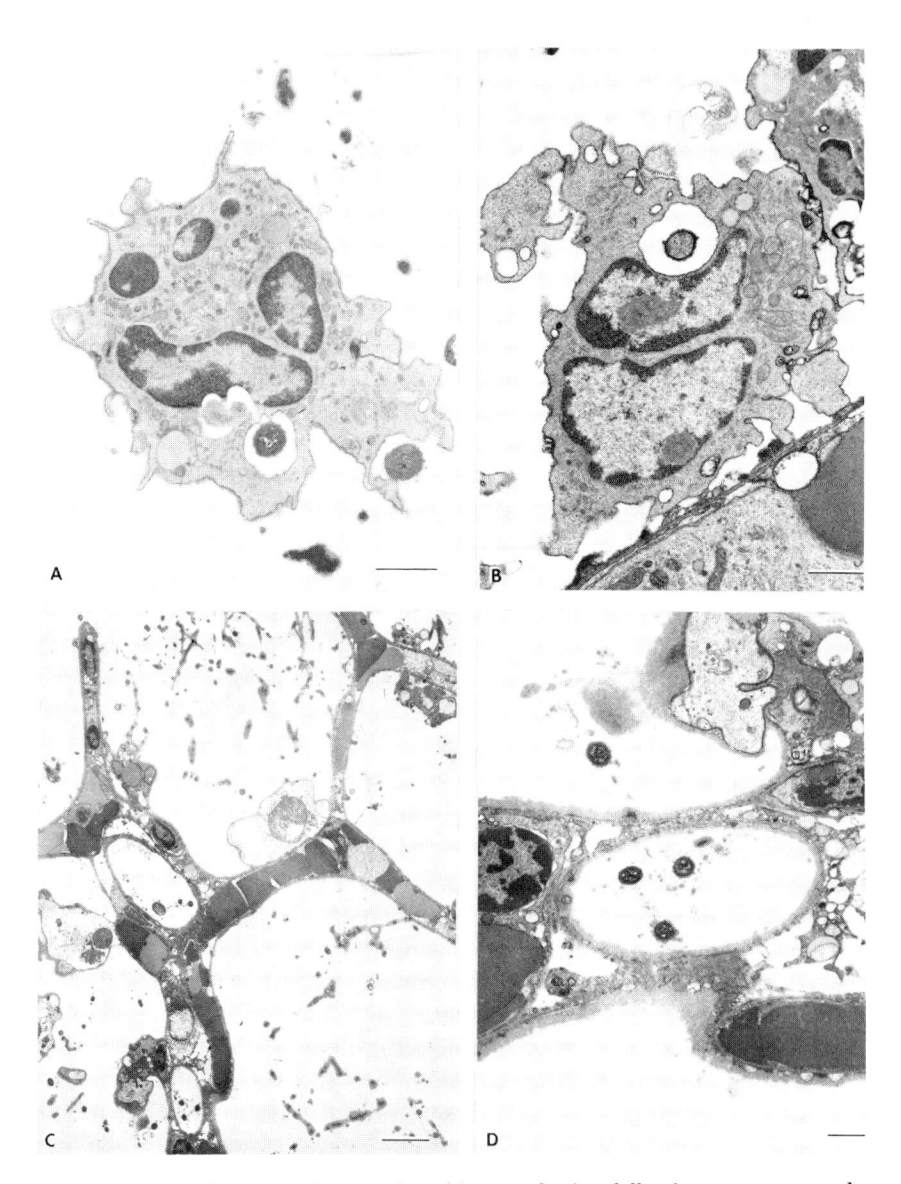

FIGURE 4.6 — Electron micrographs of lungs of mice following exposure and infection with Streptococcus pyogenes group C bacteria. (A) Polymorphonuclear leukocyte in various stages of phagocytizing bacteria; Bar = 1 μm. (B) Macrophage with engulfed bacteria; clear areas around phagocytized bacteria are indicative of the bacterial cytotoxicity. Bar = 1 μm; (C) Area of severe alveolar damage due to proliferation of bacteria. Bar = 5 μm; (D) Higher magnification of (C), showing intracellular proliferation of bacteria in alveolar septa. Bar = 1 μm.

dispersed in air. Asbestos may be serpentine chrysotile or an amphibole: actinolite, amosite, anthophyllite, crocidolite or tremolite. More than 90% of the asbestos used in the U.S. is chrysotile [Mg$_5$Si(OH)$_4$], containing small amounts of aluminum, iron and various trace elements.

Inhaled asbestos fibers are phagocytized by macrophages and taken into phagolysosomes.[10] A chemical reaction may take place between the asbestos fiber and the phagolysosomal membranes, releasing hydrolytic enzymes into the cell cytoplasm and causing cell cytotoxicity due to autodigestion. The change in membrane permeability appears to be related only to the incomplete phagocytosis of long fibers, where localized discontinuity of the cell membrane leads to liberation of enzymes, cytotoxicity and the formation of polykaryotic giant cells[11] (Figure 4.7). Cytotoxicity of asbestos fibers is a function of fiber length; fibers <5 μm are considered relatively nontoxic, while fibers >10 μm are the most toxic.

The migration of fibers through alveolar tissues and redistribution in subpleural and submesothelial tissues may be either mechanical or mediated by macrophages. Nearly 90% of inhaled asbestos fiber is cleared from the lung within two weeks after exposure; the half-life of the remaining asbestos is >100 days.[13] Fibers of <5 μm in length are rapidly cleared from the lung; fibers >10 μm length are preferentially retained. A fibrous material is seen on the surface of asbestos fibers following initial contact with macrophages. Protein adsorption into fibers is the first stage in ferruginous body formation, involving only some long fibers. An iron-containing protein coats the fibers; in rats, this protein is discontinuous, frequently occurring in the form of globules and often bizarre forms (Figure 4.7).

Chrysotile asbestos is more fibrogenic than amphibole type. Little if any pulmonary fibrosis is produced by asbestos fibers <5 μm length, while an intense pulmonary tissue reaction and fibrogenesis is seen around fibers >10 μm length. As fibrogenesis progresses, the fibrosis may spread from areas around concentrations of long fibers into diffuse lesions. The later phases of asbestosis are associated with thickening and calcification of the pleura, usually from 10–20 years after exposure.

Pleural mesothelial hyperplasia and malignant mesothelioma formation, invariably fatal, are seen after a latent period of >25 years. Mesotheliomas may develop in the pleural cavity, or in the peritoneal cavity after translocation of inhaled fibers or oral ingestion of fibers. Exposure to asbestos, no matter how brief, or whether industrial, occupationally or otherwise, is well-correlated with mesothelioma induction.[14] The risk of mesothelioma for the nonexposed general public is about 1 case per million; the risk in some asbestos mining operations has approached 100,000 cases per million.[16]

Asbestos and PAH may act synergistically as cocarcinogens in the induction of lung tumors.[15] The relative risk of developing lung carcinoma

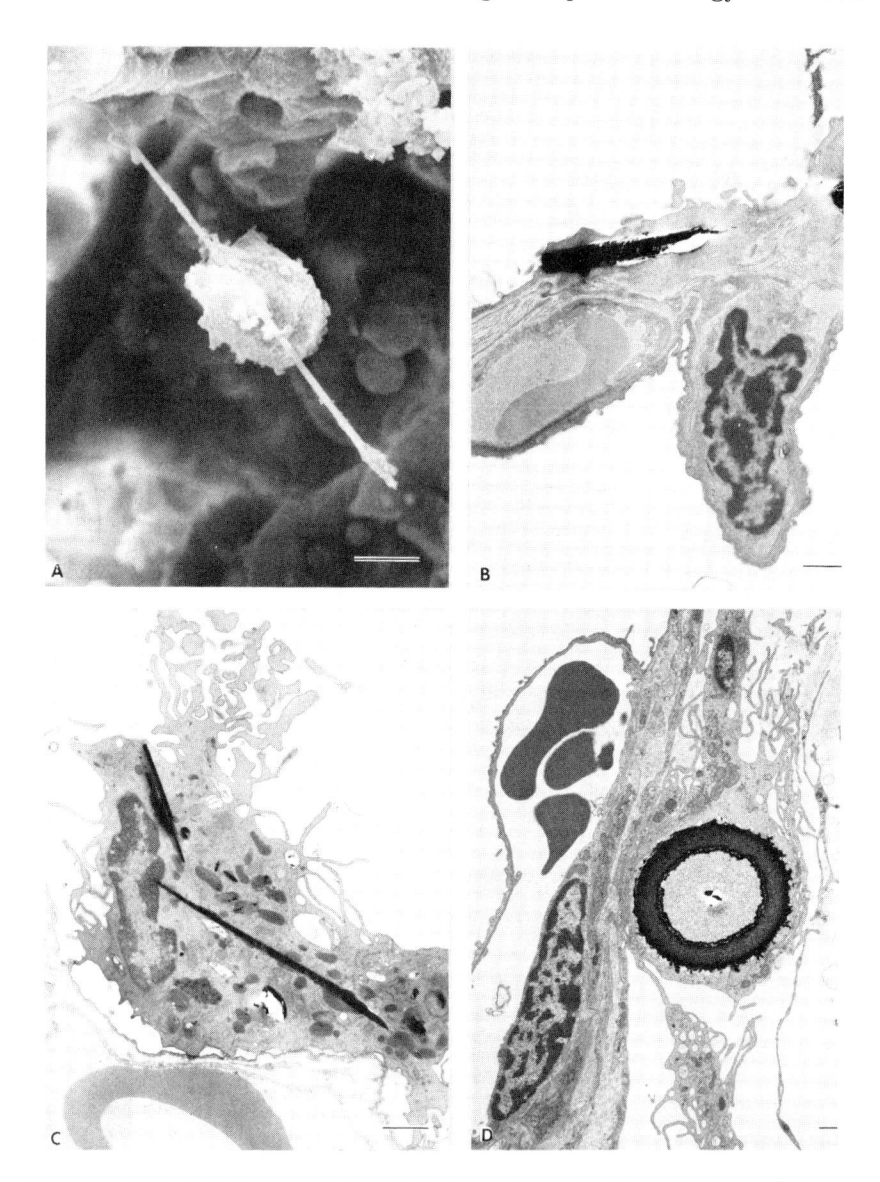

FIGURE 4.7—Cellular reactivity of the lung to crocidolite asbestos. (A) Partial engulfment of large fiber by a macrophage. Bar = 5 µm; (B) Partial engulfment of large fiber by type I alveolar epithelium. Bar = 1 µm; (C) Complete engulfment of small fibers by a macrophage. Bar = 1 µm; (D) Unusual appearing ferruginous body in alveolar septa. Bar = 1 µm.

may be as high as 80 times expected risk in asbestos workers who also smoke cigarettes.[50] Based upon studies of asbestos insulation workers in the U.S., exposure to asbestos and cigarette smoke synergistically increases the incidence of lung cancer (Table 4.4).

Silicosis

Pulmonary disease produced by inhalation of dusts was first mentioned by Agricola in 1556 in his TREATISE ON MINING. Silicosis, a fibroproliferative inflammatory disease, is produced by prolonged or heavy exposure to crystalline silica particles. The name silicosis is derived from the Latin *silex*, meaning flint, and was first applied to this disease by Visconti in 1870.[17] Crystalline SiO_2 occurs in nature as quartz, as tridymite hexagonal crystals or as cristobalite cubic crystals. In comparison, the structure of silica in silicate rocks, such as granite or sandstone, consists of tetrahydral crystals. The uncombined forms of SiO_2 are referred to as free silica or crystalline silica. Quartz, tridymite and cristobalite are all highly fibrogenic in the lung. Other forms of crystalline silica, such as keatite, coesite and stishovite, as well as opal, diatomaceous earth and fused

TABLE 4.4. Age-standardized lung cancer death rates for cigarette smoking, occupational exposure to asbestos, both or combinations thereof (E. C. Hammond, I. J. Selikoff and H. Seidman. 1979. Asbestos exposure, cigarette smoking and death rates. Ann NY Acad. Sci. 330:474-490).

Group	Exposure To: Asbestos	Cigarettes	Mortality Ratio
Control	No	No	1.0
Asbestos Workers	Yes	No	5.2
Control	No	Yes	10.9
Asbestos Workers	Yes	Yes	53.2

silica are much less fibrogenic. Cryptocrystalline forms, such as flint or chert, however, are highly fibrogenic. Silicosis is an industrial hazard in mining, quarrying, stone cutting, cleaning and grinding and in the glass, pottery, foundry and vitreous enameling industries.

Alveolar macrophages and, to a lesser degree, alveolar epithelium, phagocytize inhaled silica particles; the particles are incorporated into phagolysosomes, whose membranes react with the particles, increasing their permeability and causing leakage of lysosomal enzymes into the cells; it is this autolytic process that is a significant part of silicosis reactions (Figure 4.8). The resultant death of cells with phagocytized silica particles, the release of lysosomal, hydrolytic enzymes, cell necrosis and continuing rephagocytosis of the particles are required to induce alveolar fibrosis. In acute silicosis there is a marked accumulation of neutrophils, along with protein and fibrin, into the alveolar air spaces. A later macrophage influx results in clearance of many particles and cell debris from the alveoli. Focal concentrations of silica particles and macrophages, along with other inflammatory cells, form granulomas within a few days after exposure. A dense macrophage accumulation, followed by fibroblastic proliferation, is seen within a few weeks. By a few months, the granulomas, comprised mostly of collagen, appear nearly acellular.

Pulmonary alveolar lipoproteinosis is seen early after exposure to quartz (Figure 4.9). This process, if severe, influences the development of pulmonary fibrosis. Alveolar lipids stimulate monocyte migration to the lung and the differentiation of these cells into macrophages.[18] The lipoproteinosis reaction is also seen, although less severely, following inhalation of less toxic dusts, such as volcanic ash particles. Alveolar lipoproteinosis is a nonspecific response to continuing irritation in the lung.[19]

Three theories have been proposed for the development of silicosis: solubility, autoimmune reaction and surface reaction of cell membranes. Most experimental data indicate that the toxic effects of silica particles are due to their surface reactivity with cell membranes. The toxic effects of quartz particles in the lung can be decreased by administration of the polymer, polyvinylpyridine-N-oxide (PNO). Its protective action is explained by the binding of PNO to hydrogen bonds on the quartz surface.[20]

Heppleston describes silicotic fibrogenesis as follows: "The accumulated evidence indicates that, after ingestion of quartz, macrophages form or release a factor which enables fibroblasts to produce excess hydroxyproline. Inhaled quartz stimulates type II epithelium to generate phospholipid, sometimes in such excess that alveolar lipidosis develops and inhibits fibrosis in rats mainly by isolation of quartz particles from macrophages. Extracted lipid administered parenterally to other rats provokes a marrow monopoiesis . . . . Prostaglandins, released from stimulated alveolar macrophages, may facilitate monocytic emigration by increasing capillary permeability. Passage of monocytes from pulmonary interstitium to

4 – Inhalation Toxicology

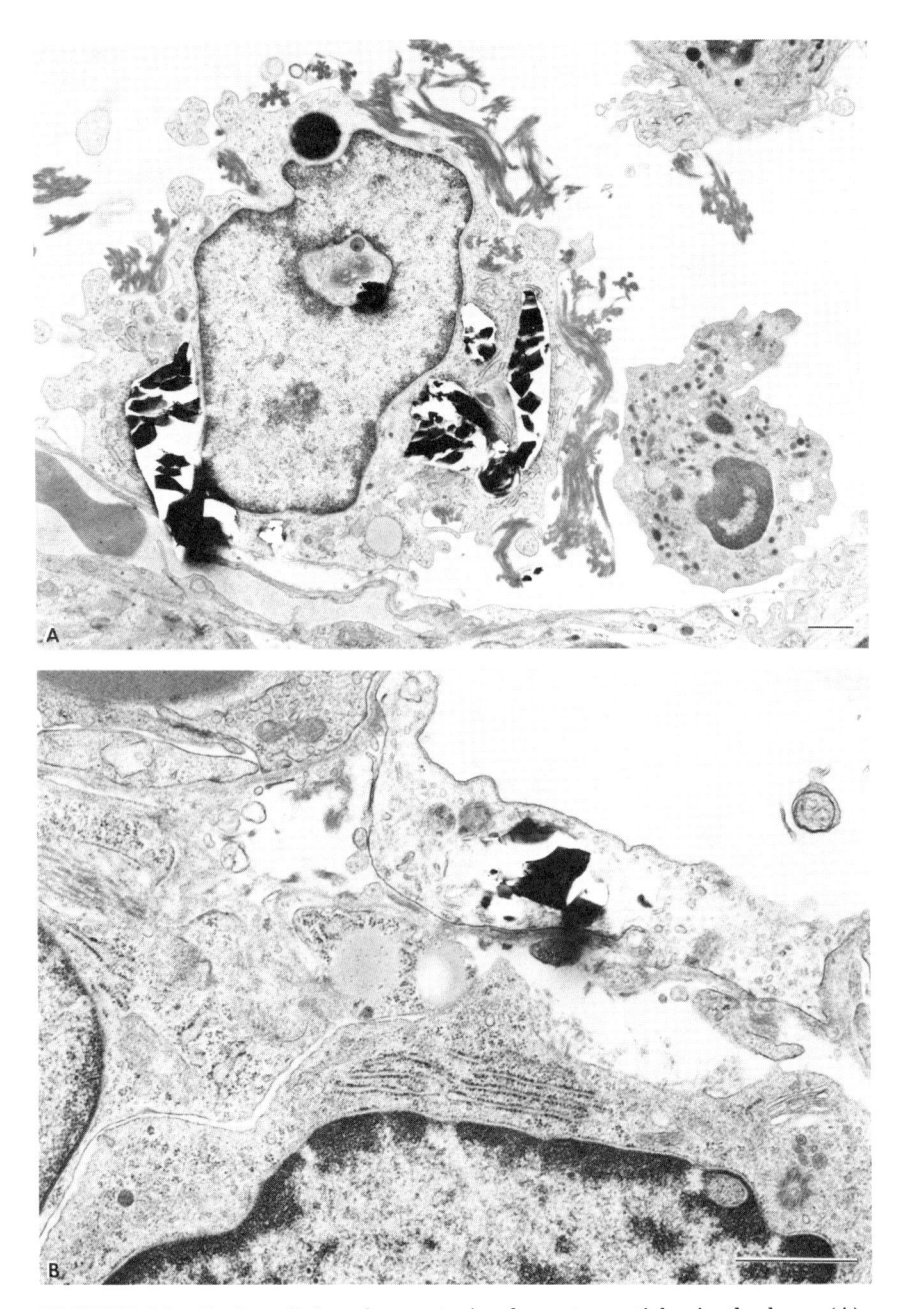

FIGURE 4.8 — Early cellular phagocytosis of quartz particles in the lung. (A) Phagocytosis of silica particles and fibrin by an alveolar macrophage after a heavy exposure. Bar = 1 µm; (B) Localization of silica particles in attenuated portion of a type I alveolar epithelial cell a few hours after exposure. Bar = 1 µm (C. L. Sanders. Unpublished).

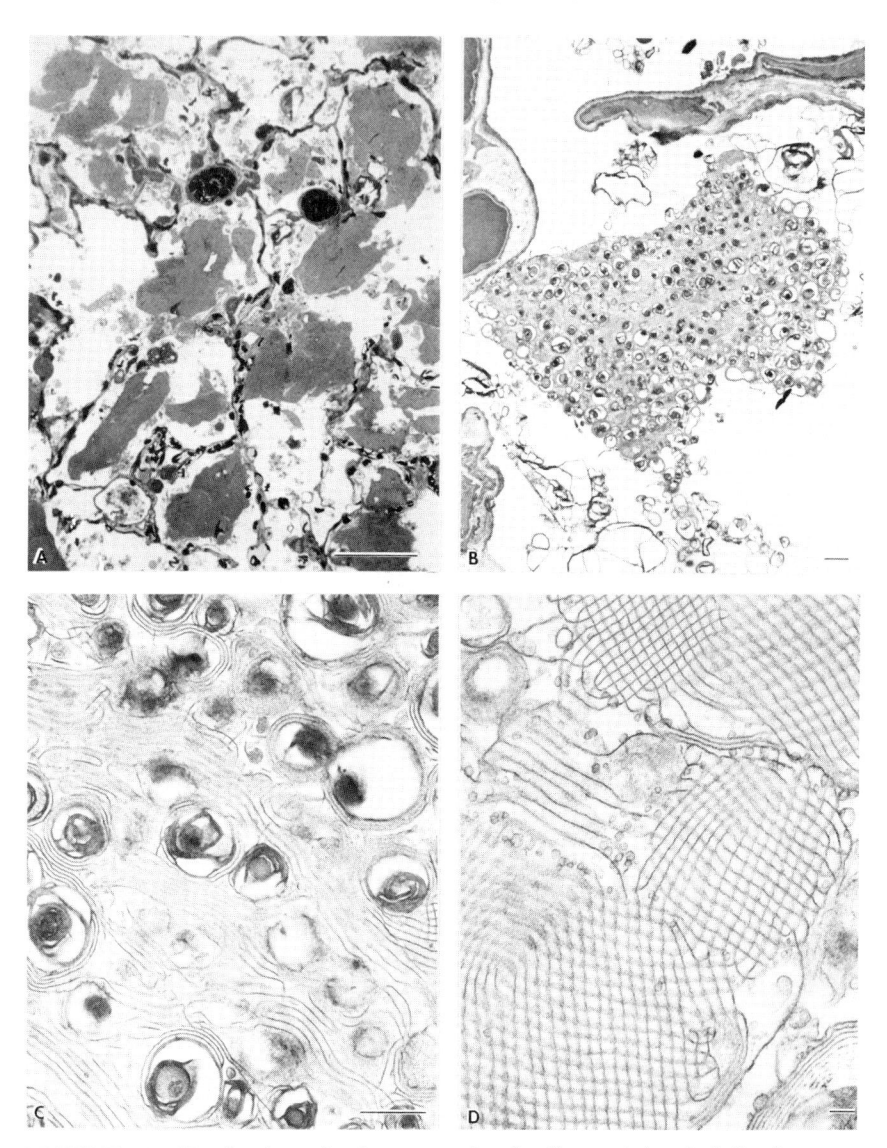

FIGURE 4.9 – Production of pulmonary, alveolar lipoproteinosis following exposure to a mixed silicate dust, Mount St. Helens volcanic ash particles. (A) Light micrograph of alveolar region. PAS stain, bar = 50 µm; (B) Electron micrograph showing lamellar bodies and membranous network material that originated from type II alveolar epithelial cells. Bar = 1 µm; (C) Higher magnification of B, showing the structure of the lamellar bodies. Bar = 0.5 µm; (D) Higher magnification of B, showing structure of the membranous network material, probably identical to surfactant. Bar = 0.1 µm (C. L. Sanders. Unpublished).

alveolar surface is likely simplified by disruption of the thin type I epithelium caused by quartz .... These interactions are integrated into a system that might be activated by agents other than silica and thus offer a wider concept of pulmonary fibrosis."[71]

Chronic silicosis occurs after protracted (20–40 years) exposure to silicogenic dusts; acute silicosis rarely occurs today. The clinical phase of chronic silicosis is associated with the abnormal radiographic appearance of the lung, fibrotic lesions, emphysema, shortness of breath and silicotuberculosis. Intractable tuberculosis tends to be a complicating factor in advanced or long-standing silicosis. Susceptibility to silicosis and infective pneumoconiosis is enhanced by cigarette smoking and consumption of ethanol. Exposure to silicogenic dusts has not been shown to cause lung cancer in humans, although several studies in animals, particularly in rodents, indicates that quartz is carcinogenic in the lung.[72] Silicosis in humans increases the carcinogenicity of cigarette smoke in the lung, probably due to the altered metabolism of smoke components by lung tissue.

Welder's Pneumoconiosis

About 1% of the U.S. working population engages in welding. High temperatures during welding operations melt the work piece and consume the wire or rod used to make the weld, producing vaporized metal fumes, vaporized flux and a complex mixture of gases and metals, including ozone, nitrogen dioxide, chromium, vanadium, arsenic, nickel, beryllium, lead and other metals.[21] In many of these cases, the air concentration of metal to which the welder is exposed exceeds the permissible occupational exposure levels for many individual pollutants.

In a study of 661 British shipyard welders, the incidence of small round opacities in lung X-rays was found to increase as a function of the number of years spent welding (Figure 4.10). Welders undergo a risk of developing opacities that is intermediate between those of coal miners and gold miners. In a study of 95 Italian welders, 92% exhibited chronic bronchitis; all showed bronchial squamous metaplasia, probably due to the presence of chromium and nickel in the welding fumes.[22] A number of epidemiological studies of welding populations has shown a statistically significant increase in lung cancer rates for welders and foundry workers which was not accounted for by differences in smoking habits; chromium, nickel and other metals in welding fumes are thought to be responsible for this increased lung cancer risk.[21,23]

Mixed-Silicate Pneumoconiosis

Mixed-silicate dusts, such as feldspar, fly ash from combustion of fossil

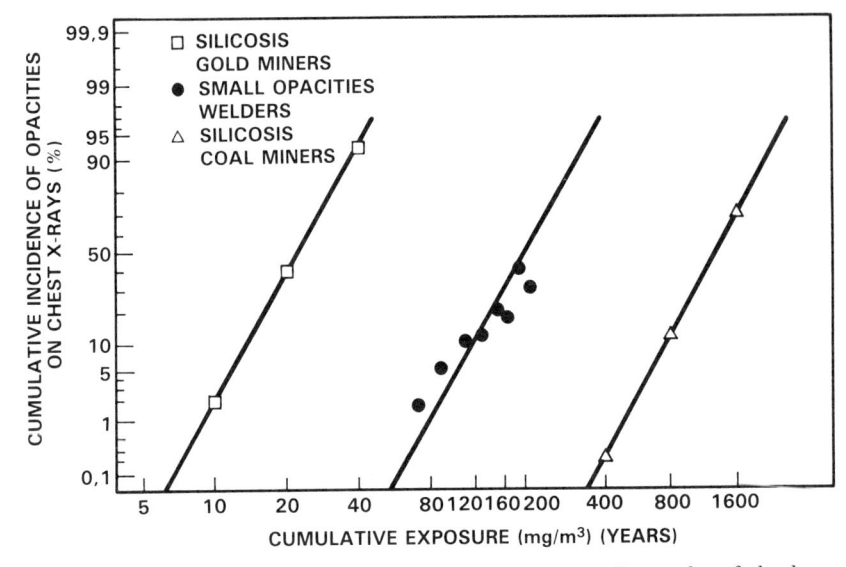

FIGURE 4.10—Incidence of small round opacities in radiographs of the lung as a function of cumulative inhalation exposure to welding fumes in British shipyard workers and following inhalation of dusts by gold miners and coal miners (R. M. Stern. 1981. Process-dependent risk of delayed health effects for welders. Environ. Health Perspect. 41:235–253).

fuels or volcanic ashes or pumice, exhibit an elemental composition (except for trace metals) that is similar to the elemental abundance seen in the region from which they were produced. The fibrogenicity of mixed-silicate dusts in the lung is limited in severity and extent by the amount of crystalline silica.[24] Simple pneumoconiosis with minimal alveolar fibrosis has been seen in experimental animals given mixed-silicate dusts, in animals residing in the San Diego Zoo, in farm workers of southern California, in Bedouins of the Negev and in pumice workers of the Lipari Islands.[25] The only effects of exposure to these dusts are loose inflammatory lesions around local concentrations of dust particles, with at most a grade I fibrosis.[26] The pulmonary response to newly formed volcanic ash appears similar to that of other mix-silicate dusts.[27] However, there is a slightly greater inflammatory, fibrotic and lipoproteinosis response than is seen with comparable doses of soil particles common to the region where the ash was deposited.[27] A pathologic scheme proposed to explain the development of pulmonary lesions from inhalation of mixed-silicate dust or ash particles is given in Figure 4.11.

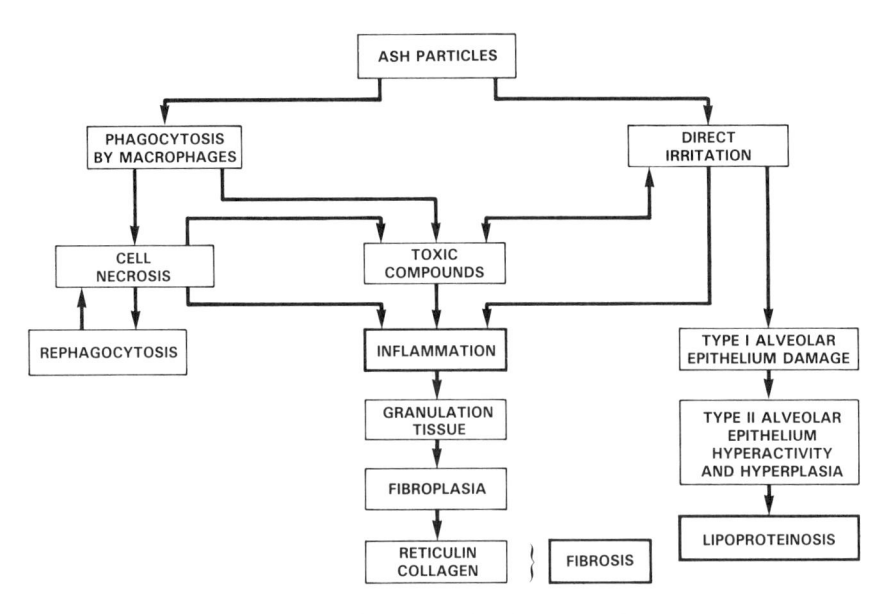

FIGURE 4.11—Pathological scheme for development of pulmonary lesions following inhalation of mixed-silicate dust or ash particles (C. L. Sanders. Unpublished).

Gases and Vapors

A gas is a state of matter in which the molecules are unrestricted by cohesive forces that cannot be condensed at any pressure (at a temperature of 20°C). In contrast, a vapor is a state of matter, normally present in a gaseous phase, which can be condensed to a liquid or a solid by pressure (at a temperature of 20°C). For practical purposes, the words gas and vapor are commonly interchanged.

Normal breathing rates will establish alveolar concentrations of an inhaled gas that has no tissue reactivity and a very low solubility, at 98% of equilibrium within two minutes; an example is helium. Gases such as sulfur dioxide and hydrogen chloride, which are very soluble in water and reactive with tissues, exhibit a rapid uptake by the upper respiratory tract with little penetration into the alveoli. Gases such as ozone and nitrogen dioxide, which have low water solubility but high tissue reactivity, much more readily penetrate to the alveoli, causing lung damage.

Gases are classified as asphyxiants or as irritants. Simple asphyxiants are physiologically inert gases, such as helium, Nitrogen, methane, or hydrogen cause oxygen deficiency due to physical displacement of oxygen, when present in the air at concentrations in excess of 20 volume percent.

Chemical asphyxiants are gases such as carbon monoxide or hydrogen cyanide, which prevent or hinder the normal uptake and transport of oxygen. Irritant gases produce inflammatory changes in tissues with which they come in contact. Sensory irritants, such as sulfur dioxide or formaldehyde, produce a burning sensation in the nose from stimulation of the trigeminal nerve endings or induce coughing by laryngeal irritation. Pulmonary irritants, such as ozone or nitrogen dioxide, cause epithelial damage, resulting in pulmonary edema, emphysema or other deep lung effects. Bronchoconstriction occurs by direct action of an irritant gas on the smooth muscle of the conducting airway, by axonal reflex or by stimulation of trigeminal-vagal nerves. Most bronchoconstrictors are also sensory irritants. Respiratory irritants are chemicals that act as a sensory irritant, bronchoconstrictor and pulmonary irritant. An example of a respiratory irritant is chlorine, which, at 50–100 ppm for 1 hour, will produce bronchitis, pulmonary edema, coughing, wheezing, nasal discharge and throat irritation.

Several gaseous organic chemicals are found in the environment in significant amounts, most of which originate from anthropogenic sources (Table 4.5).

Formaldehyde

Formaldehyde ($H_2C=O$) is a component of urea-formaldehyde, polyurethan foam insulation and has numerous other uses, including as a disinfectant, preservative of biological specimens and in paper, rubber and textile industries. Formalin is an aqueous solution that contains 40% formaldehyde. The threshold for odor recognition of formaldehyde is between 0.05 and 1.0 ppm (Table 4.6). Higher levels cause burning sensations, coughing, conjunctivitis and lacrimation. Pulmonary hypersensitivity is seen in individuals with preexisting formaldehyde-induced bronchial hypersensitivity.[64] Squamous cell carcinomas of the rat nasal cavity have been produced by inhalation exposure to 15 ppm formaldehyde vapor.[65] The current permissible exposure limit for formaldehyde, established by NIOSH, is 3 ppm. A statistically significant increase in nasal cancer has been observed in workers exposed to wood dust and formaldehyde.[66] However, no epidemiological data have shown a human carcinogenic risk from exposure to polyurethan foam insulations or from degassing of formaldehyde from foams.[67] Formaldehyde is a mutagen in bacteria, fungi and insects, induces sister chromatid exchanges in human lymphocytes and is a carcinogen in the nasal cavity of rodents. However, the possible carcinogenic risk in humans has not been clearly established.

Organic Vapors and Solvents

Organic vapors and solvents are commonly found in industrial and home environments; typical examples are gasoline, kerosene, benzene, paint removers, cleaners and degreasers. Large amounts of chloro-hydrocarbons

TABLE 4.5. Atmospheric concentration ranges of gaseous organic compounds (B. R. T. Simoneit and M. A. Mazurek. 1981. Air Pollution: The Organic Components. CRC Crit. Rev. Environ. Control 11:219–276).

Compound	Concentration Range (ng/l)	Natural (%)	Anthropogenic (%)
Methane	1000–4000	80	20
Ethane	0.05–95	–	>90
Propane	12–94	–	>90
Butane	0.01–200	–	>95
Ethylene	0.70–700		
Acetylene	0.20–230	–	>90
Propene	1–52		
Butene	1–6		
Benzene	0.03–57		>90
Formaldehyde	1–160		
Acetaldehyde	1.5–10		
Methanol	8–100		
Acetone	0.08–7		
Acrolein	1–13		
Formic acid	4–72		
Peroxyacetyl nitrate	0.10–66		
Methyl mercaptan	0.04–0.06		
Dimethyl sulfide	0.04–0.06		
Carbon disulfide	0.07–0.37		
Carbonyl sulfide	0.20–0.56		>90
Cyanogen	10–25		>90
Carbon tetrafluoride	0.10–1		
Methyl chloride	0.80–2.2		>90
Methylene chloride	0.005		>90
Chloroform	0.004–0.25		
Carbon tetrachloride	0.0004–0.26		
Fluorotrichloromethane	0.05–0.08		>90
Difluorodichloromethane	0.08–1.0		>90
1,1,1-Trichloroethane	0.03–0.4		>90

TABLE 4.6. Acute adverse effects of inhalation exposure to formaldehyde in humans (R.S.Bernstein, L. T. Stayner, L. J. Elliot, R. Kimbrough, H. Falk and L. Blade. 1984. Inhalation exposure to formaldehyde: An overview of its toxicology, epidemiology, monitoring, and control. Am. Ind. Hyg. Assoc. J. 45:778–785).

Concentration (ppm)	Effect
0.05–1.0	Odor threshold
0.05–2.0	Eye irritation
0.10–25	Nose and throat irritation
5–30	Lower airway irritation, pulmonary edema
50–100	Severe pulmonary edema, pneumonia
>100	Death

are produced in the U.S. annually: 800×10^6 kg carbon tetrachloride, 300×10^6 kg chloroform, 5000×10^6 kg 1,2-dichloroethane, 950×10^6 kg tetrachloroethylene and 500×10^6 kg trichloroethylene.

The C_1-C_4 aliphatic vapors are relatively nontoxic. Methane and ethane in natural gas produce their effects by simple asphyxiation; propane and butane produce CNS depression but no morphological damage. Among the C_5–C_8 aliphatic hydrocarbons, n-hexane produces CNS depression and polyneuropathy as a result of demyelination and degeneration of nerve axons. Prolonged exposure to gasoline or kerosene vapors can cause death due to respiratory failure from CNS depression. Aspiration chemical pneumonitis can result from the ingestion of liquid hydrocarbons, like kerosene. In general, aliphatic liquid hydrocarbon vapors produce respiratory failure only at high concentrations.[28]

Carbon tetrachloride is a formerly common solvent which produces CNS depression as well as liver and kidney damage at lower doses. Hepatic damage is due to the biotransformation of Cl_4 into the trichloromethyl radical, which causes lipid peroxidation in cell membranes. Ethanol and isopropanol potentiate the hepatotoxicity of CCl_4. Many other halogenated hydrocarbons, such as chloroform and dichloromethane produce liver and kidney damage.

Ethylene glycol, commonly used as an antifreeze agent for motor vehicles, is a renal toxicant, precipitating as the oxalate in the kidneys. Ethylene glycol commonly poisons cats and dogs when disposed on the ground, where it is ingested because of its sweet taste.

Ethanol and Methanol

Of the aliphatic alcohols, ethanol, by far the most abused as "drinking," is a CNS depressant that also causes diuresis, peripheral vasodilation, and fatty liver degeneration after prolonged usage. Immediate effects of ethanol intoxication are listed in Table 4.7. The blood ethanol level for operation of a motor vehicle, in most states, is 0.1%, wt. to volume; stupor and coma are seen at 0.3–0.4% and death at >0.5%. Over 95% of ingested ethanol is converted into acetate and incorporated into the citric acid cycle. About 30% of alcoholics exhibit hepatitis symptoms; 50% of all cases of hepatic cirrhosis are found in alcoholics. In males, the risk of liver cirrhosis

TABLE 4.7. Relationship of blood ethanol concentration and toxicity in humans.

Degree of Symptom	Clinical Symptom	Blood Ethanol Concentration	Portion of Brain
Mild	Decreased inhibitions Visual impairment Decreased reaction time Increased confidence	0.05–0.10%	Frontal lobe
Moderate	Ataxia Slurred speech Decreased motor skills Decreased attention Diplopia Altered perception Altered equilibrium	0.10–0.30%	Parietal lobe Occipital lobe Cerebellum
Severe	Vision impairment Equilibrium Stupor	0.30–0.50%	Occipital lobe Cerebellum Diencephalon
Coma	Respiratory failure	>0.50%	Medulla

triples on consumption of 21–40 g of alcohol per day and increases 600 fold with daily ingestion of 140 g of alcohol per day. A toxic intermediate in ethanol metabolism, acetaldehyde, is thought to play a central role in liver cirrhosis.[55] Ethanol is a teratogen, resulting in fetal alcohol syndrome, characterized by mental retardation and other CNS effects in offspring. It has been estimated that alcohol may be responsible for 7% of all cancer deaths among females in the U.S.[53] A significant association has been found between consumption of alcoholic beverages and cancer of the colon, rectum, lung and prostate.[54] Alcohol consumption is significantly associated with cancer of the oral cavity, pharynx and extrinsic larynx.[68] The incidence of oro-pharyngeal and laryngeal tumors is greatly increased in "heavy" drinkers and smokers above that which would be expected for only "heavy" drinkers or smokers alone.

Most methanol produced in the U.S. is now prepared synthetically by the reduction of carbon monoxide with hydrogen, rather than by the former method of wood distillation. Among other uses, methanol is a gasoline additive; new uses of methanol have been proposed for energy production.[59] Methanol is a CNS depressant that is degraded to formic acid in the body, causing acidosis and blindness because of the high sensitivity of the optic nerve.[60] Treatment of methanol poisoning involves the use of bicarbonate for the acidosis and administration of ethanol as a competitive inhibitor for biotransformation enzymes.

Benzene

Benzene is a simple aromatic hydrocarbon that is abundant in coal tar distillates and is a by-product of coke production and petroleum refining. In 1979, 11 billion pounds of benzene were commercially produced in the U.S., mostly used in the manufacture or tires and as a solvent. Benzene also has widespread use as a constituent of motor fuels and as a feedstock for the synthesis of a wide variety of aromatic compounds. Human industrial exposure to benzene occurs most often at the highest concentrations in coke production plants. Motor fuels in the U.S. contain <2% benzene while amounts in Europe range from 4–8%.[63]

Exposure to benzene can be fatal due to respiratory failure and CNS depression after a 10-minute exposure to 20,000 ppm. Toxic effects of chronic exposure to a few hundred ppm benzene are associated mostly with bone marrow damage: leukopenia, thrombocytopenia, bone marrow hyperplasia and hypoplasia, aplastic anemia and leukemia.[23] Hematopoietic effects of benzene inhalation are related not only to the exposure duration but also to the daily dose, indicating that repair of damage is more significant at lower dose-rates.[62] Alterations in erythrocytic maturation, chromosomal aberrations and abnormal bone marrow constituent proliferation all indicate an initial inhibitory effect of benzene on cell division followed often by compensatory increased proliferation.

Benzene is readily absorbed into the blood from the gastrointestinal tract and from the lungs and stored in fatty tissues throughout the body. Nearly all benzene is eventually biotransformed in the liver to phenol. A small portion of benzene is metabolized to mercapturic acid, and a very small amount is converted to benzene epoxide, a possible ultimate carcinogen.

Benzene was first related to human leukemia in 1897. Most malignancies due to benzene exposure occur in the bone marrow; these include leukemia, malignant lymphoma, myeloid metaplasia and multiple myeloma. A relationship between chronic occupational exposure to benzene and human leukemia has been established.[30] However, convincing animal data demonstrating the leukemogenic properties of benzene are lacking. An increased incidence of leukemia was observed in shoe workers in Turkey from 1967–1975; this declined when the use of benzene by that profession was phased out.[30] The latency period for benzene-related leukemia in humans is 6–14 years. The International Agency for Research on Cancer (IARC) predicts that among 1000 workers exposed to benzene for a working lifetime at 10 ppm, there would be 17 cases of Leukemia.[61] The occupational exposure standard for benzene has been lowered from 10 ppm in 1979 to the current level of 1 ppm.

Phenol

Phenol (carbolic acid) is a waste product of coal gasification, carbonization and liquefaction processes, and it is a metabolic product of benzene metabolism by the liver. In 1977, 2.4 billion pounds of phenol was produced in the U.S., mostly in synthetic processes. Phenol is a normal constituent of human tissues, occuring in urine, feces, saliva and sweat. Human background levels of phenol result from the metabolism of tyrosine and catabolism of proteins, particularly by E. coli in the gastrointestinal tract. Phenol is conjugated with sulfuric acid and glucuronic acid to yield phenyl sulfate and phenyl glucuronide, respectively. Phenol is also oxidized to carbon dioxide, 1, 4-dihydroxybenzene and o-dihydroxybenzene.

Acute phenol toxicity in experimental animals is associated with cardiovascular and neuromuscular reactions following depression of vasomotor centers. Phenol is readily absorbed through the skin, gastrointestinal tract and the lung into the blood.[31] Phenol is not a proven human mutagen, teratogen or carcinogen, but it does promote epidermal tumors in mice.[32] Human exposures to phenol have occurred through medical misapplication, suicide and industrial accidents. The use of phenol as a general disinfectant was introduced by Lister in 1860; this use demonstrated its sclerosing and irritative properties in surgical wounds. Adverse symptoms from phenol exposure in humans include local tissue necrosis, irregular pulse, darkened urine, vomiting, difficulty in swallowing, profuse perspiration, pulmonary edema, liver damage, coma, and death at high doses.[33] Hospital outbreaks of idiopathic neonatal unconjugated hyperbilirubinemia have been associated with the use of phenol as a disinfectant in nursery rooms. The OSHA (Occupational Safety and Health Administration) exposure standard for the workplace is 5 ppm (19 mg/m^3) for a 8-hour day.[33] The World Health Organization (WHO) international standard for phenol in drinking water is 1 µg/l.

Alkylbenzenes

The alkylbenzenes are, in general, obtained by extraction from coal tar derived from coking operations. Most toluene is converted to benzene; a small amount is used as a component of gasoline. Toluene vapor is readily absorbed by the lungs, and liquid toluene is rapidly absorbed by the gastrointestinal tract. Most absorbed toluene is metabolized to benzoic acid and excreted in urine as hippuric acid after hepatic conjugation with glycine.[56] Benzene-free toluene exposure affects mostly the CNS, causing dizziness, confusion and incoordination. Toluene inhalation also leads to euphoria, which attracts some individuals to "sniff" glue and paint thinners containing toluene.[57] Adverse effects, such as cancer formation, from chronic toluene exposure have not been documented.

Xylene (dimethylbenzene) is used as a solvent in the manufacture of plastics and synthetic fibers and is also a constituent of gasolines. Like toluene, absorbed xylene is preferentially taken up by adipose tissue and metabolized to methylhippuric acid. Xylene causes drying, erythema, and blistering of skin, as well as CNS symptoms, when inhaled. Benzene-free xylene is not carcinogenic.[58]

Inhalation Carcinogenesis

Lung Cancer

Lung cancer is the leading cause of death from cancer in males and the second leading cause in females in the U.S. and most other western countries of the world. In 1980, about 117,000 new cases of lung cancer were diagnosed, with about 101,000 deaths from this disease. Lung cancer is rare prior to the age of 40 and peaks in incidence at 50–60 years of age. The overwhelming etiologic agent associated with the current epidemic of lung cancer is cigarette smoking. Discussed in later sections are other factors increasing lung cancer risk; these are urban air pollution, heavy metals, carcinogenic PAH, ionizing radiation and vitamin-A deficient diets.

WHO classifies lung cancer in humans according to histological-morphological type:[35]

  I. Epidermoid carcinoma (characterized by keratinization and/intercellular bridges)
  II. Small-cell anaplastic carcinoma (so-called "oat-cell" carcinoma)
  III. Adenocarcinoma (bronchogenic originating from the bronchus, bronchiolar originating from Clara cells or bronchiolar-alveolar originating from terminal bronchioles or alveoli)
  IV. Large-cell carcinoma (undifferentiated type comprised of giant cells or clear cell types)
  V. Combined epidermoid and adenocarcinoma
  VI. Carcinoid tumors
  VII. Bronchial gland tumor (cylindroma or mucoepidermoid tumor)
  VIII. Papillary tumor of the surface epithelium
  IX. Mixed tumor (e.g. carcinosarcoma)
  X. Sarcoma (all types)
  XI. Unclassified
  XII. Mesothelioma
  XIII. Melanoma

The distribution of tumor types observed in humans varies according to lifestyle and occupational factors. A clear distinction in tumor type can be seen between smokers and nonsmokers. Only one in twenty lung tumors diagnosed in the U.S. as in a nonsmoker.[76] Most tumors in nonsmokers are adenocarcinomas; in smokers, about 40% are epidermoid carcinoma, 20% adenocarcinoma, 20% small-cell carcinoma and 15% large-cell carcinoma.

The most common site for lung cancer is in the carinae between dividing bronchi and bronchioles. The ultrastructural criteria for epidermoid differentiation include the presence of monofilament bundles, poorly developed endoplasmic reticulum and Golgi apparatus, and well-developed desmosomes. The criteria for adenocarcinoma include well-developed endoplasmic reticulum and Golgi apparatus, poorly developed desmosomes, and the presence of extracellular and/or intercellular evidence of secretions.[36]

Malignant mesothelioma is a lung tumor associated most often with asbestos exposure; the tumor arises from the mesothelial lining cells in the pleura or peritoneum, with the pleura being the site of tumor origin in 70% of all mesothelioma cases. The incidence of mesothelioma in the U.S. is currently 2.2 cases per million, with the overall risk of mesothelioma being about 300 times greater in those with asbestos exposure.[37]

Histological lung tumor types in experimental animals treated with carcinogens are comparable to those seen in humans. Bronchogenic epidermoid carcinoma is seen in rats and hamsters given intratracheal PAH. Bronchiolo-alveolar carcinomas are seen in the lung of dogs after inhalation of $^{239}PuO_2$ and in hamsters after injection of diethylnitrosamine. A good animal model for small-cell or large-cell carcinoma has not been found.[38]

Lung tumors are often metastatic by the time they are diagnosed in humans, possibly because of the often profound immunosuppression associated with this tumor. The most prevalent symptoms of lung cancer are cough, chest pain, abnormal sputum, hemoptysis, dyspnea, superior vena cava obstruction, pain in the shoulder or arm, pleural effusions, pulmonary hypertrophic osteoarthropathy, anemia and neurological signs due to brain metastases. Early detection of lung cancer by periodic radiographs or sputum exfoliative cytology have not improved survival following diagnosis. Patients with lung cancer die almost equally of local-regional disease, tumor progression and metastases. All lung tumors in humans have a poor prognosis, although survival is longer for patients with epidermoid carcinoma and adenocarcinoma. Overall, the mean survival time is 6–9 months for lung cancer patients; only 20% live >1 year while only about 5% survive 5 years after diagnosis.

Pulmonary Carcinogenesis from Organic Chemicals

The identification of organic chemical carcinogens has been based mostly on results of bioassays in animals. The critical evaluation of both epidemiological and experimental data provides a basis for classifying substances in relation to carcinogenicity (39–40). Animal models may be used for carcinogenic bioassays, as pathology reference standards for morphological characterization and diagnosis and as a model for mechanistic studies of pathogenesis.[52] The skin and lung are the most important sites of exposure to carcinogens. Mortality from skin cancer is very low; for lung cancer, it is very high.

About 1300 tons of PAH are discharged into the atmosphere annually in the U.S., primarily in urban regions. This correlates with a significantly higher incidence of lung cancer in urban areas than in rural areas of the U.S. The rate of clearance of PAH from the lung has a significant impact on lung tumor formation.[41] BaP is rapidly cleared from the lung even when given after induction of severe simple pneumoconiosis or radiation fibrosis.[42] However, PAH usually reaches the lung not as pure chemicals but adsorbed onto the surface of particles (e.g., soot) resulting from incomplete combustion of fossil fuels; this greatly increases the retention time of PAH in the lung, and greatly increases the cellular concentration at the site of particle impingement. This may explain why it is difficult to induce lung tumors in experimental animals with pure BaP, while BaP attached to "inert" particles is very effective in inducing lung tumors.

Table 4.8 lists several experimental studies that have demonstrated the respiratory carcinogenicity of a variety of chemicals or chemical mixtures; in some studies, carrier particles (such as ferric oxide, india ink, carbon, aluminum oxide or titanium oxide) were used to enhance the tumor response. Such PAH-particle interactions may explain the high lung cancer rates seen in asbestos workers and uranium miners who smoke cigarettes and in coke workers who are exposed to high levels of hematite dust. Other air pollutants, such as sulfur dioxide, may also promote PAH respiratory carcinogenesis.[43]

N-nitroso compounds are present in cigarette smoke and in a number of foods and beverages. Epidemiological studies have failed to demonstrate a correlation between exposure to N-nitroso compounds and cancer. However, the widespread use of nitrates, nitrites and nitrosatable amines makes it difficult to document such a correlation for this class of compounds. Many experimental studies in animals have shown a powerful carcinogenic effect of N-nitroso compounds in lung, liver and other tissues. The pulmonary carcinogenicity of N-nitroso compounds is enhanced by the simultaneous administration of "inert" particles in the lung.[44]

Cigarette Smoking

Until the 1870s, cigarettes were considered a minor luxury; in 1869, fewer than 2 million cigarettes were produced in the U.S., all of them hand-rolled. This all changed in 1880, when James Bonsack developed a cigarette-rolling machine that could replace 48 hand-rollers. This discovery, along with the development of a curing process for tobacco that produced a product sufficiently nonirritating to allow inhalation of the smoke, resulted in rapid, exponential growth in cigarette consumption. By 1885, 1 billion cigarettes were produced annually, increasing to 10 billion by 1910, 100 billion by 1926 and 600 billion today in the U.S.

The dense particulate aerosol formed when tobacco is smoked has a particulate weight of 400–500 mg per 10 puffs; the smoke contains about 5×10^9 particles with a diameter of 0.1–1.0 µm. The major components of cigarette smoke, by weight, are: nitrogen (50–66%), oxygen (10–15%), carbon dioxide (8–14%), carbon monoxide (3–6%) and wet particulate matter (4–9%). The particles are slightly charged and dipolar, and the pH of the smoke is 5.6–6.2. About 150 chemical compounds have been identified in the gas phase and about 2000 compounds in the particulate phase of the smoke; many have carcinogenic activity (Table 4.9). Mainstream smoke is inhaled directly into the lung, while sidestream smoke is formed between puffs, providing passive exposure to nonsmokers.

Cigarette-smoking complicates the evaluation of toxicity associated with energy production because the habit is so widespread, the toxic

TABLE 4.8. Studies that have demonstrated the induction of malignant lung tumors in experimental animals following exposure to organic chemicals (W. E. Pepelko. 1984. Experimental carcinogenesis in small laboratory animals. Environ. Res. 33:144-188).

Chemical	Date	Details of Study
Dibenzanthracene	1937	Lung implant in mice (Andervont)
Methylcholanthrene	1957	Lung implant in rats (Kuschner)
Dibenzanthracene	1957	Lung implant in rats (Kuschner)
Benzo(a)pyrene	1970	Lung implant in rats (Laskin)
Methylcholanthrene	1970	Lung implant in rats (Laskin)
Dibenzanthracene in olive oil	1949	Intratracheal in rats (Niskanen)
Dimethylbenzanthracene	1958	Intratracheal in hamsters (Della Porta)
Diethylnitrosamine	1964	Intratracheal in hamsters (Dontenwill)
Benzo(a)pyrene in between-60	1962	Intratracheal in hamsters (Harrold and Dunham)
Dimethylbenzanthracene in mineral oil	1965	Intratracheal in hamsters (Gross)
Benzo(a)pyrene in mineral oil	1965	Intratracheal in hamsters (Gross)
Benzo(a)pyrene in mineral oil	1965	Intratracheal in hamsters (Gross)
Benzo(a)pyrene + ferric oxide	1965	Intratracheal in hamsters (Saffiotti)
N-Nitroso-N-methyl-	1970	Intratracheal in hamsters (Harrold)
Methylcholanthrene	1970	Intratracheal in hamsters (Laskin)
Benzo(a)pyrene on india ink	1970	Intratracheal in rats (Shabad and Pylev)
Benzo(a)pyrene + asbestos	1966	Intratracheal in hamsters (Smith)
Methylcholanthrene	1971	Intratracheal in mice (Nettesheim)
Methylcholanthrene	1972	Intratracheal in rats (Schreiber)
Benzo(a)pyrene	1972	Intratracheal in hamsters (Feron)
Ozonized gasoline	1956	Inhalation in mice (Kotin and Falk)
Ozonized gasoline + Influenza virus	1963	Inhalation in mice (Kotin and Wiseley)
3-Nitro-3-hexane	1963	Inhalation in mice (Deichman)
Diethylnitrosamine	1962	Inhalation in hamsters (Dontenwill)
Coal tar	1963	Inhalation in mice (Horton)
Benzo(a)pyrene + Sulfur dioxide	1970	Inhalation in rats (Laskin)
Vinyl chloride	1970	Inhalation in rats (Viola)
Bis(chloromethyl)ether	1971	Inhalation in rats (Laskin)
Polyurethane	1972	Inhalation in rats (Laskin)

TABLE 4.9. List of toxic agents in cigarette smoke (U.S. Department of Health, Education and Welfare. 1979. Constituents of Tobacco Smoke. In, "Smoking and Health," pp. 14.1–14.72, DHEW NO(PHS) 79–50066, Washington, D.C.).

Chemical	Amount per Cigarette	Carcinogenicity*
Dimethylnitrosamine	1–200 ng	+
Dialkylnitrosamines	2–80 ng	+
Nitrosopyrrolidine	2–42 ng	+
Nitrosopiperdine	0–9 ng	+
Hydrazine	24–43 ng	+
Vinyl chloride	1–16 ng	+
2-Nitropropane	730–1210 ng	+
Formaldehyde	20–90 ug	+
Hydrogen cyanide	30–200 ug	–
Acrolein	45–140 ug	–
Acetaldehyde	18–1440 ug	–
Carbon monoxide	2–20 mg	–
Nitrogen oxides	10–600 ug	–
Nicotine	0.5–5.0 mg	+
Volatile phenols	20–30 mg	+
Benzo(a)pyrene	10–50 ng	+
Dibenz(a, h)anthracene	40 ng	+
Benzo(b)fluoranthrene	30 ng	+
Benzo(j)fluoranthrene	60 ng	+
Dibenz(a,j,)acridine	3–10 ng	+
Benz(a)anthracene	40–70 ng	+
Chrysene	40–60 ng	+
Benzo(e)pyrene	5–40 ng	+
2-Methylfluoranthrene	30 ng	+
N-Nitrosonornicotine	130–250 ng	+
Beta-naphthylamine	20 ng	+
Nickel compounds	0–600 ng	+
Cadmium compounds	9–70 ng	+
Polonium-210	0.03–1.3 pCi	+

*As demonstrated in animal bioassay studies.

chemicals in smoke so numerous, and the harmful effects of smoking so diverse. Cigarette-smoke lung clearance has been evaluated, using 14-C-dotriacontane labeled smoke particles. Cigarette smoke decreases the lung clearance rates of other inhaled particles, even up to a period of 6 months following cessation of smoking.[46] The particulate phase of smoke also has been shown to alter the lung metabolism of BaP.[47]

Cigarette-smoke condensates painted on the skin of mice are carcinogenic at that site. Unsuccessful efforts have been made to develop a consistent predictable, dose-response experimental model for lung-cancer induction with inhaled cigarette smoke, including bioassay studies using nose-only and whole-body exposures of animals to aged smoke, exposure of dogs through tracheostomies, and nose-only exposure of monkeys. Implantation of cigarette-smoke condensate in beeswax pellets in the lungs of rats gave a high, dose-dependent incidence of lung tumors.[48]

The standardized age-mortality correlation is considerably greater for smokers than for nonsmokers. The average male smoker loses 6.2 years of life expectancy; this increases to 8.6 years in a heavy smoker. Cessation of smoking for more than 10 years reduces lost lifespan to 0.3–2.3 years, depending upon prior smoking history. In addition to an increased lung-cancer rate, there is a two-fold increased risk of coronary heart disease, a 10-fold increased risk of emphysema, bronchitis, and peripheral vascular disease, and a two-fold increased risk of oral cancer in smokers.

A strong association exists between cigarette smoking in mothers and reduced body weight at birth of offspring, increased fetal and infant mortality and impaired physical and intellectual development. The relative risk of nonfatal myocardial infarction among women using oral contraceptives and cigarettes is about 12 times that seen in nonsmokers who use these contraceptives. In addition, nicotine in milk of lactating mothers who smoke may cause nausea, diarrhea and tachycardia in infants.[49] Many of these effects may be attributed to increased levels of carboxyhemoglobin in maternal and fetal blood.

The absolute incidence of lung cancer has greatly increased in males in the U.S., reaching a plateau in the 1970s; rates in females are currently increasing at a rapid rate. These increases in lung cancer are correlated with increasing annual per capita consumption of cigarettes (Figure 4.12). The lung cancer death rate in nonsmokers is 7.4/100,000, and 230/100,000 for smokers.[50] The risk of lung cancer in smokers who use filtered cigarettes is somewhat less than for users of nonfiltered cigarettes. The risk of lung cancer for cigar-only smokers is 2.9 the rate for nonsmokers, while pipe smokers have a rate that is 2.5 times that for nonsmokers.[75] Overall, cigarette-smoking accounts for 30% of all cancer deaths and 90% of all lung cancer deaths. Lung cancer is rare prior to the age of 40; the latent period from initiation of smoking averages about 30 years before the appearance of lung cancer.

Passive exposure to cigarette smoke is likely to be significant because people spend an average of about 80% of their time indoors. Respirable particulate levels are higher in homes of smokers than in homes of nonsmokers. Maternal cigarette smoking is associated with increases of 20 to 35% in the rates of 8 respiratory illnesses in their children.[73] Also several recent studies indicate an increased incidence of lung cancer in spouses who do not smoke but have a spouse who does smoke.[57,74]

4 – Inhalation Toxicology

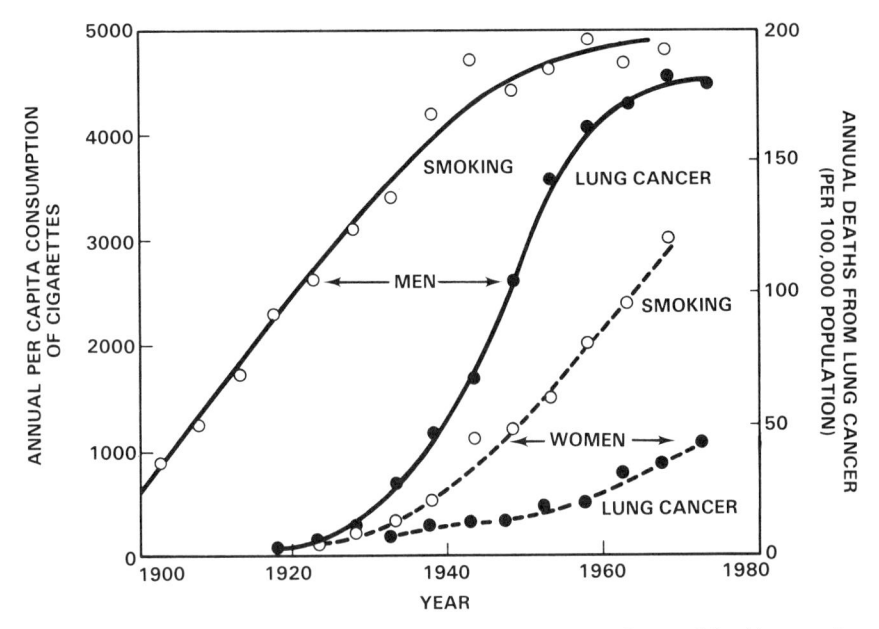

FIGURE 4.12—Relationship between cigarette consumption and incidence of lung cancer in men and women of the U.S. (J. Cairns. 1975. The cancer problem. Scientific Amer. 233:64-78).

References

1. E. R. Weibel. 1963. Morphometry of the Human Lung. Academic Press, New York, New York.
2. J. D. Brain and P. A. Valberg. 1979. Deposition of aerosols in the respiratory tract. Am. Rev. Respir. Dis. 120:1325-1373.
3. P. E. Morrow. 1966. Deposition and retention models for internal dosimetry of the human respiratory tract. Health Phys. 12:173-207.
4. P. E. Morrow, F. R. Gibb and K. M. Gazioglu. 1967. A study of particulate clearance from the human lungs. Am. Rev. Respir. Dis. 96:1209- .
5. M. O. Amdur. 1975. Air Pollutants. In, "Toxicology. The Basic Science of Poisons," Macmillan Publ. Co., Inc., New York, pp. 527-554.
6. Respiratory Disease: Task Force Report on Problems, Research Approaches, Needs. 1976. National Heart, Lung, and Blood Institute, DHEW Publication No. (NIH) 76-432. DHEW, Washington, D.C.
7. W. M. Haschek and H. Witschi. 1979. Pulmonary fibrosis—A possible mechanism. Toxicol. Appl. Pharmacol. 51:475-487.
8. D. E. Gardner and J. A. Graham. 1977. Increased pulmonary disease mediated through altered bacterial defenses. In, "Pulmonary

Macrophage and Epithelial Cells," CONF-760927, NTIS, Springfield, VA, pp. 1-21.
9. E. Goldstein, W. Lippert and D. Warshauer. 1974. Pulmonary alveolar macrophage: defender against bacterial infection of the lung. J. Clin. Invest. 54:519-528.
10. Y. Suzuki. 1974. Interactions of asbestos with alveolar cells. Environ. Health Perspect. 9:241-252.
11. K. Miller and J. S. Harington. 1972. Some biochemical effects of abestos on macrophages. Br. J. Exp. Pathol. 53:397-404.
12. J. M. G. Davis. 1979. Current concepts in asbestos fiber pathogenicity. In, "Dusts and Disease," Pathotox Publishers, Park Forest South, IL, pp. 45-49.
13. A. Morgan, J. C. Evans and A. Holmes. 1977. Deposition and clearance of inhaled fibrous minerals in the rat. Studies using radioactive tracer techniques. In, "Inhaled Particles IV," Pergamon Press, Oxford, U.K., pp. 259-272.
14. J. S. Harington. 1981. Fiber carcinogenesis: Epidemiologic observations and the Stanton hypothesis. J. Natl. Cancer Inst. 67:977-989.
15. I. J. Selikoff, W. J. Nicholson and A. M. Langer. 1972. Asbestos air pollution. Arch. Environ. Health 25:1-13.
16. M. R. Becklake. 1976. Asbestos-related diseases of the lung and other organs: Their epidemiology and implications for clinical practice. Am. Rev. Respir. Dis. 114:187-227.
17. M. Ziskind, R. N. Jones and H. Weill. 1976. State of the art. Silicosis. Am. Rev. Respir. Dis. 113:643-665.
18. A. G. Heppleston. 1963. The disposal of inhaled particulate matter: A unifying hypothesis. Am. J. Pathol. 42:119-135.
19. A. C. Allison, J. S. Harington and M. Birbeck. 1966. An examination of the cytotoxic effects of silica on macrophages. J. Exp. Med. 124:141-
20. W. R. Parkes. 1974. Occupational Lung Disorders. Butterworths, London, UK, pp. 166.
21. R. M. Stern. 1981. Process-dependent risk of delayed health effects for welders. Environ. Health Perspect. 41:235-253.
22. R. Caudarella, D. Cascella, G. Tabaroni, T. Corso and G. B. Raffi. 1979. Occupational disease of welders. I. Respiratory effects. Ital. Med. Law. 1:31-37.
23. C. E. Bates and L. D. Scheel. 1974. Processing emissions and occupational health in the ferrous foundry industry. Am. Ind. Hyg. Assoc. J. 35:452-462.
24. J. S. Fruchter et al. 1980. Mount St. Helens ash from the 18 May 1980 eruption: Chemical, physical, mineralogical and biological properties. Science 209:1116-1125.
25. R. P. Sherwin, M. L. Barman and J. L. Abraham. 1979. Silicate pneumoconiosis of farm workers. Lab. Invest. 40:576-582.

26. T. Belt and E. J. King. 1945. Special Report. Med. Res. Council (UK) 250:29.
27. C. L. Sanders, A. W. Conklin, R. A. Gelman, R. R. Adee and K. Rhoads. 1982. Pulmonary toxicity of Mount St. Helens volcanic ash. Environ. Res. 27:118–135.
28. H. H. Cornish. 1975. Solvents and vapors. In, "Toxicology. The Basic Science of Poisons," Macmillan Publ. Co., Inc., New York, pp. 503–526.
29. M. Aksoy, K. Dincol, T. Akgu, S. Erdem and G. Dicol. 1975. Haematological effects of chronic benzene poisoning in 217 workers. Br. J. Ind. Med. 28:296–302.
30. M. Aksoy. 1980. Different types of malignancies due to occupational exposure to benzene: A review of recent observations in Turkey. Environ. Res. 23:181–190.
31. H. Babich and D. L. Davis. 1981. Phenol: A review of environmental and health risks. Reg. Toxicol. Pharmacol. 1:90- 109.
32. R. K. Boutwell and D. K. Bosch. 1959. The tumor-promoting action of phenol and related compounds for mouse skin. Cancer Res. 19:413–424.
33. National Institute for Occupational Safety and Health. 1976. Criteria for a recommended standard. Occupational exposure to phenol. No. 76-196, U.S. DHEW, Cincinnati, OH.
34. D. Natusch and J. Wallach. 1974. Urban aerosol toxicity: The influence of particle size. Science 186:695–699.
35. L. Kreyberg, A. A. Liebow and E. A. Uehlinger. 1967. Histologic typing of lung tumors. International Histological Classification of Tumours No. 1, World Health Organization, Geneva.
36. E. M. McDowell et al. 1978. The respiratory epithelium. V. Histogenesis of lung carcinomas in the human. J. Natl. Cancer Inst. 61:587–606.
37. K. H. Gordon et al. 1982. Radiation therapy in the management of patients with mesothelioma. Int. J. Radiat. Oncol. Biol. Phys. 8:19–25.
38. U. Saffiotti. 1970. Experimental respiratory tract carcinogenesis and its relation to inhalation exposures. In, "Inhalation Carcinogenesis," CONF-691001, NTIS, Springfield, VA, pp. 27–54.
39. U. Saffiotti and C. C. Harris. 1979. Carcinogenesis studies on organ cultures of animal and human respiratory tissues. In, "Carcinogens: Identification and mechanisms of action," Raven Press, New York, pp. 65 - .
40. H. Reznik-Schuller and G. Reznik. 1979. Experimental pulmonary carcinogenesis. Int. Rev. Exp. Pathol. 20:211- .
41. B. Paigen et al. 1978. Genetics of aryl hydrocarbon hydroxylase in the human population and its relationship to lung cancer. In, "Polycyclic Hydrocarbons and Cancer," Vol. 2, Academic Press, New York, pp. 391- .
42. R. A. Gelman, C. L. Sanders and A. W. Conklin. 1981. Effect of lung damage caused by intratracheally instilled $_{239}PuO_2$ and volcanic ash

on the alveolar clearance of ^{14}C-benzo(a)pyrene. In, "Pacific Northwest Laboratory Annual Report for 1980 to the MOE Assistant Secretary for Environment, Part l, Biomedical Sciences, PNL-3700, pp. 95-96.
43. S. Laskin, M. Kuschner and R. T. Drew. 1970. Studies in pulmonary carcinogenesis. In, "Inhalation Carcinogenesis," CONF-691001, NTIS, Springfield, VA, pp. 321-350.
44. R. Montesano. 1970. Systemic carcinogenesis (N-nitroso Compounds) and synergistic or additive effects in respiratory carcinogenesis. Tumori 56:335-444.
45. D. Hoffmann, I. Schmeltz, S. S. Hecht and E. L. Wynder. 1978. Tobacco carcinogenesis. In, "Polycyclic Hydrocarbons and Cancer," Vol 1., Academic Press, New York, pp. 85.
46. R. E. Filipy, J. L. Pappin, D. L. Stevens and W. J. Bair. 1981. The impairment of pulmonary clearance of $_{239}$PuO$_2$ in rats by prolonged exposure to cigarette smoke. In, "Pacific Northwest Laboratory annual Report for 1980 to the Doe Assistant Secretary for Environment, Part l, Biomedical Sciences, PNL-3700, pp. 110-111.
47. W. C. Lubawy and D. G. Perrier. 1980. The effect of acute tobacco smoke on pulmonary benzo(a)pyrene metabolism. Environ. Res. 21:438-445.
48. G. E. Dagle, K. E. McDonald, L. G. Smith and D. L. Stevens. 1978. Pulmonary carcinogenesis in rats given implants of cigarette smoke condensate in beeswax pellets. J. Natl. Cancer Inst. 61:905- .
49. Smoking and Health. 1979. A Report of the Surgeon General, DHEW Publication No(PHS) 79-50066, U.S. Government Printing Office, Washington, D.C.
50. E. C. Hammond. 1966. Smoking in relation to death rates of 1 million men and women. Natl. Cancer Inst. Monogr. 19:12.
51. G. C. Kabat and E. L. Wynder. 1984. Lung cancer in nonsmokers. Cancer 53:1214-1221.
52. W. E. Pepelko. 1984. Experimental respiratory carcinogenesis in small laboratory animals. Environ. Res. 33:144-188.
53. K. J. Rothman. 1975. Alcohol. In, "Persons at High Risk of Cancer" (J. F. Fraumeni, Ed), Academic Press, New York, pp. 139.
54. E. S. Pollack et al. 1984. Prospective study of alcohol consumption and cancer. N. Eng. J. Med. 310:617-621.
55. J. P. von Wartburd and R. Buhler. 1984. Biology of Disease. Alcoholism and aldehydism: New biomedical concepts. Lab. Invest. 50:5-16.
56. T. Wilczok and G. Biekiek. 1978. Urinary hippuric acid concentration after occupational exposure to toluene. Br. J. Ind. Med. 35:330-334.
57. B. L. Weisenberger. 1977. Toluene habitation. J. Occup. Med. 19:569-570.
58. F. Gamberale, G. Annwall and M. Hultengren. 1978. Exposure to xylene and ethylbenzene. III. Effects on central nervous function. Scand. J. Work Environ. Health 4:204-211.

59. H. S. Posner. 1975. Biohazards of methanol in proposed new uses. J. Toxicol. Environ. Health 1:153–171.
60. D. R. McLean, H. Jacobs and B. W. Mielke. 1980. Methanol poisoning: A clinical and pathological study. Ann. Neurol. 8:161-167.
61. R. Snyder. 1984. The benzene problem in historical perspective. Fund. Appl. Toxicol. 4:692–699.
62. A. M. Dempster, H. L. Evans and C. A. Snyder. 1984. The temporal relationship between behavioral and hematological effects of inhaled benzene. Toxicol. Appl. Pharmacol. 76:195–203.
63. R. S. Brief et al. 1980. Benzene in the workplace. Am. Ind. Hyg. Assoc. J. 41:616–623.
64. R. S. Bernstein, L. T. Stayner, L. J. Elliott, R. Kimbrough, H. Falk and L. Blade. 1984. Inhalation exposure to formaldehyde: An overview of its toxicology, epidemiology, monitoring, and control. Am. Ind. Hyg. Assoc. J. 45:778–785.
65. J. A. Swenberg et al. 1980. Induction of squamous cell carcinomas of the rat nasal cavity by inhalation exposure to formaldehyde vapor. Cancer Res. 40:3398–3402.
66. J. H. Olsen, S. P. Jensen, M. Hink, K. Faurbo, N. O. Breum and O. M. Jensen. 1984. Occupational formaldehyde exposure and increased nasal cancer risk in man. Int. J. Cancer 34:639–644.
67. K. A. L' Abbe and J. R. Hoey. 1984. Review of the health effects of urea-formaldehyde foam insulation. Environ. Res. 35:246–263.
68. J. M. Elwood, J. C. G. Pearson, D. H. Skippen and S. M. Jackson. 1984. Alcohol, smoking, social and occupational factors in the aetiology of cancer of the oral cavity, pharynx and larynx. Int. J. Cancer 34:603–612.
69. Task Group on Lung Dynamics. 1966. Health Phys. 12:173–207.
70. B. T. Mossman, A. Eastman and E. Bresnick. 1984. Asbestos and benzo(a)pyrene act synergistically to induce squamous metaplasia and incorporation of (^{3}H)thymidine in hamster tracheal epithelium. Carcinogenesis 5:1401–1404.
71. A. G. Heppleston. 1982. Silicotic fibrogenesis: A concept of pulmonary fibrosis. Ann. Occup. Hyg. 26:449–462.
72. The Cancer Letter. 1984. Vol. 10:7.
73. J. H. Ware, D. W. Dockery, A. Spiro, F. E. Speizer and B. G. Ferris. 1984. Passive smoking, gas cooking and respiratory health of children living in six cities. Am. Rev. Respir. Dis. 129:366–374.
74. T. Hirayama. 1981. Non-smoking wives of heavy smokers have a higher risk of lung cancer: A study from Japan. Br. Med. J. 282:183–185.
75. J. H. Lubin, B. S. Richter and W. J. Blot. 1984. Lung cancer risk with cigar and pipe use. J. Natl. Cancer Inst. 73:377–381.

CHAPTER 5

ENVIRONMENTAL AIR POLLUTION

A Brief History

Ancient toxicologists described the more obvious inhalation hazards associated with mining. Pliny the Elder (23–79 A.D.) described how workers attempted to protect their faces from lead dusts with loose bags or bladders. Agricola (1494–1555) wrote a 12-volume work on mining and smelting as practiced in central Europe, including a description of many pulmonary diseases. In 1700, Bernardo Ramazzini (1633–1714), an Italian physician, published a book on the diseases of workers, including a description of the hazards of inhalation of various substances, including mercury. The term "smog" was first coined by Harold Antoine Des Voeux in 1911 as a contraction of smoke and fog. Haldane and Bancroft in this century described the toxicity of carbon monoxide which, today, causes more deaths than any other inhaled environmental pollutant.[1]

The association of heat and urban air pollution with excess mortality, particularly in the old and infirm, was noted as early as the mid-nineteenth century in England.[2] During the last 50 years, several documented acute air pollution episodes in the western world have resulted in high mortality (Table 5.1). The 1930 episodes in the Meuse Valley of Belgium, in Donora, PA in 1948, and the "killer" fogs of London in 1952, 1956 and 1962 all resulted in legislation that produced stricter air pollution standards. An eye witness account of the Donora incident is typical of the times:[3]

"The fog closed over Donora on the morning of Tuesday, October 26. The weather was raw, cloudy and dead calm, and it stayed that way as the fog piled up all that day and the next. By Thursday, it had stiffened adhesively into a motionless clot of smoke. That afternoon it was just possible to see across the street, and except for the stacks, the mills had vanished. The air began to have a sickening smell, almost a taste. It was the bittersweet reek of sulfur dioxide. Everyone who was out that day remarked on it, but no one was much concerned. The smell of sulfur diox-

TABLE 5.1. Acute air pollution episodes (E. T. Chanlett. 1973. Environmental Protection, McGraw-Hill Co., New York).

Meuse Valley, Belgium, Dec. 1-6, 1930. Area contained coke ovens, blast furnances, steel, glass, zinc and sulfuric acid plants. Inversion and air stagnation in the 15-mile river valley for 1 week caused an estimated sulfur dioxide level of 25–100 mg/m³ (10–40 ppm). Thousands became ill with coughing, breathlessness, chest pain, eye and ear irritation; 60 deaths.

Donora, PA., U.S., Oct. 27-31, 1948. Area contained a zinc smelter, wire coating mill, steel mills and sulfuric acid plants. A temperature inversion and dense fog along the horseshoe-shaped valley of the Monongahela River resulted in sulfur dioxide levels of 1.5–5.5 mg/m³ (0.5–2 ppm) along with heavy particulate and sulfuric acid mist. Of the 14,000 population in the area, 6,000 became ill; 1,400 sought medical attention and 17 died. Symptoms were coughing, sore throat, chest constriction, burning and tearing of eyes, vomiting and excessive nasal discharge.

London, England, Dec. 5-9, 1952. A highly industrialized area using coal for industry and heating homes. A "pea soup" fog and temperature inversion covered most of the U.K.; sulfur dioxide levels were as high as 3.8 mg/m³(1.4 ppm) while particulate levels reached 4.5 mg/m³. From 3,500 to 4,000 "excess" deaths were recorded in the week of Dec. 5–12 from chronic bronchitis, bronchopneumonia and heart disease.

London, England, January, 1956. Extended heavy fog conditions similar to the 1952 episode resulted in 1,000 excess deaths; also resulted in passage by Parliment of the Clean Air Act.

London, England, Dec. 5-7, 1962. Severe fog and temperature inversion caused sulfur dioxide levels to be higher than those of 1952; particulate levels were lower. An alert and emergency medical care plan caused excess deaths to be only about 700.

---

ide, a scratchy gas given off by burning coal and melting ore, is a normal concomitant of any durable fog of Donora. This time it merely seemed more penetrating than usual."

Although there is an indication in the older literature of the possible role of mixture of pollutants and additive interactions on health,[1] the importance of mixtures in inhalation exposures is only beginning to receive much attention. Just how complex a polluted environment may be is in-

dicated by Howard Lewis in his book, "With Every Breath You Take" published in 1965:
"Chatanooga, Tennessee, has a population of only 130,000, and surrounding Hamilton County has but 108,000 more. Yet residents of this community endure in their air this barrage: smoke from wire salvaging; smoke and odor from burning junked cars and tires; odor from a rendering plant; dust from a cement plant; odor from a food company; smoke from a brickworks; dust from a glass company; smoke from burning wood wastes; ammonium chloride from galvanizing operations; emissions from heat treating and annealing; acids and alkalis from electroplating; fumes from a brass foundry; fluorides from an aluminum foundry; acid mist from a battery plant; ferrosilicon, chrome, manganese, and tars from an electrometallurgical plant; odors and tars from a roof-material producer; solvents, varnish, and oxides of nitrogen from a paint company; paint spray; solvents from enameling operations; odors from a tannery; dust and odor from the burning of varnish and insulation; and smoke from burning coal and wood wastes for fuel."[3]

The picture, while highly graphic, is incomplete; Lewis did not mention the oxides of nitrogen, sulfur and carbon and the metals and PAH associated with fossil-fuel combustion nor the pesticides and herbicides drifting in from nearby agricultural regions.

General Aspects of Air Pollution

Industrialized societies generate a wide variety of air pollutants, some of which are known or suspected carcinogens (Tables 5.2 and 5.3). Levels of air pollutants in rural or nonindustrialized regions of the U.S. are considerably less than in urban regions. Anthropogenic contributions account for about a fifth of the total particulate burden in the air and for substantially greater proportions of specific pollutants. An estimated 5×10^7 kg of vapor-phase chemicals and 10^6 kg particulate matter are released daily into the atmosphere in the U.S. as vehicle exhaust, emissions from power plants, emissions from industrial sources and from cigarette smoke.[10] Anthropogenic hydrocarbons tend to exhibit higher toxicity than do natural emissions of hydrocarbons, which occur mostly as exudations from plants.

In addition to particulates, a variety of "benchmark" chemicals are in polluted air. Among these are numerous hydrocarbon species (particularly PAH), oxides of nitrogen (NO_x, mostly as NO and NO_2), ozone (which, combined with NO_x and other minor oxidants, gives a measure of total oxidant level), oxides of sulfur (SO_x, mostly SO_2), oxides of carbon (CO and CO_2) and a variety of trace metals. Atmospheric levels of these pollutants show large diurnal and seasonal variability and are also greatly affected

TABLE 5.2. Major stationary sources of air and water pollution (Hittman Associates, Inc. 1979. Environmental assessment report: Solvent refined coal [SRC] systems. U.S. Department of Commerce, PB–300 383, NTIS, Springfield, VA.)

Power plants	Coke-oven batteries
Coal-cleaning plants	Sulfur-recovery plants
Kraft pulp mills	Carbon-black plants
Portland cement plants	Primary lead smelters
Primary zinc smelters	Fuel-conversion plants
Iron & Steel mill plants	Sintering plants
Primary aluminum ore reduction plants	Secondary metal-production facilities
Primary copper smelters	Chemical process plants
Municipal incinerators	Fossil-fuel boilers
Hydrofluoric acid plants	Petroleum storage and transfer facilities
Sulfuric acid plants	Taconite ore-processing facilities
Lime plants	Glass-fiber-processing plants
Phosphate rock-processing plants	Charcoal-production facilities

by meteorological conditions. Typically, polluted air contains 400–700 times more carbon monoxide, 200 times more NO_x, 25 times more ozone and 1000 times more SO_x than unpolluted air. Examples of the wide variety of chemicals emitted from one source are given in Table 5.4 for stack emissions from a fossil-fuel power plant.

Attempts have been made to assess the associations between environmental air pollution and public health problems in epidemiological studies such as CHESS (Community Health and Environmental Surveillance System); studies in Los Angeles County emphasize oxidants, CO, SO_x and particulates and sensitive human health indicators as part of that program.

TABLE 5.3. Concentrations of known and suspected carcinogens in urban and rural atmospheres (D. F. S. Natusch. 1978. Potentially carcinogenic species emitted to the atmosphere by fossil-fueled power plants. Environ. Health Perspect. 22:79–90).

		Concentration	
Chemical	Range	Urban Air Average	Rural Air Average
Inorganic gases, μg/m³			
SO_x	20–1200	70	0.1–5
NO_x	50–400	100	2–6
O^3	20–400	100	20–100
Hg	.001–.20	.007	–
Inorganic Particulates, ng/m³			
Arsenic	2–130	10	<0.5–5
Asbestos	10–100	20	–
Beryllium	<0.2–8	5	–
Cadmium	4–250	10	–
Cobalt	0.5–15	2	<0.5–2
Chromium	5–120	15	<1–10
Copper	10–4000	60	1–280
Iron	1000–2000	1400	10–1000
Nickel	10–1000	100	<10
Lead	500–3000	1500	10–100
Selenium	<1–10	1	–
Uranium	0.01–2	0.2	–
Vanadium	50–2000	500	<1–50
Alkanes, ng/m³			
n-Pentane	1–40	15	–
2-Methylbutane	5–60	25	–
Alkenes, ng/m³			
2-Butene	1–5	5	–
1,3-Butadiene	1–5	2	–
Propene	1–20	6	–

(continued)

Table 5.3 (continued)

Chemical	Range	Urban Air Average	Rural Air Average
Aldehydes and ketones, ng/m³			
Formaldehyde	5–100	20	0.5–5
Acrolein	<1–20	5	–
Nitrosamines, ng/m³			
Dimethyl-nitrosamine	20–100	–	–
Peroxides, ng.m³			
Peroxyacyl-nitrates	2–30	–	–
Aromatic Hydrocarbons, ng/m³			
Benzene	5–90	20	–
Toluene	10–100	40	–
1,2-Dimethyl-benzene	5–100	40	–
Polyaromatic hydrocarbons, ng/m³			
Anthracene	0.5–700	1	–
Benzo(a)pyrene	1–50	10	–
Benzo(e)pyrene	0.1–50	5	–
1,2-Benzanthracene	1–70	20	–
Coronene	0.2–50	1	–
Chrysene	0.5–200	5	–
Pyrene	0.2–50	10	–
Polycyclic nitrogen compounds, ng/m³			
Acridine	0.1–0.5	–	–
carbonitrile	0.02–0.1	–	–
Tetraethyl lead, ng/m³	50–2000	75	–
Benzene-soluble organics, ng/m³	1000–20000	7000	200–3000

(continued)

TABLE 5.4. Known air pollutants emitted by the stack of fossil fuel plants with a concise toxicological evaluation [N. I. Sax. 1982. Toxicological effects of fossil fueled power plant pollutants. Dangerous Properties of Industrial Materials Report 2(1):5-15].

Air Pollutant—Agents and Toxic Effects

Acids—Nitric, sulfuric, hydrochloric and organic acids as irritants to eyes, skin and mucous membranes; 1–2 ppm tartaric or citric acids will etch teeth in 6 months.

Alcohols—Methanol, ethanol, isopropanol and propanol fumes act as eye and mucous membrane irritants and are CNS depressants.

Aldehydes—Acrolein, crotonaldehyde and formaldehyde are irritants to eyes, skin and mucous membranes; suspected carcinogens.

Aluminum—Aluminum powder pneumoconiosis at high doses.

Ammonia—Gaseous irritant, caustic and lethal at high concentrations.

Amines—Aromatic amines are suspected human carcinogens.

Antimony—Highly toxic element that is an irritant at low concentrations.

Arsenic—Extremely toxic and carcinogenic element. Causes a variety of skin and other disorders, such as "blackfoot."

Asbestos—Causes asbestosis, pleural calcification and mesothelioma.

Barium—Caustic and highly toxic as soluble salts.

Benzo(a)pyrene—Common "benchmark" PAH from pyrolysis of fossil fuels; carcinogen.

Beryllium—Extremely toxic element, causing berylliosis; carcinogenic.

Benzene—Very toxic to hematopoietic system; leukemogen.

Bromine—Moderate irritant of mucous membranes.

Cadmium—Extremely toxic metal; target tissues are lung, liver, kidney and testis.

Carbon bisulfide—Highly toxic CNS and hematopoietic toxin.

(continued)

Table 5.4 (continued)

Carbon dioxide – Reduces oxygen availability by volume substitution.

Carbon monoxide – Highly toxic gas that lowers tissue oxygen levels due to binding with heme in erythrocytes.

Chlorinated hydrocarbons – Unpredictable acute toxicity; narcotic at high concentrations; both aliphatic and aromatic classes are suspected carcinogens.

Chlorine – Highly toxic common air pollutant that combines in air for form HCl; 1000 ppm fatal.

Chloral hydrate – Poisonous drug and carcinogen in animals.

Chromium – Dermatitis in trivalent form; carcinogen in nasal, laryngeal and pulmonary regions of the respiratory tract.

Cobalt – Toxic to myocardium and experimental carcinogen.

Copper – Metal fume fever and cirrhosis of liver and pancreas; allergic irritant of skin and eyes.

Creosols – Highly toxic and corrosive to skin, eyes and mucous membranes.

Dimethylacetamide and dimethylformamide – moderately toxic irritants and experimental teratogens.

Dimethylsulfate – High toxicity with levels in fly ash as high as 830 ppm; potential carcinogen.

Dimethyl sulfide and diphenyl sulfide – moderately toxic irritants.

Ethyl acrylate – Acrid, penetrating odor and moderately toxic irritant.

Fluorine and fluorides – Powerful irritants; chronic exposure causes osteoporosis and experimental teratogenesis.

Gold – Moderate toxicity and powerful allergen.

Hafnium – Toxic in the liver.

Hydrochloric acid – Extreme irritant to skin, eyes and mucous membranes.

Hydrogen sulfide – Rotten-egg odor at low concentrations; highly tox-

(continued)

Table 5.4 (continued)

ic gas, irritant and asphyxiant at >100 ppm.

Iron dust – Irritant pneumoconiotic dust causing mild pulmonary fibrosis; eye irritant; causes siderosis.

Ketones – High volatility of some causes neurotoxicity.

Lead – Highly toxic metal, stored in bone and toxic to CNS, kidney and bone marrow.

Lithium – Oxide is powerful irritant; neurotoxin.

Manganese – Highly toxic metal; chronic exposure leads to CNS toxicity with characteristic sensory and motor effects.

Mercaptans – Powerful sulfur-containing odorants; offensive odor that may lead to nausea.

Methacrylates – Low toxicity but suspected experimental teratogen.

Mercury – Highly toxic element, causing stomatitis, excessive salivation, gingivitis and loosening of teeth, psychological changes and CNS damage; confimred teratogen and carcinogen.

Molybdenum – Moderately toxic metal; may interfere with copper metabolism.

Nickel – Carcinogen in nasal cavity, paranasal sinuses and lungs; dermatitis or "nickel itch."

Niobium – Highly toxic salts, causing severe but reversible damage to skin and eyes on contact.

Nitriles – Exhibit toxicity similar to cyanide; examples, acrylonitrile and organic cyanides.

Nitrobenzene – Highly toxic, forming methemoglobin, causing cyanosis.

Oxides of Nitrogen – NO_x only slightly irritating in upper respiratory tract but powerful oxidant in alveolar region, leading to edema, dyspnea and death at high concentrations; forms acids with water in air.

Ozone – Not emitted directly from stack but formed in atmosphere after exposure to chemical milieu and sunlight; powerful oxidant with disagreeable odor. Damage to lung like NO_x.

(continued)

Table 5.4 (continued)

Phenols – Toxic to skin, kidney and lung; experimental cocarcinogen.
Phosphorus – Extremely irritating as oxides; chronic exposure leads to necrosis of mandible, weakness and generalized debilitation.

Phosphine – Highly toxic gas with pulmonary, CNS and hematopoietic effects.

Phosgene – "Choking gas" of WW I, due to pulmonary edema and irritation of the upper respiratory tract.

PAH compounds – Many are proven mutagens, teratogens and carcinogens.

Rubidium – Moderate irritant to eyes, skin and mucous membranes.

Samarium – Toxicity limited to interference with blood coagulation.

Selenium – Highly toxic metal in lung, liver, kidney and bone marrow; anti-oxidant and antagonist of cadmium toxicity at low concentrations.

Scandium – Toxicity like that of boron.

Silver – Moderately toxic metal.

Styrene – Vinyl benzene or styrene monomer are powerful lacrimators and irritants to eyes; teratogen in animals.

Sulfur dichloride – Penetrating odor; irritating to lung, skin and mucous membranes. Decomposes in water to form HCl, thiosulfuric acid and sulfur.

Sulfur oxides – Sulfur dioxide, sulfurous acid, sulfuric acid, sulfides and sulfates cause irritation to eyes and lungs, mostly in the upper airways.

Tin – Salts are irritants; alkyl tin compounds are highly toxic in skin; tin oxide produces tin pneumoconiosis.

Toluene – Moderately toxic common solvent.

Uranium – A radiological and chemical toxin, principally in lung and kidney.

Vanadium – Irritant in eye and lung.

Zinc – Some compounds astringent, corrosive and emetic; gastrointestinal irritant at high concentrations.

Standards for Airborne Contaminants

Numerous agencies have established standards for air pollutants based upon the potential for adverse health effects or damage to the environment. The past 10 to 20 years have seen a spate of federal air pollution control laws and regulations. Premier among these has been the Clean Air Act of 1970, which provided the Environmental Protection Agency (EPA) with the power to adopt and enforce air-pollution regulations. EPA then promulgated the National Primary and Secondary Ambient Air Quality Standards (AQS), setting maximum ambient concentrations for various pollutants (Table 5.5). AQS standards were set for the general population, considering the impacts of exposure to the most sensitive groups of the population, such as the very young and old. The Clean Air Act Amendments of 1977 included comprehensive new requirements for the prevention of significant air quality deterioration in areas with air quality cleaner than minimum national standards.

In addition, the Resource Conservation and Recovery Act requires that solid wastes comply with stringent air standards. The Toxic Substances Control Act (TOSCA) regulates the disposal of specific hazardous substances, such as nickel catalyst, used in conversion processes. The Occupational Safety and Health Act (OSHA) regulates air contaminant exposures in the workplace. Threshold Limit Values (TLV) set by the American Conference of Governmental Industrial Hygienists for occupational exposure to various airborne materials are based on continuous exposure for 8 hours daily and 5 days per week. TLV values are not necessarily indicators of toxicity and do not take into account sensitive groups in the populations or interactions of pollutants with other agents that may enhance their toxicity, although they do consider such factors as eye and respiratory tract irritation. Short-Term exposure Limits (STL) and Emergency Exposure Limits (EEL) are also set for brief exposures under unusual conditions. All standards are set at exposure levels that are not expected to cause permanent human health problems (Figure 5.1). TLV limits provide nearly complete protection; there is not a fine line between safe and dangerous concentrations. EEL limits, on the other hand, provide a fine line between safe and dangerous concentrations, having no intentional safety factor; reversible but frank injury may result. STL limits provide for a reasonable margin of safety; discomfort but no injury may result from exposure.

The concentration of an air pollutant is normally expressed in either mg/m^3 or in parts per million (ppm); the translation from one to the other units can be accomplished by the following equation:

$$mg/m^3 = ppm \times molecular\ weight\ 24.5$$

TABLE 5.5. Examples of environmental—AQS and occupational—TLV standards.

Substance	Air Concentration, ug/m³	
	AQS	TLV
Particulates	75 (annual)	–
	260 (24 hours)	–
Sulfur oxides	80 (annual)	13
	365 (24 hours)	–
Carbon monoxide	10,000 (8 hours)	55,000
	40,000 (1 hour)	–
Nitrogen dioxide	100 (annual)	9,000
Ozone	240 (1 hour)	200
Arsenic	–	50
Nickel	–	1,000
Beryllium	–	2
PAH (benzene-soluble)	–	200
Hydrogen sulfide	–	10,000
Lead	–	150
Vanadium	–	50

Polycyclic Aromatic Hydrocarbons

Polycyclic aromatic hydrocarbons (PAH) compounds are widely distributed in the soil, air and water. Natural production of atmospheric hydrocarbons is mostly from plant exudations and is estimated at $75-200 \times 10^9$ kg/year. The National Academy of Sciences[18] has emphasized the possible contribution of plants to the total PAH carcinogenic burden in the environment. The more volatile tetracyclic hydrocarbons are present mostly in gaseous form, while pentacyclic hydrocarbons tend to associate primarily with

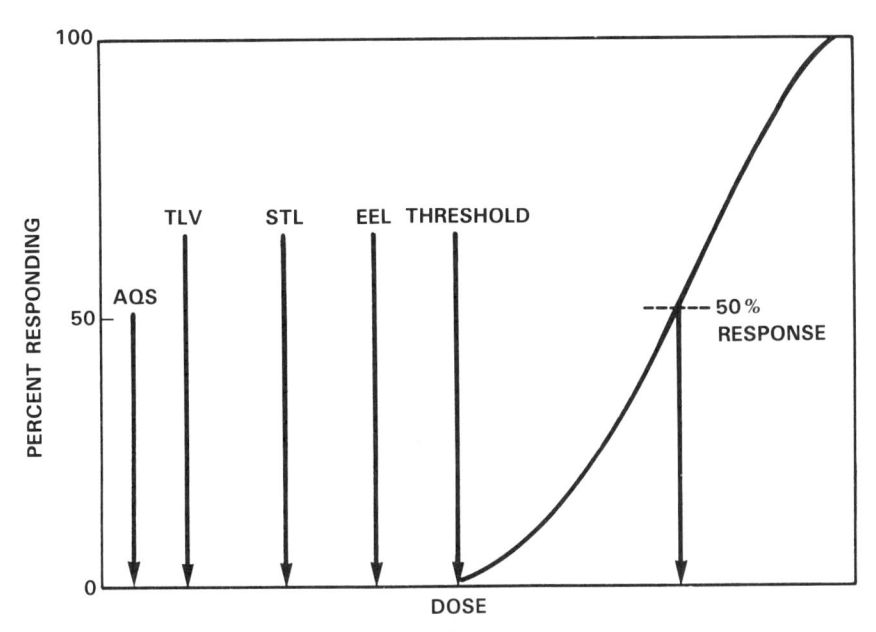

FIGURE 5.1—Relationship of dose exposure to airborne levels of toxic compounds and the setting of various exposure standards. Air quality standard (AQS) is set by the EPA for the general population. Threshold Limit Value standard is set by the American Conference of Governmental Industrial Hygienists for a 40-hr-per-week exposure; the same group sets Short-Term exposure limits (STL), a maximal limit for periods not to exceed 1 hour. The Emergency Exposure Limit (EEL), set by the Committee on Toxicology of the National Academy of Sciences, is a short-term exposure limit that is thought not to cause disability or interfer with an emergency task. The TLV is the minimum exposure dose that produces significant adverse effects.

respirable particles. Most of the PAH of potential health concern are adsorbed onto respirable particles in the environment.[4]

Concern over the adverse health effects of PAH as well as the observed higher incidence of lung cancer in urban and industrialized regions of the U.S. has produced considerable interest in determining levels of PAH; those species most studied are fluoranthene, benz(a)anthracene, chrysene, benzo(k)pyrene, benzo(e)pyrene, benzo(b)fluoranthene, benzo(j)fluoranthene, benzo(k)fluoranthene, anthracene, benzo(g,h,i)perylene, indeno(1,2,3-c,d)-pyrene, coronene and benzo(a)pyrene (BaP). The most commonly studied is BaP. About 90% of BaP air emissions are from stationary sources of coal combustion, including coal refuse fires, residential coal use and coke production. Less than 0.5% of total BaP emissions come from coal-fired power plants (Table 5.6). BaP levels in cities that have coke oven facilities are significantly greater than in cities that do not have such facilities. The

TABLE 5.6. Sources of benzo(a)pyrene emissions in the United States from 1971–1973 (R. H. Ross. 1977. Environmental interactions. In, "Environmental, Health, and Control Aspects of Coal Conversion: An Information Overview," Vol. 2, ORNL/EIS–95, Energy Research and Development Administration, Washington, D.C., pp. 6–1 to 6–131).

	BaP Emissions (tons/year)	
Source	U.S.	Worldwide
Heating & power generation		
Coal	431	2376
Oil	2	5
Gas	2	3
Wood	40	220
Subtotal	475	2604
Industrial processes		
Coke production	192	1033
Catalytic cracking	6	12
Subtotal	198	1045
Refuse & open burning		
Enclosed incineration	34	102
Open burning		
Coal refuse fires	340	680
Forest & Agriculture	140	420
Other	74	148
Subtotal	588	1350
Vehicles		
Trucks & Buses	12	29
Automobiles	10	16
Subtotal	22	45
Grand Total	1283	5044

facilities. The annual global emission of BaP, based upon data obtained from 1966–1969, is estimated at 5,000 tons; the U.S. contribution may be about 1,000 tons.[29]

The level of BaP is a very rough estimate of the total carcinogenic PAH level in the air but is not a good indicator of most other PAH compound concentrations. BaP air levels in urban regions range from a low of 0.1 ng/m^3 in the spring in residential New York City to high values of over 20 ng/m^3 in the autumn, observed in Nashville, Tennessee.[4] Air levels of BaP are considerably higher near freeways than in residential areas of a city. The highest recorded BaP level in the U.S. was for Altoona, PA in 1967, 29.5 ng/m^3.[50] One of the highest exposure levels to BaP occurs in tavern air as a result of passive exposure to cigarette smoke. Such levels reach 30–140 µg/m^3.[5] In one Czechoslovakian study, it was found that air of a beer hall was concentrated with BaP to levels up to 100 times that of city air.[18] The most significant mode of PAH or BaP removal from air is by photooxidation by sunlight. PAH also react with ozone, NO_x and SO_x. An example is the ozonolysis of benz(a)anthracene which undergoes oxidation of a double bond on exposure to ozone.[18] PAH are primarily adsorbed onto the surfaces of particles, particularly by hydrogen bonding to soot particles. The average uptake of BaP in drinking water is 20 ng/l; it is over 100 µg/kg in some smoked foods.

"... PAH formed by high temperature processes, whether from natural open burning and volcanic eruptions or from man-induced combustion reactions, are all emitted into the atmosphere, and thus are subject to the same dynamic forces which govern the movement, transport and fallout of aerosols generally, Because a significant portion of PAH, adsorbed onto the aerosols, will decompose by photooxidation while still in the atmosphere, either stationary or in motion, their fallout at greater distances from the source ... will be relatively very limited .... The degradation of PAH in the atmosphere by photooxidation will also continue to some extent when they have settled back on earth and water surfaces, for as long as they are exposed to sunlight. However, some PAH will be degraded by soil bacteria and aquatic organisms. PAH, while adsorbed on particles or in solution, may ... remain there stable for extremely long periods, given absence of light and anaerobic conditions."[29]

Some studies have indicated a positive correlation between BaP air levels and lung-cancer mortality, while others have failed to demonstrate such a correlation.[6] Such correlations are complex in attempting to identify any one causative agent. For example, death rates from lung cancer for smokers in urban areas is 25% to 125% higher than for smokers in rural areas (Table 5.7). A quantitative estimate of the relationship between lung cancer rates and atmospheric BaP concentrations was attempted by the Committee on Biological Effects of Atmospheric Pollutants of the National Academy of Sciences.[58] Using the comparison between urban and

TABLE 5.7. Comparison of lung cancer mortality in urban and rural areas of the United States; data are standardized for age and smoking history and are expressed as number of deaths from lung cancer per 100,000 population (L. B. Lave and E. P. Seskin. 1970. Air pollution and human health. Science 169:723–733).

| Deaths in Smokers ||| Deaths in Nonsmokers ||| Location |
Urban	Rural	Ratio	Urban	Rural	Ratio	
101	80	1.26	36	11	3.27	California men
189	85	2.23	50	22	2.27	England and Wales
-	-	-	38	10	3.80	Northern Ireland
100	50	2.00	16	5	3.20	American men

rural lung cancer rates and urban (6.6 µg /1000 m³ BaP) and rural (0.4 µg/1000 m³ BaP) concentrations, they concluded that a 100% increase in lung cancer death rate is associated with an 6.2 µg/1000 m³ BaP increase, using data from the most polluted urban area and the most nonpolluted rural area in the U.S.

Aromatic Amines

The aromatic amines represent a large and growing group of valuable chemicals in industry and by-products of fossil fuel conversion and combustion. Aromatic nitro and amino compounds, such as aniline, nitrites and organic nitrates generate methemoglobin; methemoglobin is the chemical analogue of hemoglobin in which the iron of heme has become oxidized. Oxygen bound to methemoglobin is very tightly attached and not available to tissues. Normal level of methemoglobin in humans is 2%. Cyanosis developes at methemoglobin levels of > 15%; however, recovery is possible even when levels reach 75%.[60]

The epidemiology of aromatic amine carcinogenesis is essentially the epidemiology of bladder cancer of industrial origin;[58] among the more important carcinogenic chemicals are 1- and 2-naphthylamines, benzidine, 4-biphenylamine, 3-3′-dichlorobenzidine and diphenylamine. Among these, 2-naphthylamine, benzidine and 4-biphenylamine are established human bladder carcinogens. The carcinogenic potential of aromatic amines has

5 – Environmental Air Pollution   121

certain structural requirements:[1] the type of aromatic ring system;[2] the position of the amino group on the ring;[3] the presence of other substitutes.[59] Biological data on many of these compounds are inadequate to assess their carcinogenic activity.

Particles

In the late 1800s, scientists began to realize that disease was transmitted by particles or aerosols in the air. Following Tyndall's publication "Essays on the Floating Matter of the Air" in 1882 and Arnold's "Untersuchungen uber Staubinhalation und Staubmetastase" in 1885, the toxicological importance of aerosols was better appreciated. Natural sources are the origin of most particles (Table 5.8); these include, erosion of soils, forest fires, volcanic debris, sea salt and large amounts of sulfates, nitrates, ammonia salts and hydrocarbon exudations from plants. Anthropogenic particles exhibit greater toxicity than natural particles and account for 15–20% of total particle levels (Table 5.9). They tend to be concentrated in urban areas; significant differences in particle levels are noted in rural regions of the U.S.

Atmospheric suspended particle levels are the cumulative result of the following mechanisms:[18]

1. Growth or change in particles as a result of chemical reactions on the particle surface
2. Change in particles by attachment and adsorption of trace gases and vapors
3. Change caused by collisions of particles undergoing Brownian motion
4. Change caused by collision of particles in turbulent air flow
5. Gain or loss in particles caused by diffusion or convection from nearby regions
6. Loss by gravitational settling; proportional to particle size and density
7. Removal at earth's surface by impaction, interception, Brownian motion and turbulent flow
8. Loss or modification by rainout within clouds
9. Loss by washout below cloud level

Residence times of particles <5 µm in diameter exceed 100 hours in the atmosphere in dry climates. Washout by rainfall can rapidly remove particles from the air. For example a light rainfall of 8 hours duration will remove 25% of the particles from the air while a heavy rainfall of <4 hours duration will remove 75% of atmospheric particles below the level of the clouds.

Respirable-sized particles are continually emitted from buildings, farms and industries in the form of soot, fly ash, sulfates, metals, rubber, asphalt

TABLE 5.8. Estimated emissions into the air of particles smaller than 20 microns radius from natural and man-made sources (Report on the Study of Man's Impact on Climate. 1971).

Type of Particle		Amount
Natural		
Soil and rock debris		100–500
Forest fires and slash-burning debris		3–150
Sea salt		300
Volcanic debris		25–250
Sulfates from H_2 emissions		130–200
Ammonium salts from NH3		80–270
Nitrate from NO_x		60–430
Hydrocarbons from plant exudations		75–200
	Subtotal	773–2200
Man-made		
Particles by direct emissions		10–90
Sulfates from SO_2		130–200
Nitrates from NOx		30–35
Hydrocarbons		15–90
	Subtotal	185–415
	Total	958–2615

and various other emissions. The overall amount of anthropogenic particles has fallen in the last few decades because of air pollution control devices such as electrostatic precipitators, which creates an electrostatic charge in the stack of an industrial facility, charging the particles, which then collect on an oppositely charged area within the stack (Figure 5.2). The efficiency of electrostatic precipitators has risen from about 75% in 1940 to over 95% today. Even with 99% efficiency, large amounts of respirable particles escape into the air; they also are of greater toxicological concern because they are smaller and penetrate deeper into the lung. Thus, by the use of electrostatic precipitators, visibility and appearance around an industrial plant can be greatly improved, but the invisible long-term health hazards to surrounding populations may not be as significantly reduced.

TABLE 5.9. Estimates of global particulate organic carbon (POC) from anthropogenic sources for 1973–1974 (B. R. T. Simoneit and M. A. Mazurek. 1981. Air Pollution: The Organic Components. CRC Rev. Environ. Control 11:219–276).

Source	World Production or Consumption, 10^6/y	Total Particulate Emissions, 10^6/y	POC Emissions,* 10^6/y
Coal	3069	–	–
Power production	1290	14.2	1.4
Industry	770	20.0	2.0
Domestic and commercial fuel	400	4.0	0.4
Cleaning refuse	860	1.8	0.2
Coke	910	2.4	0.2
Carbon black	4	0.3	0.1
Cement	696	6.5	0.3
Pig iron and crude steel	1220	11.0	–
Ferroalloys	12	0.7	–
Copper Smelting	9	1.5	–
Al,Pb,Zn products	22	1.2	–
Lime production	119	3.8	0.2
Nitric acid production	30	0.2	–
Phosphate fertilizer	23	1.2	0.3
Chemical wood pulp	91	0.9	0.2
Incineration	630	5.4	0.9
Noncommercial fuel	1940	9.7	1.9
Cotton ginning	14	0.2	0.1
Wheat handling	360	9.0	3.6
Petroleum refining	2650	0.4	–
Petroleum combustion			
Gasoline	613	1.2	0.2
Kerosene	78	0.1	–
Fuel oil	592	0.6	0.1
Residual oil	959	1.3	0.1
Aircraft jet fuel	106	–	–
Natural gas	1.3×10^{12}m^3	0.4	0.1
Agricultural burning	1000	8.5	1.6
Totals	18,444	100.0	13.9

* Diameter < 1μm

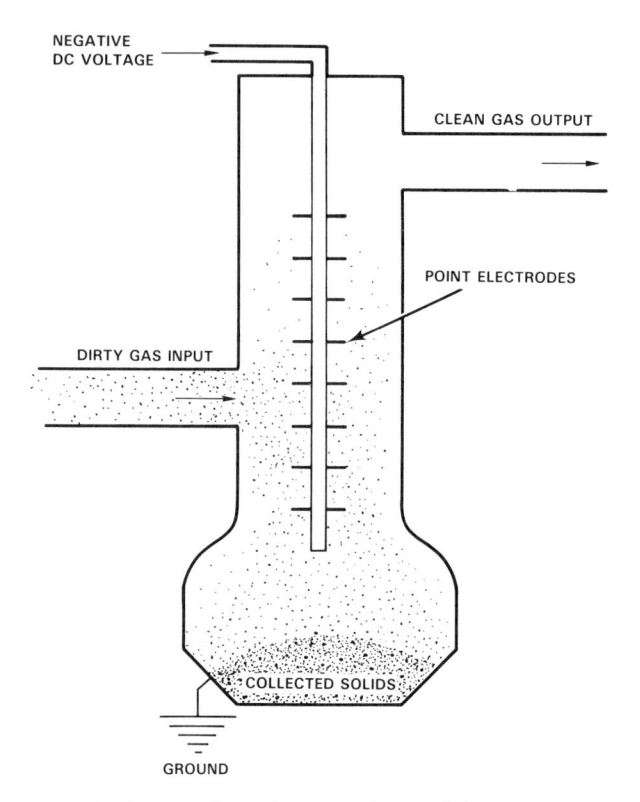

FIGURE 5.2 – Basic design of an electrostatic precipitator.

Several factors determine injury to the respiratory tract from inhaled particles:

A. Quantity of particles inhaled
   a. Concentration in the atmosphere
   b. Particle size
   c. Duration of exposure
   d. Respiratory rate and volume
B. Quantity of particles retained
   a. Physicochemical properties of the particles
   b. Anatomical considerations
   c. Clearance efficiencies
C. Site of action of the particles
   a. Site and magnitude of deposition
   b. Solubility
   c. Toxicity of particle constituents

Toxicity from inhaled particles may be mediated through physical properties of the particles, as with radionuclides or quartz, or through chemical action, as with PAH. Loading of pulmonary macrophages with phagocytized particles will slow their overall clearance rates from the lung. Toxic gases and chemicals may be absorbed onto particle surfaces and carried into tissues following inhalation. Phagocytized particles may alter lysosomal membrane permeability, releasing hydrolytic enzymes into the cell cytoplasm. Macrophages may also secrete powerful oxidant radicals, such as superoxide (O_2^-) and hydroxide, and increase production of hydrogen peroxide, particularly after phagocytosis of particles. An inducible enzyme, superoxide dismutase, protects cells against superoxide radical formation.[8]

Human exposure to high particulate levels is associated with increased incidences of asthma, pneumonia and bronchitis as well as lung cancer. Inhaled particles may promote lung-cancer development even though they are deposited at a different time or route than the pulmonary carcinogen.[9] Uptake and retention of chemical carcinogens adsorbed onto the surface of particles may be enhanced by the particles themselves.

Nitrogen Dioxide

The oxides of nitrogen (NO_x) are a complex series of compounds, including nitrous oxide (N_2O), nitric oxide (NO), nitrogen dioxide (NO_2), nitrogen trioxide (N_2O_3), nitrogen tetroxide (N_2O_4) and nitrogen pentoxide (N_2O_5). The term NO_x is used to indicate the sum of NO and NO_2. The other nitrogen compounds are either inert or present in such small quantities in the air as to not contribute significantly to air pollution. All the oxides of nitrogen react in air so that the principal remaining form is NO_2. The end product of UV irradiation in polluted atmospheres containing hydrocarbons, NO_x and ozone is collectively termed photochemical smog (Figure 5.3). In unpolluted air, the most abundant NO_x compound is N_2O (mean level of 0.5 ppm), produced by the microbial decomposition of nitrates, ammonia and amino acids.[11] Average values in North America are 4 ppb for NO_2 and 2 ppb for NO.[52]

Bacterial action is responsible for the major production of global NO_x, which is mostly NO. Microbial action produces at least 10 times more NO_x than formed by anthropogenic sources.[52] The source of NO_x compounds in polluted air is mostly the burning of fossil fuels, particularly gasoline in vehicles. In 1975, the U.S. total production of NO_x was 24 million tons; of this amount, 10.7 million tons came from vehicle exhaust, 6.8 million tons from coal and oil-fueled boilers, and 5.6 million tons from stationary industrial sources. NO_x emissions from automobile exhausts range from 0.1 to 2.0 g/mile, depending on vehicle maintainance, engine capacity, type of fuel and use of antipollution devices. Other sources of NO_x are welding operations, combustion of natural gas in homes, and cigarette smoke.

NO produced by motor vehicles from the reaction of N_2 and O_2 in the combustion process undergoes atmospheric photochemical reactions with hydrocarbons and UV light:

$$NO + RO_2 \rightarrow NO_2 + RO.$$

NO_2 is also produced by oxidation of NO with ozone:

$$NO + O_3 \rightarrow NO_2 + O_2.$$

The conversion of NO into NO_2 is fairly slow, resulting in a characteristic diurnal peak in NO and NO_2 levels, with NO levels peaking during morning rush hours and then being consumed to form peak NO_2 levels later in the day (Figure 5.3). After NO disappears, ozone starts to accumulate, peaking after NO_2 during the day.

NO_x is also a source of peroxyacetyl nitrates (PAN), which are strong lacrimators:

$$RCOO + NO_x \rightarrow RCOONO_x.$$

PAN levels peak only after the formation of high levels of NO_2.

Nitrous acid (HNO_2) and nitric acid (HNO_3) are produced in varying amounts in the atmosphere and in the humidified air of the deep lung following exposure to NO_x. The relative insolubility of NO_2 in water allows

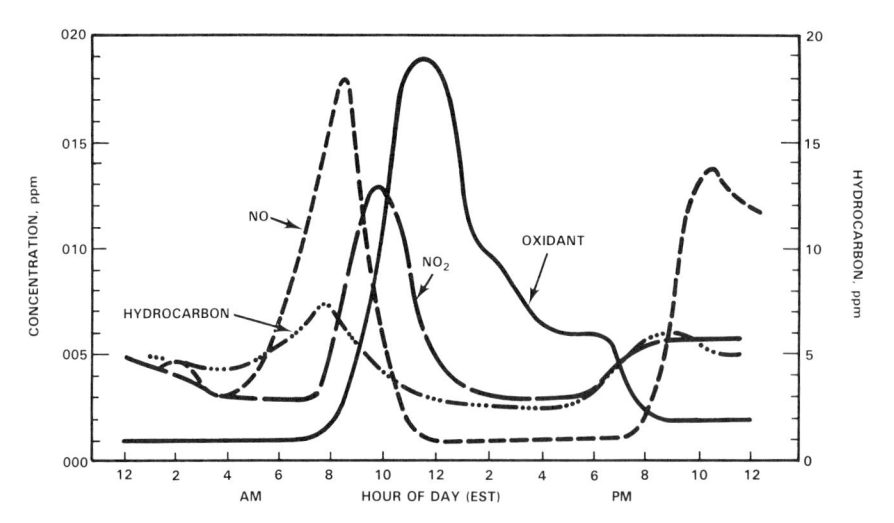

FIGURE 5.3—Typical daily air concentrations of nitrous oxide, nitrogen dioxide, total oxidant and hydrocarbon levels in photochemical smog of Los Angeles County in the 1960s (U.S. Department of Health, Education, and Welfare. 1970. Air Quality Criteria for Photochemical Oxidants. National Air Pollution Control Administration Publication AP-63, Washington, D.C.).

NO_x to pass through the airways of the upper respiratory tract to the alveoli, where it is converted to acids which are highly irritating and damaging to lung tissues. Nitrites and nitrates are formed from the reactions of NO_x in solution with metals. Nitric acid may be adsorbed onto the surface of soot particles in the air, resulting in deeper pentration of the acid into the lung.

The formation of N-nitrosation compounds, such as nitrosamines, results from atmospheric reactions between nitrogenous compounds and NO_x and nitrous acid. Elevated nitrosamine levels, seen in the atmosphere of several eastern U.S. cities, are of particular concern because of the known carcinogenic potency of nitrosamines. Nitrosamines are also found in cigarette smoke and in some cooked foods and industrial processes.[11]

Studies of rhesus monkeys indicate a rapid transfer of inhaled $^{13}NO_2$ from the alveolar air to the blood, where high level of pulmonary radioactivity is maintained for 10–30 minutes, in contrast to zeon inhalation, which is "washed out" of the lung during this time. A one-pack-a-day cigarette smoker may absorb about 3.0 gram NO_x daily.

Nitrogen dioxide as an oxidant is less damaging to the lung than is ozone. Both ozone and nitrogen dioxide are relatively insoluble in water and are irritants in the deep lung. Both oxidants cause lipid peroxidation in the lung, which are antagonized by vitamin E and other anti-oxidant compounds. However, the oxidation of unsaturated fatty acids by nitrogen dioxide is mechanistically different than by ozone. Remarkably stable free radicals are formed after nitrogen dioxide exposure, from the oxidation of lecithin, a constituent of cell membranes and lung surfactant.

Nitrous oxide has a strong association with iron in hemoglobin (six orders of magnitude greater than the association for oxygen), which results in the formation of detectable levels of methemoglobin and depression of the oxygen-hemoglobin dissociation curve after exposures of 10–30 ppm NO for several hours. However, the interaction of nitrogen dioxide with heme is not an important toxic factor, so far as acute toxicity is concerned;[11] long-term effects are unknown.

Nitrogen dioxide is particularly damaging to alveolar macrophages, seriously impairing phagocytosis, interferon production and antibactericidal capability. Mice exposed to levels of nitrogen dioxide as low as 0.5 ppm experienced a significantly higher mortality after challenge with Klebsiella pneumoniae or Streptococcus pyogenes. Prolonged impairment of pulmonary clearance of inhaled particles is seen following exposure to nitrogen dioxide levels that produce permanent histological lesions in the lungs; lung clearance was transiently impaired with nitrogen dioxide-induced, reversible histopathological lesions in the lung.[14]

Acute exposure to nitrogen dioxide results in dyspnea, bronchospasm, cough, headache, tachycardia and chest pain. Delayed symptoms following somewhat lower exposures include pulmonary edema following a symptom-free interval. Bronchitis and bronchiolitis, with persistent cough, bronchiolitis obliterans or progressive deterioration and pneumonia, may follow subacute exposure to nitrogen dioxide. Bronchiolitis obliterans results from the dysplastic regeneration of bronchial epithelium with an entrapped intrabronchiolar exudate and fibroplasia, occurring at 2–6 weeks after exposure and often resulting in permanent lung damage.[11] In one area of Chattanooga, TN, exposed to an annual mean nitrogen dioxide level of 0.083 ppm from an explosives factory emission, a high incidence of acute bronchitis and decreased resistance to acute pulmonary infections were noted in the surrounding population. Increased airway resistance and exacerbated symptoms in individuals with pre-existing bronchitis and asthma are seen following short-term exposures to more than 1 ppm nitrogen dioxide.[12]

Endothelial damage and early pulmonary edema are seen in experimental animals exposed to 30–40 ppm nitrogen dioxide. Exposure of rats to 15–25 ppm nitrogen dioxide caused rapid destruction of cilia and focal hyperplasia of nonciliated bronchiolar epithelium. Bronchiolitis obliterans appeared after 16 weeks exposure; after 10 weeks exposure to 15 ppm, rats recovered. Cuboidal metaplasia of the alveolar epithelium and replacement of damaged type I cells by proliferating type II cells was seen in rats at 2 days after exposure to 15–20 ppm.[13] Prolonged exposure to nitrogen dioxide results in interstitial fibrosis and centrilobular emphysema. There is a marked variation among species in sensitivity to nitrogen dioxide: the hamster is resistant, and the mouse, rabbit, monkey and dog are sensitive to nitrogen dioxide effects. Nitrogen dioxide-induced pulmonary fibrosis tends to continue, even after termination of exposures.[13]

Bronchospasm has been seen in human volunteers after a 15-minute exposure to 5 ppm nitrogen dioxide, due to histamine release in the lung from degranulating mast cells. Increased bronchoconstriction was seen in susceptible human asthmatic patients after inhalation of as little as 0.1 ppm nitrogen dioxide for 1 hour. However, results of most human exposure studies suggest that nitrogen dioxide at concentrations common in polluted ambient air (0.5 ppm and below) has little or no direct effect on pulmonary function, even in sensitive populations like asthmatics (37–38). Histopathological effects of nitrogen dioxide inhalation in experimental animals after short to prolonged exposures to exposures levels of 0.25 to 3.0 ppm are listed in Table 5.10. A proposed explanation for the development of pulmonary injury following inhalation of nitrogen dioxide is given in Figure 5.4.

The recommended maximum atmospheric concentration of NO_x for occupational exposure is 5 ppm. The mean annual national AQS for nitrogen dioxide is 0.05 ppm. Los Angeles is currently the only urban region in the U.S. that regularly exceeds the AQS standard.

TABLE 5.10. Histopathological effects of inhaled nitrogen dioxide in experimental animals (R. F. Bils and B. R. Christie. 1980. The experimental pathology of oxidant and air pollution inhalation. Int. Rev. Exp. Pathol. 21:195–293).

Conc.(ppm)	Species	Effects
0.25	Rabbit	Irreversible lung collagen changes
0.40	Guinea pig	Increased alveolar capillary permeability
0.50	Rat	Lung mast cell degranulation
0.50	Mouse	Alveolar wall thickening
0.50	Guinea pig	Decreased lung phospholipids
0.80	Mouse	Pulmonary alveolar edema
0.80	Rat	Terminal bronchiolar hypertrophy
1.0	Rat, dog Rabbit, hamster	Focal interstitial pneumonia
1.0	Monkey	Slight emphysema
1.1	Guinea pig	Emphysema, bronchiolar damage
1.5	Mouse	Desquamative bronchitis
2.0	Rat	Increased lung weight due to edema, increased collagen formation and bronchiolar hyperplasia
2.0	Guinea pig	Type II cell hypertrophy and hyperplasia
2.0	Monkey	Bronchial-bronchiolar hyperplasia and metaplasia

Ozone

Ozone is not formed in significant quantities near the earth's surface by natural environmental interactions in unpolluted air. UV radiation dissociates O_2, producing atomic oxygen which reacts with oxygen to form ozone, mostly at high altitudes by the following reactions:

$$O_2 \,(UV) \rightarrow 2O\cdot$$

$$O + O_2 \rightarrow O_3$$

However ozone accumulation is increased at low altitudes when hydrocarbons (particularly olefins and substituted aromatics) and NO_x enter the photochemical smog cycle (Figure 5.3). Hydrocarbons in photochemical

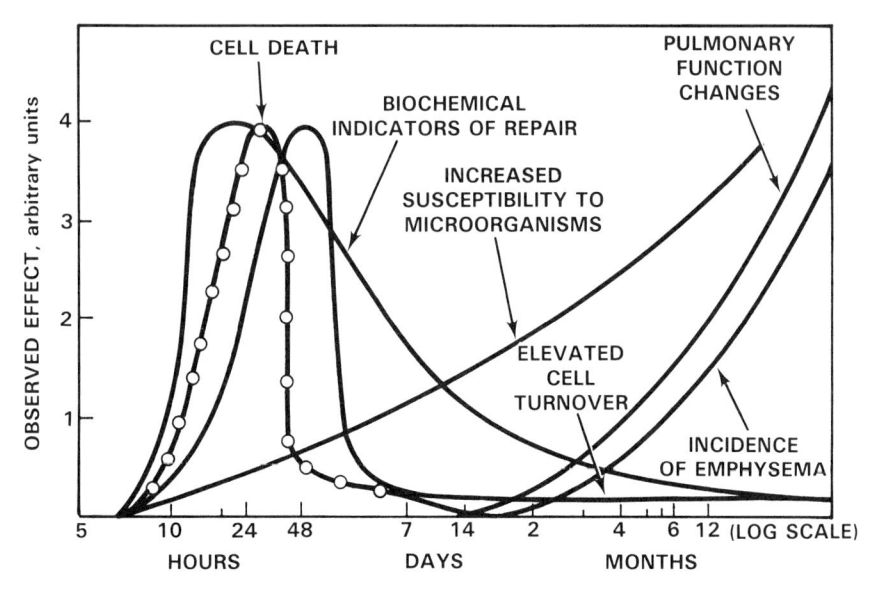

FIGURE 5.4 — Probable temporal sequence of damage and repair in the lung following a single, short-term exposure to nitrogen dioxide (D. B. Menzel. 1980. Pharmacological mechanisms in the toxicity of nitrogen dioxide and its relations to obstructive respiratory disease. In, "Nitrogen Oxides and Their Effects on Health," Ann Arbor Science, Ann Arbor, MI, pp. 199–216).

smog are oxidized to form oxygen radicals which react with oxygen to form ozone; ozone then oxidizes NO to form NO_2 with a subsequent rapid buildup in the air of ozone and NO_2.

Ozone, a highly toxic, biologically reactive gas, is a major component of photochemical smog. Ozone levels as high as 0.6 ppm are periodically encountered in Southern California on high-smog days. The "yellow alert" level for photochemical oxidants is 0.1 ppm for a 1-hour exposure; the "red alert" level is 0.25 ppm for 2.75-hours; and the "emergency alert" level is 0.37 ppm for 2.75-hours. Ozone levels as high as 0.6 ppm are also found in cabins of high-altitude passenger airplanes; the ozone enters the cabin from the outside air through the ventilation system.

Molecular oxygen (O_2) acts as an oxidant by virtue of its univalent reduction during cellular respiration to reactive metabolites (e.g., superoxide anions, hydroxyl radicals and hydrogen peroxide). Preexposure to ozone has been found in some animal species to markedly increase the tolerance to hyperoxia, indicating cross-tolerance between the atmospheric oxidants, ozone and oxygen.[39,55]

Acute exposure to ozone causes pulmonary edema and epithelial necrosis and induces characteristic lesions in the terminal bronchioles and centroacinar alveoli.[16] Species differences in ozone toxicity are due in part to differences in anatomy of the terminal airway systems, particularly its length. In animals, the LC_{50} values for ozone range from 6 ppm to 20 ppm; rats are the most sensitive, and dogs the least sensitive. In contrast, the LC_{50} value for nitrogen dioxide ranges from 60 ppm to 100 ppm.

A marked serous and fibrinous inflammatory exudate in the alveoli, along with alveolar interstitial edema and acute bronchiolitis, are seen in squirrel monkeys after several hours exposure to 2–4 ppm ozone. A variety of pulmonary and systemic effects of short-or long-term exposures to ozone have been described (Table 5.11). Long-term exposure is associated

TABLE 5.11. Health effects from ozone inhalation exposure in humans.

Conc. (ppm)	Effects
0.05	Threshold for headache
0.10	For 1-hour, causes slight increase in airway resistance
0.15	Threshold for eye discomfort
0.26	Threshold for cough
0.10–0.30	Eye irritation, decreased athletic performance. Increased airway resistance in smokers with pre-existing bronchitis and emphysema.
0.30	Threshold for chest discomfort in young adults
0.25–0.56	For 2-hours, causes chest tightness during exercise
0.37–0.75	For 2-hours, causes decreased vital capacity, increased residual volume, and a dose-dependent alteration in exercise ventilatory patterns
0.50	For 2 to 6-hours, causes marked changes in pulmonary mechanics and blood gases in subjects with asthma. Decreased compliance and tidal volume and increased respiratory frequency in all others. Decreased visual acuity in dark-adapted room, increased sphering and osmotic fragility of erythrocytes, and decreased phagocytosis by neutrophils. Chromosome aberrations in circulating lymphocytes.

with chronic bronchitis, bronchiolitis, pneumonitis and emphysema. Partially reversible interstitial fibrosis is also the result of chronic ozone exposure.[18]

The presence of unsaturated fatty acids in cell membranes makes the cell susceptible to oxidant attack by ozone. The initial reaction can be expressed thus:

$$RX + X \cdot \rightarrow R \cdot + XH$$

where R· is the alkyl radical of a fatty acid; it is then propagated:

$$R \cdot + O_2 \rightarrow RO_2 \cdot.$$

$$RO_2 \cdot + RH \rightarrow R \cdot + RO_2H$$

where RH is the fatty acid, $RO_2 \cdot$ is the hydroperoxyl free radical, and RO_2H is a fatty acid hydroperoxide. The molecular basis for ozone toxicity is therefore attributed to free radical formation and peroxidation of the cell membrane. In addition, ozone causes ozonolysis of olefinic bonds in unsaturated fatty acids. Some of the effects of ozone inhalation can be duplicated by administration of fatty acid ozonides. The free radical chain reaction can be terminated by an anti-oxidant (AH):

$$RO_2 + AH \rightarrow RO_2H + A \cdot$$

The anti-oxidant at highest concentration in the body is vitamin E or alpha tocopherol. Mortality in rats deficient in vitamin E increases after exposure to ozone but decreases with vitamin E supplements in the diet.

The infectivity of bacterial pathogens is enhanced by ozone exposure due to inhibition of phagocytosis and lysosomal-enzyme-related digestion of bacteria and to direct destruction of alveolar macrophages. Several studies of populations living in Southern California have shown an increased incidence of eye irritation, chest discomfort, cough and decreased athletic performance with increasing oxidant exposure levels (Figure 5.5). Patients with angina appear to be more susceptible to ozone toxicity than are individuals without any evidence of heart disease; extensive exercise may increase pulmonary and cardiovascualr symptoms in such individuals.[56]

Oxidants are not mutagenic, teratogenic or carcinogenic despite the ability of ozone to induce chromosomal aberrations in circulating lymphocytes. However, both ozone and nitrogen dioxide may increase the formation of nitrosamines in the atmosphere or alter the pulmonary metabolism of inhaled PAH, thus influencing the carcinogenic potential of other inhaled carcinogens.[17]

Sulfur Oxides

Anthropogenic sulfur dioxide emissions are about 100 million tons per year

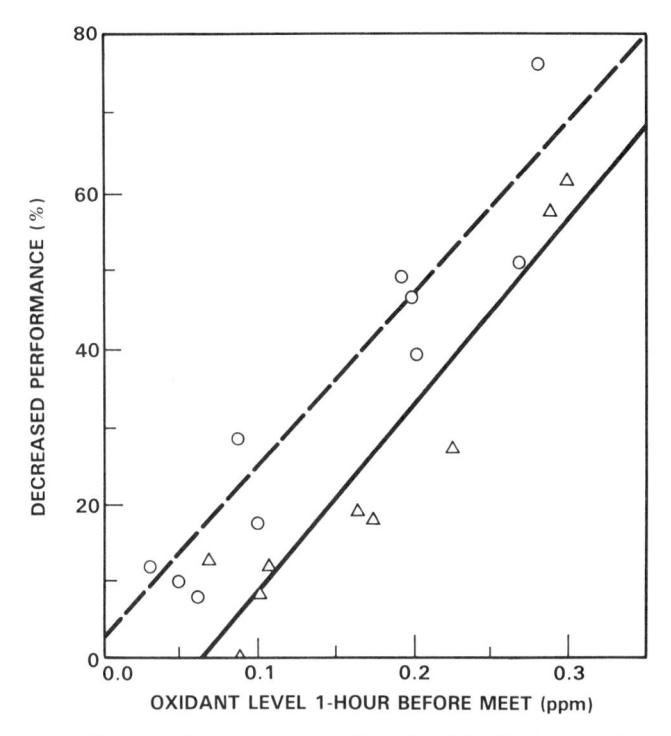

FIGURE 5.5 — Relationship between oxidant level in the hour prior to an athletic event and percent of team members with decreased performance; solid line for 1959–1961 and broken line for 1962–1963 (W. S. Wayne, P. F. Wehrle and R. E. Carroll. 1967. Oxidant air pollution and athletic performance. J. Am. Med. Assoc. 199:901–904).

while natural sources, such as volcanoes, contribute only about 1.5 million tons per year.[51] Sulfur dioxide is capable of travelling considerable distances in the atmosphere. While in the air SO_2 may:

1. React with other gases, with or without photochemical precipitation
2. Be absorbed by water droplets
3. Be chemically absorbed onto dry metal oxide particles
4. Be adsorbed onto particles containing solutions of metal salts that convert SO_2 to sulfates.

The oxides of sulfur (sulfur dioxide, SO_2; sulfur trioxide, SO_3; sulfuric acid, H_2SO_4; and sulfates, $SO_4^=$) originate from the combustion of coal, from sulfuric acid plants and from metalurgical processes using sulfur-containing ores. The major emission from these sources is sulfur dioxide, a pungent, suffocating and irritant gas that is highly soluble in water. Sulfur dioxide undergoes a variety of photochemical and catalytic reactions in the atmosphere, forming sulfuric acid and sulfates, principally ammonium sulfate, from the reaction of sulfuric acid with atmospheric ammonia. In fog or water droplets, SO_2 becomes sulfurous acid (H_2SO_3) which is rapidly oxidized to sulfuric acid (H_2SO_4) by dissolved oxygen. Most atmospheric SO_2 and sulfate is removed by rainout or washout due to the high solubility of sulfur oxides in water (e.g., SO_2, 10.8 g/100 g water at 20°C).

Sulfur dioxide, along with NO_x produced during the combustion of coal and other fossil fuels, has been implicated in the production of acid precipitation. The pH of rain water in unpolluted regions of the U.S. ranges from 5.5 to 5.8, while in the polluted northeastern industrial regions the range is 4.1 to 4.3. Poorly buffered aquatic environments in the northeast are particularly sensitive to acid precipitation. Sulfates, a major component of particles in many regions of the U.S., are derived from atmospheric reactions of sulfur dioxide, sulfuric acid and metals which promote the conversion of sulfur dioxide to sulfuric acid, which then forms sulfate salts.

Sulfuric acid is the most important sulfur compound, and most commonly used industrial chemical, with 34,000,000 short tons sold or used in the U.S. in 1976.[57] Most sulfuric acid is used in making fertilizers; other uses are in petroleum alkylation, iron and steel pickling, uranium leaching and processing and manufacture of alcohols, other acids, pulp and paper, rayon, explosives and storage batteries. Sulfuric acid mist, derived from industrial applications and from the oxidation of atmospheric sulfur dioxide, is very hygroscopic and a strong acid.

Typical sulfur dioxide levels in urban or industrialized regions range from 0.01 ppm to 0.08 ppm, with higher levels found near focal sources of pollution, such as coal fired-power plants. The majority of inhaled sulfur dioxide is absorbed by the upper respiratory tract at high exposure concentrations, while deeper penetration into the lung is observed at low sulfur dioxide concentrations. In rabbits, about 95% of sulfur dioxide is absorbed by the upper respiratory tract at 100 ppm, 50% at 0.5 ppm, but only 5% is absorbed at 0.1 ppm. In humans, at typical environmental sulfur dioxide levels, about 50% of inhaled sulfur dioxide penetrates into the deep lung. Sulfur dioxide is rapidly absorbed from the air in the respiratory tract and evenly distributed throughout the body tissues; the half-life of intratracheally instilled ^{35}S-sulfate in rat lung is only 35 minutes.

Species vary greatly in their sensitivity to sulfur oxides: the rat is the most resistant and guniea pig the most sensitive animal. However, there is evidence suggesting that humans are even more sensitive than the guinea pig. The toxicity of sulfur dioxide is enhanced when inhaled with a particulate aerosol.[19] Although health effects in humans are ranked according to ambient levels of sulfur dioxide or sulfuric acid (Tables 5.12 and

TABLE 5.12. Health effects in humans following inhalation of sulfur dioxide.

Conc. (ppm)	Effects
0.0035–0.1	Excessive acute respiratory disease, including chronic bronchitis
0.008–0.05	Increased respiratory infections in asthmatics
0.02	Community mortality is 2% greater than expected
0.03–0.05	Aggravation of pre-existing chronic bronchitis
0.04	Increased severity of illness during influenza epidemics
0.08	Increased frequency of asthmatic attacks
0.3	Taste threshold for detection
0.5	Odor threshold for detection
0.3–6.4	Increased light sensitivity
1.6	Increased bronchoconstriction
1–10	Increased airway resistance and decreased compliance
7	Increased nasal and laryngeal resistance
10	Lacrimation and rhinorrhea, pulmonary edema and epithelial necrosis. Fatal if exposure is prolonged.

5.13), actual exposure is to a mixture of sulfur dioxide, sulfuric acid, sulfates and particles in the environment.

The odor and taste thresholds for sulfur dioxide range from 0.3 to 0.5 ppm. High exposures in humans causes bronchoconstriction, resulting in increased airway resistance and upper airway inflammation. Inhaled sulfuric acid is 3–5 times more damaging than equal concentrations of sulfur dioxide. A histamine-related bronchoconstriction, which is partially antagonized by atropine, is seen in humans after exposure to sulfur oxides. The all-or-none response is similar to many asthmatic attacks seen in humans. Response to inhaled sulfur dioxide is more readily reversible than the response to inhaled sulfuric acid mist. Irritancy decreases with

TABLE 5.13. Health effects in humans following inhalation of sulfuric acid.

Conc. (mg/m³)	Effects
0.003–0.02	Excessive acute respiratory disease
0.35–5.0	Bronchoconstriction
0.6–2.0	Transient increased sensitivity to light
>1.0	Increasing incidence of chronic bronchitis
5.0	High level of discomfort and irritancy in upper respiratory tract following short-term exposures
>20	Lethal for prolonged exposures

an increase in particle size of the mist, indicating that the site of mist deposition is of prime importance in toxicity.

Sulfur oxides decrease mucociliary flow and upper-airway particle clearance from the lung in the dog, monkey and humans, but increases these parameters in the rat;[20-21] the reasons for this species difference are not known. The abnormal elimination of mucus produced by the tracheobronchial epithelium is associated with the development of chronic obstructive pulmonary disease, especially with chronic bronchitis. Significant alterations in mucociliary clearance are seen in donkeys at sulfuric acid exposures that are only 10% of the current TLV.[20] Chronic bronchitis in humans is a disease whose incidence is correlated with sulfur dioxide levels in the air or with cigarette smoking. A high incidence of chronic bronchitis is also seen in occupational exposures at sulfuric acid levels >1 mg/m³. During the 1950s, a significant correlation between sulfur dioxide levels and increased mortality from bronchitis was seen in the U.S.[22] Sulfur dioxide is a promoter of BaP pulmonary carcinogenesis in rodents.[54]

Epidemiological studies have shown associations between mixtures of sulfur dioxide and particles and increased incidences of pulmonary infections.[41] Reduction of sulfur dioxide and particulate levels in England and the U.S. correlate with a reduction in chronic bronchitis.[23] Temporary and reversible increases in airway resistance are seen at particulate levels of 250 μg/m³ and at sulfur dioxide levels of <300 μg/m³, when both are present together in the air. These data indicate an adverse health effect for combined exposures, presumably due to changes in sulfur dioxide deposition in the lung. Epidemiologic studies do not indicate that the concentration of sulfates is a more important air pollution variable than total particulates suspended as smoke. The current WHO recommended population exposure limits are 100–150 µg/m³ for sulfur dioxide as a 24-hour mean and 40–60 µg/m³ smoke as an annual mean.[40]

The national Air Pollution Control Administration listed the following conclusions:[53]

"At concentrations of about 1500 µg/m³ (0.52 ppm) of sulfur dioxide (24-hr average) and suspended particulate matter measured as a soiling index of 6 COH or greater, increased mortality may occur . . . . At concentrations of about 750 µg/m³ (0.25 ppm) of sulfur dioxide and higher (24-hr mean), accompanied by smoke at a concentration of 750 µg/m³, increased daily death rate may occur . . . . At concentrations of about 500 µg/m³ (0.11 ppm) of sulfur dioxide (24-hr mean), with low particulate levels, increased hospital admissions of older persons for respiratory disease may occur; absenteeism from work, particularly with older persons, may also occur . . . . At concentrations of about 715 µg/m³ (0.25 ppm) of sulfur dioxide (24-hr means), accompanied by particulate matter, a sharp rise in illness rates for patients over age 54 with severe bronchitis may occur . . . . At concentrations of about 600 µg/m³ (0.21 ppm) of sulfur dioxide (24-hr mean), with smoke concentrations of about 300 µg/m³, patients with chronic lung disease may experience accentuation of symptoms . . . . At concentrations ranging from 105 to 265 µg/m³ (0.037 to 0.092 ppm) of sulfur dioxide (annual mean), accompanied by smoke concentrations of about 185 µg/m³, increased frequency of respiratory symptoms and lung disease may occur . . . . At concentrations of about 120 µg/m³ (0.046 ppm) of sulfur dioxide (annual mean), accompanied by smoke concentrations of about 100 µg/m³, increased frequency and severity of respiratory diseases in school children may occur . . . . At concentrations of about 115 µg/m³ (0.040 ppm) of sulfur dioxide (annual mean), accompanied by smoke concentrations of about 160 µg/m³, increases in mortality from bronchitis and from lung cancer may occur."

Hydrogen Sulfide

Hydrogen sulfide is a colorless gas with a characteristic odor of rotten eggs; the olfactory threshold in humans is 0.1–0.2 ppm (Table 5.14). Natural gas is among the richest sources of H_2S; concentrations as high as 42% are seen in some Wyoming gases. Hydrogen sulfide is occasionally encountered in coal mining due to the action of steam on sulfides; it is also found during coke production. Other sources of environmental hydrogen sulfide are volcanic gases, geothermal wells, the Kraft paper pulp process and by bacteriologic action on a variety of substrates. Pollution recovery

TABLE 5.14. Health effects in humans following inhalation of hydrogen sulfide (W. M. Gafafer [ed]. 1964. Occupational Diseases: A Guide to Their Recognition. U.S. Government Printing Office, Washington, D.C.).

Concentration (ppm)	Effects
0.2	Detectable odor
10.	TLV for daily 8-hour exposure
150.	Olfactory nerve paralysis
250.	Pulmonary edema after prolonged exposure
500.	Systemic symptoms after 0.5–1.0 hour exposure
1000.	Respiratory collapse imminent
5000.	Immediate death

processes in industry remove, at best, only about 80% of emitted hydrogen sulfide.[24]

About a third of inhaled hydrogen sulfide is absorbed in the undissociated state; the remainder exists as the hydrosulfide ion (HS-). A sulfide oxidase system in the liver and kidney metabolizes HS- to thiosulfate. However, the majority of hydrogen sulfide is nonenzymatically metabolized at low to moderate exposure levels. Sulfide reacts with ferric ions in hemoglobin and myoglobin to form stable but dissociable complexes, such as sulfhemoglobin and sulfmethemoglobin.

Toxicity from inhaled hydrogen sulfide is associated with acute respiratory distress. Low levels of hydrogen sulfide are continuously being produced in the gastrointestinal tract; hydrogen sulfide, in the ppb range, in exhaled air is also the source of malodor.[14] Exposure of guinea pigs to 220 ppm hydrogen sulfide for 22 days resulted in mild pulmonary inflammation; in vitro rabbit alveolar macrophages were killed by exposure to 55 ppm hydrogen sulfide for 24 hours.[25] Exposures of 220 ppm hydrogen sulfide for 7 days are not toxic to reproductive ability or embryogenesis in rats.[26]

Hydrogen sulfide gives an offensive odor at 3–5 ppm, which is considerably below its TLV of 10 ppm. The threshold exposure level for serious eye injury is 70–140 ppm. At 210–350 ppm, the olfactory apparatus is rapidly paralyzed, neutralizing the sense of smell as a warning of acute exposure. Acute pulmonary edema and apnea occur at 400–700 ppm.

Respiratory paralysis and death occur in humans at exposure levels of >700 ppm.[24] At high concentrations, hydrogen sulfide activates carotid body chemoreceptors, causing respiratory depression similar to that seen with hydrogen cyanide. The cytochrome oxidase system appears to be the point of greatest biochemical sensitivity to hydrogen sulfide.[27]

During a thermal inversion, an accidental release of hydrogen sulfide for 25 minutes at a petrochemical plant in Poza Rice, Mexico in 1950, resulted in 22 deaths and 320 hospitalizations of nearby residents. The fatalities resulted from respiratory paralysis. Management of acute hydrogen sulfide inhalation includes respiratory assistance and oxygen administration. Lasting neurological changes due to severe cerebral hypoxia from systemic hypotension and hypoxia occur in some survivors of acute hydrogen sulfide exposure; these may include acoustic neuritis and dysarthria.

Carbon Monoxide

Carbon monoxide (CO) is a colorless, odorless, and tasteless gas that gives no warning of its presence until clinical effects appear. Ten times more carbon monoxide is produced in nature than by anthropogenic sources, mostly from oxidation of methane; other natural sources are kelp in oceans, fires and volcanos. Anthropogenic carbon monoxide production occurs largely from the combustion of fossil fuels, particularly from motor vehicle exhausts.[28] The atmospheric levels of carbon monoxide have increased about 100% in the U.S. during the past 40 years. The current rate of increase in atmospheric carbon monoxide levels is about 6% per year.[42] The greatest direct individual source of carbon monoxide exposure in humans is cigarette smoke, which contains about 40,000 ppm carbon monoxide.

Carbon monoxide is endogenously formed from the breakdown of hemoglobin in the liver. For each molecule of hemoglobin degraded to urobilinogen, one molecule of carbon monoxide is formed. Drugs such as phenobarbital and diphenylhydantoin induce heme biosynthesis and subsequent heme catabolism, increasing endogenous carbon monoxide formation. Patients with hemolytic anemia exhibit carbon monoxide formation of about 10 times normal.[29]

Health effects from inhalation of carbon monoxide may be due to a reduction in the oxygen-carrying capacity of the heme molecule in red blood cells, since carbon monoxide has a 240-fold greater affinity for heme than does oxygen. In the presence of carbon monoxide, the oxyhemoglobin dissociation curve is shifted to the left, so that the amount of oxygen available to tissues is diminished. Myoglobin also has a high affinity for carbon monoxide. At a 10% HbCO level, about 30% of myoglobin in cardiac muscle is combined with carbon monoxide. Normal carboxyhemoglobin levels in blood are 0.3-0.7% from endogenous sources. HbCO levels in patients with hemolytic anemia may approach 6%; levels in heavy cigarette smokers may be as high as 15%. Levels in nonsmoking steel workers at the end of a work shift range from 1.5-4.0%; in smoking steel workers, they may reach 20%. HbCO levels in rural residents are typically half those in urban residents.[29]

Exposure to a constant carbon monoxide level results in a rapid, predictable increase in HbCO; a doubling of baseline HbCO is seen after 30 minute exposure to 500 ppm CO and in 4 hours after 100 ppm CO. Equilibrium HbCO levels occur after 7-8 hours continuous exposure to 50 ppm CO. The HbCO half-time in blood is about 4-6 hours, irrespective of degree of hemoglobin binding with carbon monoxide.

Acute effects of increasing carbon monoxide exposure include a sequence of: headache, dizziness, lassitude, flickering before the eyes, ringing in the ears, nausea, vomiting, palpitations, pressure on the chest, muscular weakness, collapse, coma and death. Subtle physiological, behavioral, motor and intellectual changes are seen following exposures as low as 5-10 ppm CO (Table 5.15). Exposure of monkeys, baboons, beagles, rats and mice to 370-460 ppm CO for 168 consecutive days resulted in high HbCO levels but not in death, nor did the exposure cause evidence of histopathological damage in the lung.[30]

The body adapts to hypoxic stress by increasing cardiac output and blood flow to critical tissues such as brain and myocardium. If the capacity of the body to adapt to increasing HbCO is exceeded, tissue hypoxia results. Increased cardiac output, ventilation and coronary blood flow and reduced myocardial oxygen have been observed in humans with 4-13% HbCO levels. Individuals with cardiovascular disease are particularly susceptible to increasing HbCO levels. Significant increase in angina pain and decrease in exercise tolerance are noted in patients with advanced coronary disease at HbCO levels as low as 5%.[31]

Continuous, chronic exposure to carbon monoxide, resulting in HbCO levels as low as 5%, causes damage to the cardiovascular system, including increasing the rate of cerebrovascular accidents, decreasing auditory threshold sensitivity, increasing neuroretinitis and causing optic nerve atrophy. The significance of these effects at low continuous exposure levels is controversial.[43]

Carbon Dioxide

Carbon dioxide (CO_2) is formed from oxidation of carbonaceous material by reaction with oxygen in the air. A steady, continuous increase in atmospheric carbon dioxide levels has been seen during this century, primarily from the increased burning of fossil fuels.

TABLE 5.15. Summary of health effects in humans associated with inhalation exposure to carbon monoxide. Individuals with advanced cardiovascular disease show aggravated symptoms, such as angina pain; levels as low as 20% HbCO may be fatal to individuals with advanced pre-existing disease. Short exposures to >50,000 ppm CO may be fatal due to cardiac arrhythmia before HbCO levels are markedly elevated.

Blood Levels of Carboxyhemoglobin (HbCO) as % Saturation	Effects
0.3–0.7	Normal range for endogenous production
1–5	Increased blood flow
5–10	Changes in myocardial metabolism, diminished visual perception, manual dexterity and ability to learn
10–20	Headache, impaired coordination, altered visual evoked response and electroencephalograph, and impaired psychomotor functions
20–30	Throbbing headache and nausea
30–40	Severe headache, nausea, vomiting and syncope
>50	Coma, convulsions; lethal if untreated or prolonged exposure

Symptoms of carbon dioxide toxicity are due to oxygen deprivation, becoming evident when air oxygen levels fall below 12%. Up to 30% CO_2 in air can be temporarily tolerated, provided there is an adequate oxygen supply. Headache and rapid breathing occur at 8% oxygen; unconsciousness and death occur at oxygen levels of <5%. Respiratory acclimatization is seen at up to 3% CO_2; impaired performance in submarine personnel is found, even at oxygen levels of 21%, due to carbon dioxide-linked respiratory acidosis.[44]

Accidents involving exposure to high air levels of carbon dioxide have been encountered in fermentation situations (silos, grain elevators, brewery vats, sugar beet reduction basins) and in the use of dry ice (frozen carbon dioxide) in poorly ventilated spaces. Reversible and irreversible cerebral and retinal damage have been seen in those who survive carbon dioxide-induced comas.[45,46] Current ceiling occupational exposure limits for carbon dioxide is 3%.

Ammonia

Ammonia is a colorless gas with an easily discernible, pungent odor at concentrations of <1 ppm. It is a ubiquitous constituent of soil, air and water and an integral part of the ecological nitrogen cycle. More than 99.5% of atmospheric ammonia is produced by natural biological processes, mostly from microorganism decomposition of organic matter. The main anthropogenic sources of ammonia are combustion processes, feedlots, fertilizer plants, petrochemical industries and household uses.[32] The emission factors for ammonia from the combustion of fossil fuels are listed in Table 5.16; the emission for coal combustion is twice that for oil combustion and nearly 5,000 times greater than for natural gas combustion.[33] The emission factor for ammonia for gasoline or diesel-powered motor vehicles is about 2 lb/ 1,000 gallons.

Ammonia is a constituent of many atmospheric aerosols, and reacts with nitric and sulfuric acids at high humidities to form nitrates and sulfates, respectively. Neutralization of acids by atmospheric ammonia is an important factor in decreasing the toxicity of both ammonia and acid components.[34] Levels of particulate ammonium sulfate and nitrate have reached 35 µg/m³ in polluted regions of the U.S., with nearly all particles in the respirable range. Ammonium sulfate and nitrate particles account for a significant fraction of the total particulate burden found in the air of the eastern U.S.; ammonium nitrate is seen mostly in the photochemical smog of southern California. Ammonia is also thought to react with soot particles formed during combustion of fossil fuels to produce potentially toxic complex particulates.

Ammonia gas is detectable by its odor at concentrations exceeding 1 ppm, produces irritation of the throat at 400 ppm, immediate irritation of the eyes at 700 ppm and severe coughing at 1,700 ppm. It is fatal after a short exposure at >5,000 ppm. Continuous exposure of humans to 500 ppm is fatal after 1 hour.[32] Bats who live in caves have developed physiological tolerance mechanisms, which allow them to live in environments with ammonia levels as high as 5,000 ppm; the ammonia is derived from microbial decomposition of bat manure in the caves.

About 95% of inhaled ammonia is absorbed by the nasopharyngeal region of the respiratory tract. Endogenous ammonia results from liver disease (hepatic coma associated with cirrhosis or primary or metastatic liver tumors), uremia and constipation. The cause of hepatic coma is thought

TABLE 5.16. Amounts of ammonia discharged following combustion of fossil fuel (S. Miner. 1969. Preliminary air pollution survey of ammonia. A literature review. National Air Pollution Control Administration Publ. APTD 69-25. Raleigh, NC, U.S. DHEW, 39 p.).

Combustion Source	Emission Factor
Coal	2 lb/ton
Fuel oil	1 lb/1000 gal
Natural gas	0.4 lb/10^{6}ft^3
Bottle gas (Butane)	1.7 lb/10^{6}ft^3
Propane	1.3 lb/10^{6}ft^3
Wood	2.4 lb/ton
Forest fire	0.3 lb/ton

to be interference with the blood-brain barrier dynamics from increasing levels of ammonia in the blood; this increases the excitability of nervous tissues in the brain.

Ammonia burns the skin and eyes at exposure levels of about 300 ppm; exposure for a few hours causes upper respiratory tract irritation and early signs of developing hepatic coma. Chronic exposures at 25–250 ppm increases the infectivity of pulmonary pathogens. Exposure of guinea pigs to 250 ppm for 22 days caused a mild, reversible inflammation of the respiratory tract and chronic inflammation with intraluminal calcification of renal cortical tubules.[35] Exposure of rats for 6 weeks to 25–250 ppm ammonia increased the respiratory symptoms associated with murine respiratory mycoplasmosis.[36] The potential carcinogenicity of ammonia-soot interactions is important because of the possible formation of carcinogenic compounds on particle surfaces.

Hydrogen Cyanide

Hydrogen cyanide (HCN) is found in a variety of situations, including blast-furnace gas in the manufacture of illuminating gas; HCN is also produced by the burning of nitrocellulose. Hydrogen cyanide is a rapidly acting, toxic chemical that inactivates cytochrome oxidase by the reaction of CN$^-$ with iron in the enzyme, blocking cellular respiration. This leads to cellular hypoxia. The blood in the venous system remains arterial in color, producing the characteristic cherry-red skin color of acute cyanide poisoning. Once absorbed, hydrogen cyanide or the cyanide ion is detoxified, mostly in kidney and liver, by conversion to the relatively nontoxic thiocyanate ion (SCN⁻). The reaction is enhanced by the presence of nitrite and thiosulfate:

$$\text{Cytochrome oxidase} + CN^- \rightarrow$$

$$\text{Cytochrome oxidase cyanide complex}$$

$$NaNO_2 + \text{hemoglobin} \rightarrow \text{Methemoglobin}$$

$$\text{Methemoglobin} + CN^- \rightarrow \text{Cyanomethemoglobin}$$

$$Na_2S_2O_3 + NaCN + \cdot \rightarrow NaSCN + Na_2SO_4$$

Tobacco smoke is the major source of human hydrogen cyanide exposure. Hydrogen cyanide yields are widely ranging among cigarette brands (from 5 to 241 μg per cigarette) with a mean of 39 μg per cigarette. There is also good correlation between tar content and HCN content of cigarettes.[47] Hydrogen cyanide is a potent ciliotoxic agent that leads to irritation and infections of the lung. Dermatitis is frequent in workers chronically exposed to cyanide solutions. Behavioral and central nervous system effects have been seen in workers chronically exposed to cyanide; symptoms include headache, weakness, giddiness, vomiting, alterations in taste and smell and dyspnea. The current permissible limit for hydrogen cyanide and cyanide salts in the U.S. is 5 mg CN/m^3 air.[49]

References

1. J. N. Stannard. 1980. Breathing is an old habit. In, "Pulmonary Toxicology of Respirable Particles," CONF-791002, NTIS, Springfield, VA., pp. 616–624.
2. A. Macfarlane. 1978. Daily mortality and environment in English conurbations. II. Deaths during summer hot spells in greater London. Environ. Res. 15:332–341.
3. H. R. Lewis. 1965. With Every Breath You Take. Crown Publishers, Inc., New York, p. 42.
4. E. J. Baum. 1978. Occurrence and surveillance of polycyclic aromatic hydrocarbons. In, "Polycyclic Hydrocarbons and Cancer," Vol. 1, Academic Press, New York, pp. 45–70.
5. A. I. Schmeltz, D. Hoffmann and E. L. Wynder. 1975. The influence of tobacco smoke on indoor atmospheres. Prevent. Med. 4:66.
6. P. Stocks. 1960. On the relations between atmospheric pollution in ur-

ban and rural localities and mortality from cancer, bronchitis and pneumonia with particular reference to 3:4 benzopyrene, beryllium, molybdenum, vanadium and arsenic. Br. Med. J. 14:397.
7. P. J. Buell and J. E. Dunn. 1967. Relative impact of smoking and air pollution on lung cancer. Arch. Environ. Health 15:291.
8. G. E. Hatch, D. E. Gardner and D. B. Menzel. 1980. Stimulation of Oxidant production in alveolar macrophages by pollutant and latex particles. Environ. Res. 23:121–136.
9. J. R. Lakowicz. 1979. Can human exposure to airborne and injected particulates increase our risk from exposure to environmental carcinogens. J. Environ. Pathol. Toxicol. 2:1433-1437.
10. T. J. Hughes, E. Pellizzari, L. Little, C. Sparacino and A. Kobler. 1980. Ambient air pollutants: Collection, chemical characterization, and mutagenicity testing. Mutat. Res. 76:51-83.
11. T. L. Guidotti. 1978. The higher oxides of nitrogen: Inhalation toxicology. Environ. Res. 15:443–472.
12. C. M. Shy and G. J. Love. 1980. Recent evidence on the human health effects of nitrogen dioxide. In, "Nitrogen Dioxide," p.291.
13. R. F. Bils and B. R. Christie. 1980. The experimental pathology of oxidant and air pollution inhalation. Int. Rev. Exp. Pathol. 21:195–293.
14. D. A. Creasia. 1981. Pathogenesis of nitrogen dioxide induced respiratory lesions in reference to respiratory clearance of inhaled particulates. J. Toxicol. Environ. Health 8:857-871.
15. D. B. Menzel. 1976. Oxidants and human health. J. Occup. Med. 18:342-345.
16. H. E. Stokinger. 1965. Ozone toxicity. Arch. Environ. Health 10:719-731.
17. M. O. Amdur. 1975. Air pollutants. In, "Toxicology: The Basic Science of Poisons," Macmillan Publishing Co., Inc., New York, pp. 527–554.
18. National Academy of Sciences (NAS). 1972. Particulate polycyclic organic matter. Committee on Biologic Effects of Atmospheric Pollutants. National Academy of Sciences, NAS Printing Office, Washington, D.C.
19. J. H. Ware, L. A. Thibodeau, F. E. Speizer, S. Colome and B. G. Ferris. 1981. Assessment of the health effects of atmospheric sulfur oxides and particulate matter: Evidence from observational studies. Environ. Health Perspect. 41:255–276.
20. R. B. Schlesinger, M. Halpern, R. E. Albert and M. Lippmann. 1979. Effect of chronic inhalation of sulfuric acid mist upon mucociliary clearance from the lungs of donkeys. J. Environ. Pathol. Toxicol. 2:1351-1367.
21. Y. Alarie, W. M. Busey, A. A. Krumm and C. E. Ulrich. 1973. Long-term continuous exposure to sulfuric acid in Cynomolgus monkeys and

guinea pigs. Arch. Ind. Health 18:407–414.
22. D. O. Anderson and B. G. Ferris. 1965. Air pollution levels and chronic respiratory disease. Arch. Environ. Health 10:307.
23. M. O. Amdur. 1974. The long road from Donora: The 1974 Cummings memorial lecture. Am. Ind. Hyg. Assoc. J. 35:589–597.
24. Subcommittee on Hydrogen Sulfide. 1979. Hydrogen Sulfide. University Park Press, Baltimore, MD, 183 p.
25. A. V. Robinson. 1982. Effect of in vitro exposure to hydrogen sulfide on rabbit alveolar macrophages cultured on gas-permeable membranes. Environ. Res. 27:491–500.
26. F. D. Andrew, R. A. Renne and W. C. Cannon. 1980. Reproductive toxicity testing for effects of hydrogen sulfide in rats. In, "Pacific Northwest Laboratory Annual Report for 1979 to DOE, Pt. 1, Biomedical Sciences, PNL-3300, NTIS, Springfield, VA., pp. 276–278.
27. F. Hays. 1972. Studies of the effects of atmospheric hydrogen sulfide in animals. Ph.D. Thesis, University of Missouri-Columbia, MI.
28. H. E. Stokinger and D. L. Coffin. 1967. Biologic effects of air pollutants. In, "Air Pollution," Vol. 1, Academic Press, New York, pp. 445–544.
29. M. J. Suess. 1975. The environmental load and cycle of polycyclic aromatic hydrocarbons. Presented in, "Environmental, Health, and Control Aspects of Coal Conversion: An Information Overview," Vol. 2, ORNL/EIS-95, Energy Research and Development Administration, Washington, D.C., p. 6.
30. J. Theodore, R. D. O'Donnell and K. C. Back. 1971. J. Occup. Med. 13:242.
31. S. M. Ayres, S. Giannelli and H. Mueller. 1970. Myocardial and systemic responses to carboxyhemoglobin. Ann. N.Y. Acad. Sci. 174:268–293.
32. Subcommittee on Ammonia. 1979. Ammonia Committee on Medical and Biologic Effects of Environmental Pollutants. University Park Press, Baltimore, MD, 384 p.
33. R. K. Evans et al. 1968. Energy-systems design survey. Special Report. Power 112:S1-S48.
34. T. V. Larson, D. S. Covert, R. Frank and R. J. Charlson. 1977. Ammonia in human airways: Neutralization of inspired acid sulfate aerosols. Science 197:161–163.
35. R. A. Renne and K. E. McDonald. 1981. Subacute inhalation toxicology of hydrogen sulfide and ammonia in rodents. In, "Pacific Northwest Laboratory Annual Report," Pt. 1, Biological Sciences, PNL-3700, NTIS, Springfield, VA., p 240.
36. J. R. Broderson, J. R. Lindsey and E. J. Crawford. 1976. The role of environmental ammonia in respiratory mycoplasmosis in rats. Am. J. Pathol. 85:115–130.
37. M. T. Kleinman et al. 1983. Effects of 0.2 ppm nitrogen dioxide on

pulmonary function and response to bronchoprovocation in asthmatics. J. Toxicol. Environ. Health 12:815–826.
38. M. J. Hazucha et al. 1983. Effects of 0.1 ppm nitrogen dioxide on airways of normal and asthmatic subjects. J. Appl. Physiol. 54:730–739.
39. R. M. Jackson and L. Frank. 1984. Ozone-induced tolerance to hyperoxia in rats. Am. Rev. Respir. Dis. 129:425–429.
40. G. Ericsson and P. Camner. 1983. Health effects of sulfur oxides and particulate matter in ambient air. Scand. J. Work Environ. Health 9:1–51.
41. C. Aranyi et al. 1983. Effects of subchronic exposure to a mixture of O_3, SO_2 and $(NH_4)_2SO_4$ on host defences of mice. J. Toxicol. Environ. Health 12:55–71.
42. M. A. K. Khalil and R. A. Rasmussen. 1984. Carbon monoxide in the earth's atmosphere: Increasing trend. Science 224:54–56.
43. WHO: Environmental Health Criteria. 13: Carbon monoxide. Geneva, World Health Organization (1979).
44. NIOSH: Criteria for a Recommended Standard. Occupational exposure to carbon dioxide. Washington, D.C., National Institute of Occupational Safety and Health (NIOSH 76-194). 1976.
45. A. Freedmay and D. Sevel. 1966. The cerebro-ocular effects of carbon dioxide poisoning. Arch. Ophthalmol. 76:59–65.
46. R. F. Johnston. 1959. The syndrome of carbon dioxide intoxication: It's etiology, diagnosis and treatment. Univ. Michigan Med. Bull. 25:280–292.
47. W. S. Rickert, J. C. Robinson, N. E. Collishaw and D. F. Bray. 1983. Estimating the hazards of "less hazardous" cigarettes. III. A study of the effect of various smoking conditions on yields of hydrogen cyanide and cigarette tar. J. Toxicol. Environ. Health 12:39–54.
48. S. H. El Ghawabi et al. 1975. Chronic cyanide exposure: A clinical, radioisotope, and laboratory study. Br. J. Ind. Med. 12:215–219.
49. NIOSH: Criteria for a Recommended Standard. Occupational exposure to hydrogen cyanide and cyanide salts. Washington, D.C., National Institute for Occupational Safety and Health (NIOSH 77-100). 1976.
50. Environmental Protection Agency (EPA). 1975. Scientific and technical assessment on particulate polycyclic organic matter (PPOM). EPA-60016-75-001.
51. W. W. Kellogg, R. D. Cadle, E. R. Allen, A. L. Lazarus and E. A. Martell. 1972. The sulfur cycle. Science 175:587–596.
52. E. Robinson and R. C. Robbins. 1970. Gaseous nitrogen compound pollutants from urban and natural sources. J. Air Pollut. Control Assoc. 20:303–306.
53. National Air Pollution Control Administration (NAPCA). 1970. Air quality criteria for sulfur oxides. AP-50, Washington, D.C., NAPCA.
54. S. Laskin, M. Kuschner and R. T. Drew. 1970. Studies in pulmonary

carcinogenesis. In, "Inhalation Carcinogenesis," CONF-691001, NTIS, Springfield, VA, pp. 321-351.
55. R. M. Jackson and L. Frank. 1984. Ozone-induced tolerance to hyperoxia in rats. Am. Rev. Respir. Dis. 129:425-429.
56. H. R. Superko, W. C. Adams and P. W. Daly. 1984. Effects of ozone inhalation during exercise in selected patients with heart disease. Am. J. Med. 77:463-472.
57. B. Meyer. 1977. Sulfur, Energy, and Environment. Elsevier Scientific Publ., New York, NY.
58. H. G. Parkes. 1976. In, "Chemical Carcinogens," American Chemical Society Monograph No. 173, American Chemical Society, Washington, D.C., pp. 462-480.
59. E. Kriek. 1979. Aromatic amines and related compounds as carcinogenic hazards to man. In, "Environmental Carcinogens," Elsevier/North-Holland Biomedical Press, Amsterdam, Holland, pp. 143-164.
60. O. Bodansky. 1951. Methemoglobin and methemoglobin-producing compounds. Pharmacol. Rev. 3:144-196.

CHAPTER 6

TOXICOLOGY OF METALS

General Principles

Metals have a unique property, in contrast to most organic compounds, in that they are not catabolized in the body or by organisms in the biosphere but cycle mostly as metal ions or simple organometallic compounds, often concentrating in segments of the biosphere. For example, cadmium is concentrated up to 250,000-fold from water levels in marine invertebrates, and lead is concentrated up to 100,000-fold in freshwater invertebrates. Following environmental deposition in sediments, most heavy metals (mercury, lead, cadmium, zinc, copper, nickel, cobalt, manganese and chromium) are chemically bound and require heat and a low pH to convert them to soluble forms.[46] Sediments constitute a major sink for metals. The toxicity of most metals is related to the amounts ingested or inhaled. Oral administration of metals results in processing by digestive enzymes, acid, natural chelates, bacteria and intestinal mucosal cells. Absorption of metals across the intestinal epithelium is influenced by:

- Physical-chemical characteristics of the metal
- pH
- Microbial metabolism of the metal
- Presence of natural chelates in the diet
- Interactions among metals
- Mechanisms of metal transport through the epithelium
- Amount of metal
- Age at time of exposure to the metal.

Most metals are absorbed most effectively in the mid-jejunum. Absorption is highest in newborns, rapidly decreasing with age; absorption may be 100 times greater in the newborn than in the adult.

Absorption of inhaled metals from the lung is influenced by dose, physical-chemical form and interactions of the metal with lung tissues. Lung clearance of inhaled metals is not necessarily indicative of whole-body clearance (Figures 6.1 and 6.2). For example, cadmium and arsenic are rapidly cleared from the lung but avidly retained in other body tissues. The urine and feces are the most common routes of excretion for ingested or inhaled metals. Metals may be found in urine as simple ions, water-soluble salts, complexes with amino acids or as chelates. Urine is the major route of excretion for Be, Cd, As^{+5}, Sb^{+5}, V as vanadate and methylated Se compounds. Feces are the major route of excretion for Cu, Zn, Hg, Sn, Pb, Zr, As^{+3}, Sb^{+3}, V, Se, Cr, Co and Ni.[1]

Metals may bind to amino acids, forming bonds with proteins or nucleic acids, or they may bind to biological membranes, causing altered cell permeability. Some metals are capable of forming bonds with simple, carbon-containing molecules. Biotransformation of metals alters their toxicity; for example, methylmercury is considerably more toxic than is inorganic mercury, particularly in the central nervous system, because of the ability of methylmercury to pass through the blood-brain arrier. Nutritional factors also alter metal toxicity.

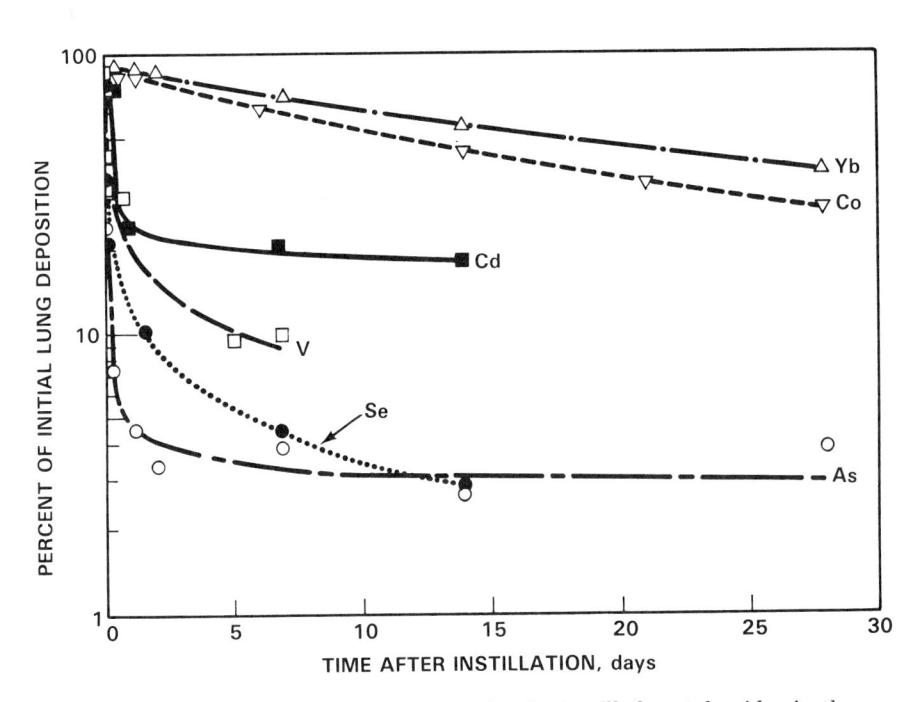

FIGURE 6.1—Lung clearance of intratracheally instilled metal oxides in the rat (K. Rhoads and C. L. Sanders. Unpublished).

Several detoxification mechanisms are available in the body to minimize metal toxicity. These mechanisms are most effective in chronic exposures since saturation of the detoxification mechanism(s) rapidly occurs at high doses of metal. Detoxification mechanisms include:

- Redox or hydrolysis reactions
- Incorporation into nontoxic compounds (e.g., cadmium into metallothionein)
- Sequestering into subcellular structures (lysosomes or portions of bone)
- Binding to natural chelates that enhance excretion
- Interactions with other metals (e.g., Zn decreases Cd toxicity)

The dose-response curve for most metals indicates a complex relationship over a wide dose range. Small quantities of most metals, even As, Cd, and Se, are essential for optimal health, but larger quantities produce toxic effects. Deficiencies in these metals leads to disease, which is of a

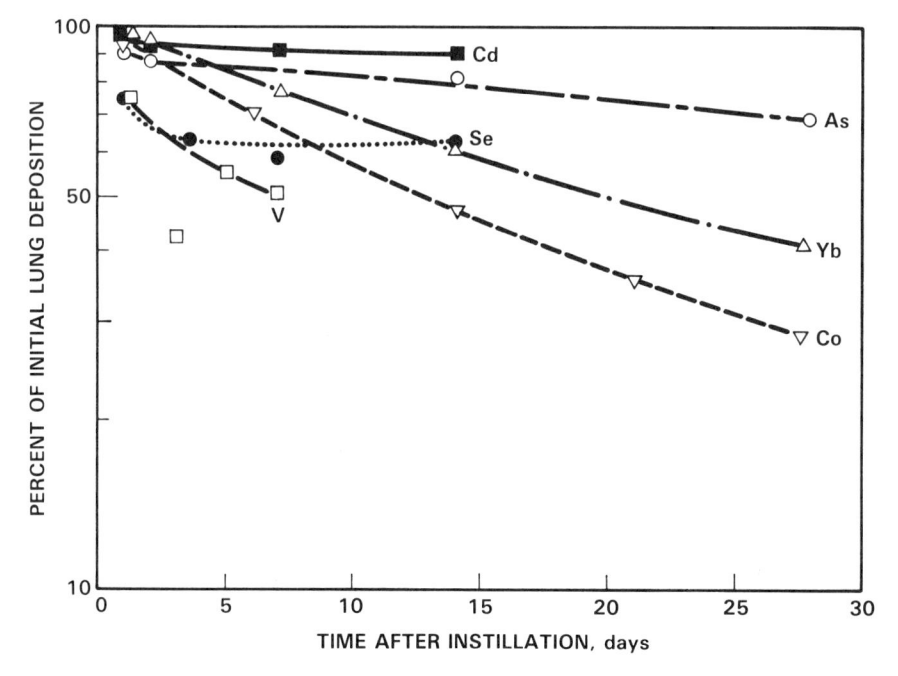

FIGURE 6.2 — Whole-body clearance of intratracheally instilled metal oxides in the rat (K. Rhoads and C. L. Sanders. Unpublished).

different nature than that caused by excess levels. Acute toxicity of metals is often determined in acute lethality studies giving the LD_{50}, LC_{50}, or pT_{50} values. The toxicity of these metals is best expressed by the ratio of the beneficial to the toxic dose; the lower the ratio, the greater the toxicity. Unfortunately, this analysis gives no hint as to the long-term toxic effects of the metal.

Several short-term bioassays for examining the toxicity of metals in mammalian systems have been developed. These in vitro assays are useful in providing estimates of relative or comparative metal toxicity. In one test, the 20-hour survival of human lung fibroblasts or rabbit alveolar macrophages was examined.[2,3] Another test examined the clonal growth of VERO monkey kidney cells cultured in the presence of metal ion[4], in which the relative plating efficiency (RPE) is used to compare treated and control cultures. The ranking of metals in this system is different from the toxicity ranking used for other bioassays (Figure 6.3). The sensitivity of the VERO bioassay is up to 4,000 times greater than fibroblast or macrophage assays.[4] These in vitro tests are useful only in screening of various compounds and providing data for experimental design of in vivo tests.

Metals with low TLV values are often teratogenic and/or carcinogenic in animals or in humans (Table 6.1). Except for Se, all identified metal carcinogens have electronegativity values ranging from 1.4 to 1.8, but not all metals in this midrange have been shown to be carcinogens.

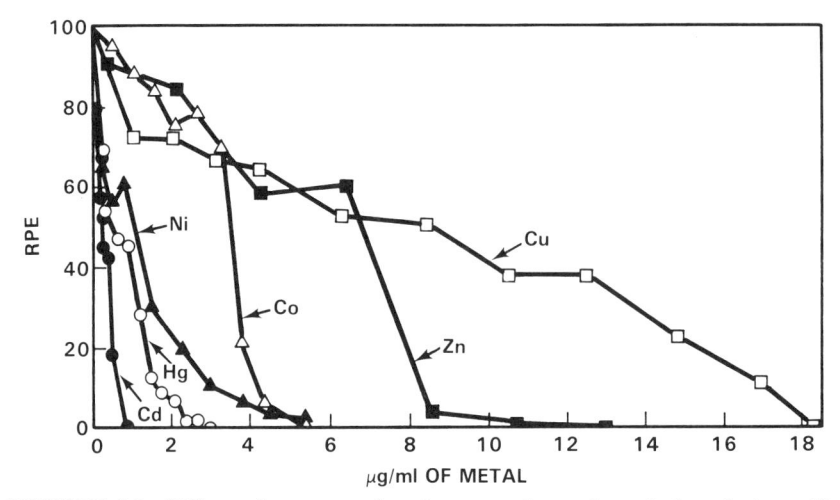

FIGURE 6.3—Effect of amount of various metals on the relative plating efficiency (RPE) of monkey kidney cells (M. E. Frazier and T. K. Andrews. 1979. In vitro growth assay for evaluating toxicity of metal salts. In, "Trace Metals in Health and Disease," Raven Press, New York, pp. 71–81).

TABLE 6.1. Some toxicological characteristics of metals. Suspected carcinogen (S), proven carcinogen (P) (T. D. Luckey and B. Venugopal. 1977. Metal Toxicity in Mammals. 1. Plenum Press, New York, pp. 129–160).

Metal	Annual World Production Metric Tons	TLV, mg/m³	Carcinogenicity Animals	Carcinogenicity Humans	Required For Health
As	52,000	0.2–0.5	S	P	+
Be	7,200	0.002	P	S	
Cd	17,000	0.05–0.2	P	S	+
Co	20,000	0.1	P	S	+
Cr	5,100,000	0.1	P	P	+
Cu	6,000,000	0.2			+
Fe	720,000,000	5.0			+
Hg	9,900	0.01–0.1			
Mn	18,000,000	5.0	S		+
Ni	480,000	1.0	P	P	+
Pb	3,200,000	0.15	P		+
Pd	105	–	P		+
Sb	65,000	0.5			
Se	1,250	0.2	P		+
Sn	230,000	2.0–10	P		+
V	10,300	0.05–0.5			+
Zn	5,300,000	5.0	P		+
Zr	380,000	5.0	P		

Hexavalent chromium, which is reduced in the cell to the trivalent state, is the most potent mutagen of the metals. The information available allows a classification of metals according to their mutagenic potential as follows: Cr > Be > As > Ni > Hg > Cd > Pb. Taking into account the number of persons potentially exposed, metals can be classified according to their carcinogenic risk as follows: As > Cr > Ni Be > Pb = Cd = Hg, the latter three metals being almost noncarcinogenic in humans. Very little information is available on the teratogenicity of metals in humans. Based on studies in animals, mostly rodents, the risk of malformations seems greatest for cadmium, followed by arsenic, nickel and chromium.[5]

Increased incidence of lung cancer has been found in workers involved in the mining, production or processing of ores or chemicals containing arsenic, chromium or nickel. Inconclusive associations have been noted between cancer development in the lung or other tissues and occupational exposures to Al, Be, Cd, Fe, and Pb compounds, although compounds of

Al, Be, Cd, Co, Cr, Cu, Fe, Mn, Ni, Pb, Pt, Ti, and Zn have been reported to induce carcinomas in the lung or kidney of animals.[5]

Conversely, several metals have been shown to exhibit anticarcinogenic properties: Zn protects against PAH skin carcinogenesis, Cu protects against hepatocarcinogenesis from methoxy dyes, and Se protects against skin carcinogenesis from benzo(a)pyrene or colon carcinogenesis from dimethylhydrazine.[5]

Aluminum

Aluminum is the most common metal found in the earth's crust (8.1%) and has a variety of industrial uses due to its low density and high tensile strength. Foods contain an average of 10–50 mg Al/kg; total daily intake in humans amounts to about 80 mg. Gastrointestinal uptake of Al is < 10% of that ingested; the lung contains more because it is being continuously exposed to aluminosilicate dusts.[6]

Inhalation of aluminosilicate dusts leads to simple pneumoconiosis at high exposure levels. Pulmonary deposition of either aluminosilicate dusts or aluminum salts induces alveolar lipoproteinosis. Fibrotic lung disease has been documented in a few workers exposed to high levels of powdered aluminum metal. Toxic symptoms from aluminum exposure are mostly related to the depletion of phosphate stores by the formation of insoluble phosphate complexes with Al in the gut. Thus, Al is used clinically to treat hyperphosphatemia associated with renal disease.

The main health hazard associated with aluminum production from bauxite ore is not from the metal but from fluoride exposure. A major health hazard is associated with the electrolytic reduction of bauxite (aluminum oxide) where fluorspar (CaF_2) and AlF_3 are used to alter the melting point and current efficiency. An increase in lung cancer was seen in aluminum reduction workers. Recent studies have also indicated neurotoxicity from Al exposure; significant differences were noted in neurological and behavioral functions between groups with high serum Al levels (504 ng/ml) and those with low serum Al levels (387 ng/ml).[7] A relationship has been found between Alzheimer's disease and increased amount of Al in neurofibular tangles of the brain. Aluminum intoxication has also been implicated in the pathogenesis of dementia, microcytic anemia and osteomalacia in patients undergoing maintainence dialysis.[63]

Antimony

Antimony, a common constituent of alloys, is used in fireproofing various materials, in metal type, and in glass and ceramic manufacturing. Daily environmental uptake by humans from all sources ranges from 10 µg to 1000 µg antimony. Little is known about the fate of inhaled or ingested antimony in humans. It is excreted primarily in the feces for trivalent compounds and primarily in the urine for pentavalent compounds. Skin, lung, liver, adrenals and kidneys contain higher concentrations of antimony than other organs.

Chemical pneumonitis with limited interstitial fibrosis was produced in guinea pigs after inhalation exposure to large amounts of antimony trioxide. Degenerative lesions in the myocardium, liver and kidneys were also noted in animals exposed to antimony compounds. Human toxicological data for antimony compounds are derived from occupational exposures and from adverse effects of pharmacological therapy for tropical diseases. Inflammatory pulmonary lesions were seen following acute inhalation exposure; simple pneumoconiosis, emphysema and obstructive lung damage followed chronic inhalation exposure; and skin and liver damage resulted in some schistosomiasis patients treated for long periods with antimony compounds.[8]

Arsenic

Inorganic arsenic compounds are used extensively in glass, textile, pigment, tanning and antifouling paint manufacturing. In addition, arsenicals have been used as pesticides and herbicides and have a long history in the therapy of cancer and syphilis. High arsenic exposures are also associated with copper, lead and zinc smeltering operations. Serious environmental exposures to arsenic may also occur from the burning of high-arsenic coals. Arsenic ranks 20th in elemental abundance in the earth's crust and 12th in abundance in the human body, with a body burden of about 18 mg As. About 1.5 million American workers, in 80 occupations, are potentially exposed to arsenic compounds.

Arsenic is a metalloid with valences of -3, $+3$ and $+5$, forming both cationic and anionic trivalent arsenites and pentavalent arsenates. Airborne arsenic is usually in the form of As_2O_3. Arsenites are more toxic than arsenates; arsine gas (AsH_3) is the most toxic inorganic arsenic compound.

Uptake of ingested or inhaled inorganic arsenic is rapid, while percutaneous uptake is limited and slow. The whole-body retention of intratracheally instilled ^{74}As in rats is 60 days.[9] In humans, the whole-body half-life of inorganic arsenic is 10 hours, but for methylated arsenic compounds following ingestion, it is 30 hours.[10] High levels of arsenic are found in the hair, nails and skin. Inorganic arsenic is converted to methylated compounds in the body and excreted, mostly in urine; methylcobalamin acts as a major methyl group donor in the biotransformation process. In humans, ingested arsenic acid is mostly excreted in urine: 51% as dimethylarsenic acid, 21% as methylarsenic acid and 27% as inorganic arsenic (Figure 6.4). Because inorganic arsenic compounds are more toxic than organic compounds, the methylation process serves as a detoxification mechanism. The TLV for airborne As is between 0.3–0.5 mg/m^3 in most countries, with ceiling limits as low as 2 µg/m^3 suggested in the U.S. by NIOSH.

Acute lethality is seen in rabbits after parenteral administration of 4 mg/kg for As_2O_3 and 6 mg/kg for As_2O_5. Inhaled arsine breaks down in the blood into inorganic arsenic, which causes anemia due to hemolysis, associated with binding to erythrocytes. Sudden hemolysis, bloody urine, jaundice and abdominal complaints are characteristic of acute arsenic poisoning. Arsenic ions inhibit mitochondrial heme biosynthetic enzymes, ALA synthetase and heme synthetase, resulting in increases in urinary porphyrins.[11] Both pentavalent and trivalent forms of arsenic increase the amount of uroporphyrin over coproporphyrin; hepatic porphyria occurs at relatively low concentrations of arsenic exposure prior to the onset of histopathological signs of frank hepatotoxicity.[64]

Several studies have shown an increase in chromosomal aberrations in cultured lymphocytes in industrial, agricultural or medical groups exposed to arsenic compounds; even, in some cases, decades following the last exposure.

Chronic exposure to arsenic compounds may lead to hyperkeratosis, portal hypertension and disturbances of the peripheral vascular and nervous systems.[49] The skin is commonly affected by exposure to inorganic arsenic. Eczematoid eruptive lesions on the palms of the hand and soles of the feet are indications of focal arsenic contact. Melanosis, and leukodermal hyperkeratosis leading to skin cancer are seen after chronic arsenic exposures. Endemic arsenic poisoning is seen in Cordoba, Argentina, where concentrations of As in drinking water ranges from 0.9–3.4 mg/l. Certain areas in Taiwan have high As levels in drinking water, which causes blackfoot disease, a peripheral extremity vascular disorder resulting in gangrene. A dose-response relationship between the incidence of blackfoot disease and the duration of exposure has been documented.[12]

Occupational exposure to arsenic compounds is strongly related to increased numbers of lung cancers; however, limited animal studies have

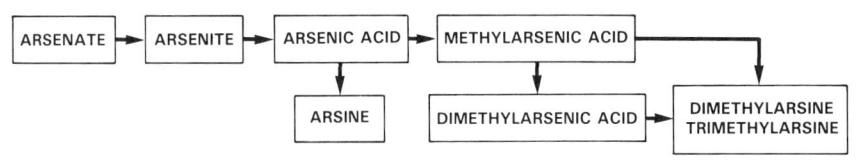

FIGURE 6.4 — Biotransformation of arsenic compounds mediated by enzymes in mammalian tissues and/or microorganisms.

failed to demonstrate the carcinogenicity of arsenic compounds. Mortality due to lung cancer was three-fold greater than normal in copper-smelter workers exposed to arsenic trioxide.[13] Other epidemiological data indicates increases in both lung and skin cancer following prolonged arsenic exposure.

Beryllium

The U.S. consumption of beryl ore, the principal source of beryllium, reached 9,000 short tons in 1970. Beryllium phosphors were used in fluorescent tubes starting in the 1930s, but this was terminated in 1949, when chronic and acute beryllium disease was found in fluorescent lamp workers.[14] Toxicologically significant amounts of beryllium enter the body almost entirely by inhalation. Inhaled soluble forms of beryllium are rapidly absorbed from the lung and excreted mostly in the feces. The lung retention of high-fired, insoluble beryllium oxide is 325 days in the rat. Absorbed beryllium is stored mostly in the bone but also transiently, in the liver, kidney and lung. Toxicity to pulmonary macrophages containing phagocytized beryllium particles results in impairment of alveolar clearance.[15]

Soluble beryllium salts are cutaneous sensitizers, resulting in eczematous contact dermatitis, indistinguishable from irritant contact dermatitis. Ionic beryllium applied to the skin binds to alkaline phosphatase and nucleic acids in epidermal cells, inhibiting phosphatase, ATPase and phosphoglucomutase activities.

Inhalation of beryllium dust produces acute pneumonitis at high doses and chronic berylliosis at low or prolonged exposures. Berylliosis is not considered a form of pneumoconiosis because beryllium produces local as well as systemic toxicity; the term "beryllium disease" is more appropriate. Patients with chronic beryllium disease show emphysema, diffuse interstitial fibrosis and a sarcoid-like response resembling Boeck's sarcoidosis with scattered granuloma formations.

Beryllium disease has an immunological basis with antigenic reactions, resulting in an increased circulating level of gamma globulins and a delayed hypersensitivity reaction. The expression of beryllium disease is highly variable among individuals. In some cases, latent beryllium disease "flares up" during times of stress, such as pregnancy, infection or surgery. Antiinflammatory steroid therapy is often effective in limiting, but not reversing, beryllium disease.

The Beryllium Case Registry was established in 1952 to collect data on the epidemiology, diagnosis, clinical features, course and complications of beryllium disease. Of the 887 cases of beryllium disease reported up to 1977, 631 have been classified as chronic and 256 as acute.[16]

Soluble beryllium compounds give a consistent, dose-related tumor response in several animal species. Beryllium sulfate, low-fired beryllium oxide, beryllium hydroxide, beryllium phosphate and beryl ore dust have all been found to be carcinogenic in the lungs of rats and/or monkeys. Beryllium exposure in humans is clearly related to the development of non-neoplastic beryllium disease. Several recent epidemiological studies of beryllium workers also indicate a high rate of lung-cancer mortality. The data, however, are not sufficient to prove that beryllium is a human carcinogen.[17]

Cadmium

World production of cadmium increased from 14 tons in 1900 to 17,000 tons in 1974, markedly increasing potential human occupational and environmental exposure. Cadmium is an industrial product or by-product of considerable environmental health concern. Cadmium enters the U.S. environment at the rate of 2,000 to 5,000 tons per year; of this amount, 20% arises from zinc mining and smelting, 30% from the industrial use of cadmium, and the remaining 50% comes from fertilizers, sewage-sludge disposal and the combustion of municipal refuse and fossil fuels.

Air levels of cadmium have often exceeded 0.1 $\mu g/m^3$ in the vicinity of lead and zinc mines and smelters. Average urban air contains 0.05 $\mu g/m^3$ compared to average levels of 0.005 $\mu g/m^3$ in rural air. Occupational exposure levels of about 100 $\mu g/m^3$ have been correlated with health problems in cadmium workers.[18]

Daily uptake of cadmium by Americans averages 3 μg, mostly from diet. A similar amount of cadmium is also absorbed from the lung daily by heavy cigarette smokers. Cigarette-smoking results in significantly higher levels of cadmium in the liver and kidney. So-called "reference man" contains 30 mg cadmium, 50% of which is in the liver and 33% in the kidneys, mostly in the cortical portion. In the kidney, the critical tissue for cadmium-induced health effects, irreversible damage occurs when cadmium levels in the cortex exceed 200 $\mu g/g$ wet weight.[21] The levels of chronic cadmium exposure necessary to induce significant renal damage presently exceeds by only 5–8 times the levels presently encountered in the average American. Any further increase in cadmium exposure, e.g., by cigarette smoking, will reduce this already low margin of safety before adverse effects are seen. Cadmium is tenaciously retained in the body and is slowly excreted, mostly in the feces. Urinary cadmium levels, however, are good indicators of chronic, low-level cadmium exposure.[19,20] At constant environmental exposure levels, the cadmium concentration in the renal cortex increases almost linearly from ages 5 to 50 and remains constant or decreases at later ages. Because of its unusually long retention time in the body, particularly in the renal cortex, cadmium's adverse health effects are considered irreversible.

Cadmium is transported by an inducible, metal-binding protein, metallothionein, which selectively binds cadmium, zinc and several other divalent cations. Metallothionein is a low-molecular-weight protein that is synthesized in several organs, particularly in the liver following cadmium exposure. Cadmium is deposited for the most part, in the liver and kidney where it is bound to metallothionein. Intraveneously injected cadmium-metallothionein complex is rapidly taken up directly by the renal cortex, while inorganic cadmium is first taken up by the liver and kidney and combined with metallothionein. Feeding supplemental zinc significantly decreases the cadmium content of muscle, liver, blood and kidney.[20] Polyaminocarboxylates provide the best protection against a lethal dose of cadmium salt when given immediately after exposure to cadmium; 2,3-dimercaptopropanol (BAL) and sodium diethyldithiocarbamate were able to produce a significant reduction in cadmium content of the kidney.[66]

Cadmium effects the reabsorption of low-molecular-weight proteins by the proximal tubules of the kidney, resulting in urinary proteinuria, one of the earliest signs of cadmium poisoning. Morphological changes in the tubule epithelium accompany progressively increasing amounts of cadmium, resulting in increasing amounts of protein, glucose, amino acids and phosphate in the urine. Although experimental hypertension has been produced in animals exposed to cadmium, possibly due to kidney damage, there is no conclusive data that cadmium exposure causes hypertension in humans. An imbalance in calcium and phosphorus metabolism in combination with cadmium exposure results in osteoporosis and osteomalacia in Japanese populations; this disease, termed "Itai-Itai," is seen specifically in individuals deficient in dietary calcium and vitamin D.[21]

Acute pneumonitis with pulmonary edema resulting from the destruction of type I alveolar epithelium is seen within 24 hours in animals after inhalation of a high level of cadmium aerosol. The acute lethal dose for inhaled cadmium oxide, which is very soluble, in rats is 20 µg/g wet weight of lung tissue; this can be mitigated by pretreating the animals with selenium or thiol compounds. Recovery from the pneumonitis is rapid, with survivors displaying a type II alveolar epithelial hyperplasia during the first two weeks after exposure (Figures 6.5 and 6.6).

Cadmium causes damage in the testes of experimental animals by means of an initial interstitial edema, followed by hemorrhage, necrosis of spermatogenic cells, destruction of seminiferous tubules, atrophy of testes and finally, in some species, by Leydig cell tumor formation.[22,65] Cadmium exposure in animals has also been associated with damage to the kidney, liver and lung. Cadmium is also embryotoxic, teratogenic, mutagenic, and carcinogenic, and may cause hyperglycemia, reduced immunocompetency and anemia due to its interference with iron metabolism. The metal is a suspected carcinogen in the lung and prostate gland of humans. Cadmium sulfate, given subcutaneously in rodents, causes sar-

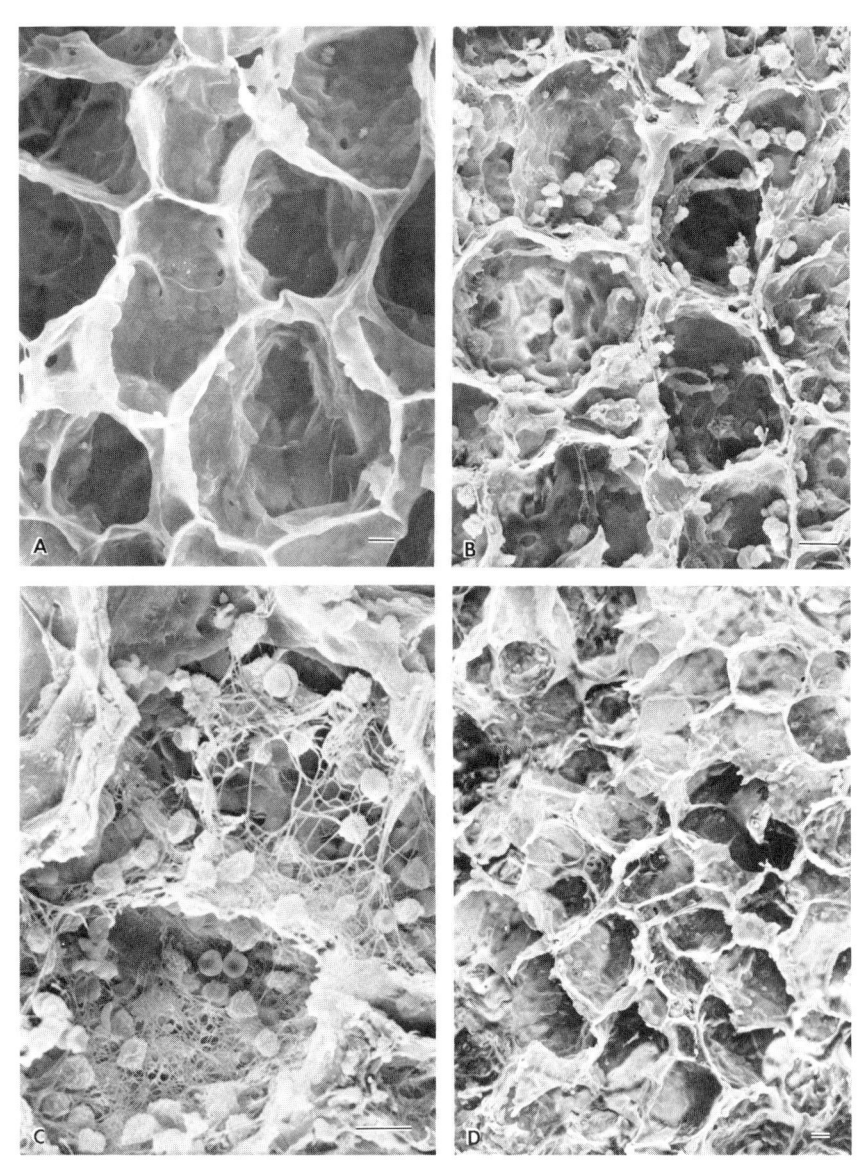

160   Toxicological Aspects of Energy Production

FIGURE 6.5—Scanning electron micrographs of rat lung following deposition of 20 µg cadmium oxide. (A) unexposed control; (B) 24 hour after cadmium exposure, showing roughened alveolar surfaces and macrophage accumulation; (C) 48 hour after cadmium exposure, with macrophages, erythrocytes and a fibrin-like precipitate present in alveoli; and (D) 2 weeks after cadmium exposure, with partial repair and removal of alveolar edudate. Bar in all photographs = 10 micrometer (C. L. Sanders. 1982. Alveolar clearance of inhaled metal oxides. Pacific Northwest Laboratory Annual Report for 1981 to the DOE Office of Energy Research, Part 1. Biomedical Sciences, PNL-4100 PT1, Richland, Washington, pp. 143-146).

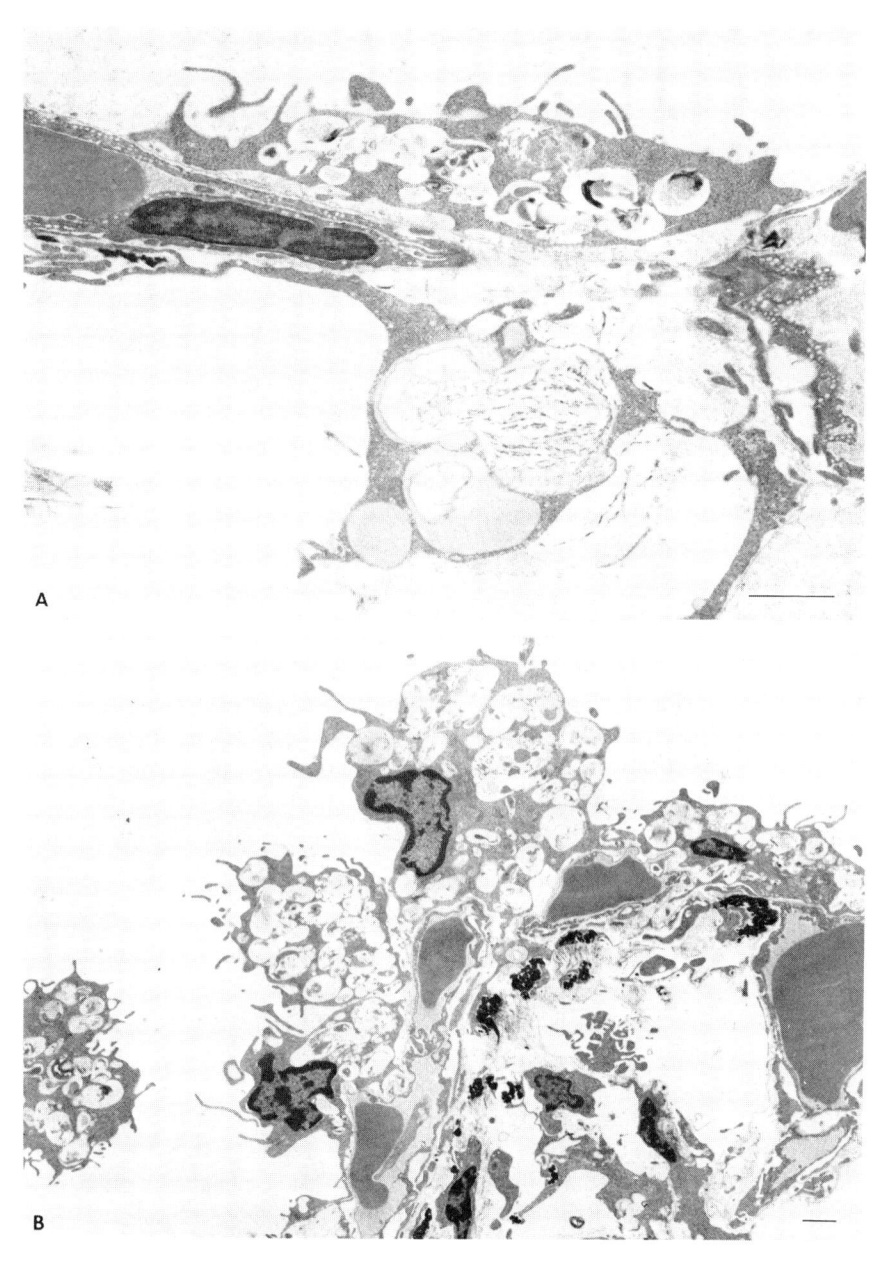

FIGURE 6.6—Transmission electron micrographs of alveolar epithelial damage at 24 hours after pulmonary deposition in rats of 20 µg cadmium oxide. (A) Type I alveolar epithelium; (B) Type II alveolar epithelium. Bar in both photographs = 1 micrometer. (C. L. Sanders. Unpublished).

coma formation. Cadmium oxide has not been shown to be a carcinogen in the kidney or liver of experimental animals.[21,51] A recent chronic inhalation study showed that cadmium chloride is a potent carcinogen in rats, inducing up to 70% incidence of lung tumors at the highest exposure level.[52]

Chromium

Chromium is used in the metallurgical and chemical industries. The daily environmental intake of trivalent chromium in air, water and food ranges from 0.05–1.0 mg.[23] Environmental exposure is mostly to trivalent chromium compounds, whereas occupational exposure is mostly to hexavalent (chromate) compounds. About 10% of the chromium in food is absorbed by the gastrointestinal tract, but less than 2% of chromium is absorbed when it is given by itself. Chromates are more readily absorbed from the lung than are trivalent chromium compounds, and trivalent compounds are excreted from the body faster than are chromates. Liver, spleen, bone marrow and kidney concentrate chromium following ingestion; the lungs and hair concentrate inhaled chromium.

Chromates are considerably more toxic than are the trivalent compounds. Effects of chromates include skin ulceration, accompanied by irritative and allergic dermatitis and asthmatic reactions.[23] Small-cell lung carcinoma is seen in chromate workers, but lung cancer incidence in workers exposed to trivalent chromium compounds is not demonstrably higher than for non-exposed populations. Chromates induce adenocarcinoma in the lungs of exposed animals but do not induce bronchogenic carcinomas.[25]

Cobalt

Most U.S. usage of cobalt is in steel manufacturing. Cobalt is an essential element; the daily requirement is about 0.12 µg, for the biosynthesis of cyanocobalamine. Daily uptake of cobalt in a typical diet ranges from 5–50 µg, with levels in drinking water ranging from 0.1–5.0 µg/l. Inhaled cobalt oxide is very soluble in the lung; less than 1% of initial alveolar deposition remains in the lungs of hamsters by 10 days after exposure (Figure 6.7). Syrian hamsters exposed to high levels of cobalt oxide aerosol die from pulmonary edema and myocardial lesions. Use of cobalt as an antifoam agent in beer has led to an outbreak of cardiomyopathy in heavy beer drinkers; estimated daily intake was 10 mg cobalt from beer.[26] Cobalt compounds have not been shown to be mutagenic, teratogenic or carcinogenic in humans.

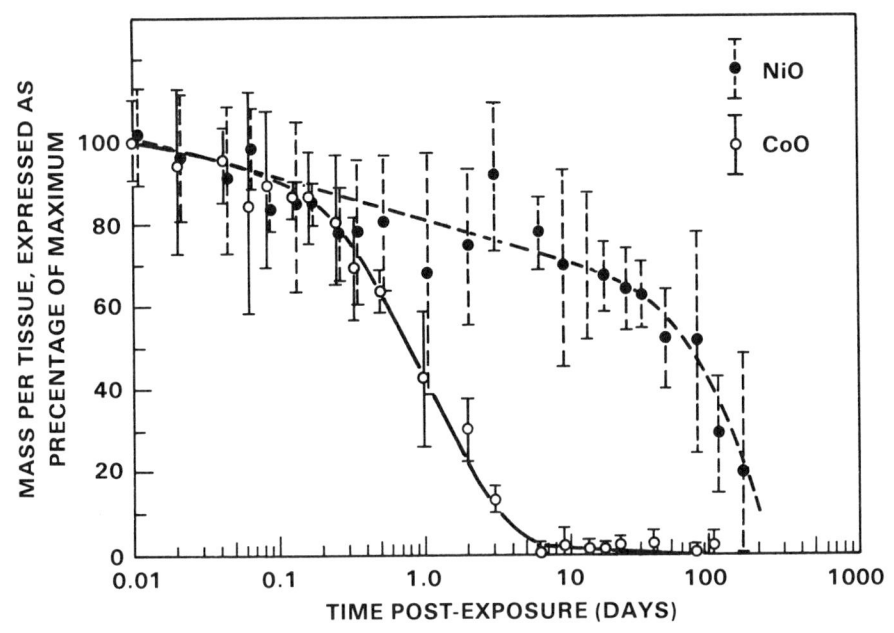

FIGURE 6.7—Lung clearance of inhaled NiO and CoO from Syrian hamsters (A. P. Wehner and D. K. Craig. 1972. Toxicology of inhaled NiO and CoO in Syrian golden hamsters. Am. Ind. Hyg. Assoc. J. 33:146–155).

Copper

Copper is used primarily in electrical, plumbing and heating equipment. Estimated daily environmental uptake of copper in the U.S. varies from 1–3 mg, mostly from food. About half of the copper ingested in food is absorbed by the gastrointestinal tract and transported, mostly bound to albumin, to the liver, where it is carried, primarily as a high-molecular-weight protein, ceruloplasmin, to other tissues. Copper is an integral part of several enzymes, e.g., cytochrome oxidase and superoxide dismutase. The biological retention half-time for parenterally administered copper is about 4 weeks. Normal copper levels in liver are 5–10 mg/kg; copper is excreted mainly in bile.[27]

Excessive copper intake, or copper imbalance as is seen in genetically related Wilson's disease, leads to liver degeneration, hepatic granulomas and greenish lung nodules. These sequelae are seen in "vineyard sprayer's lung" in France following inhalation of copper sulfate during spraying of grape vines. Copper sulfate is also a powerful emetic, used clinically in children at doses of about 250 mg. Acute inhalation of copper dusts may cause metal fume fever and acute upper respiratory tract inflammation. No chronic diseases have been associated with low-level, chronic exposure to copper.[27]

Iron

Iron is an abundant metal, comprising about 5% of the earth's crust, used for the production of raw steel. Environmental uptake of iron by humans is mostly in food; a typical U.S. diet provides about 6 mg Fe/1000 kcal. Iron is an essential nutrient which combines with protoporphyrin to form hemoglobin and is a component of myoglobin, catalases, cytochromes and peroxidases. Daily iron requirements by humans range from 10 to 30 mg, depending on age, health status and sex. Iron-deficiency anemia is one of the most common clinical conditions in humans.

About 10–20% of ingested iron is absorbed by a healthy person (more if iron-deficient) from the GI tract. About 70% of absorbed iron is bound to hemoglobin and 10% to myoglobin. Excess iron is stored, as ferritin (soluble in water) and hemosiderin (insoluble in water), mostly in the liver, spleen and bone marrow. Iron is excreted at a rate of about 1 mg/day. Considerable iron may be lost through bleeding (0.5 mg Fe/ml blood); about 20 mg Fe is lost during a normal menstrual cycle.[28]

Severe poisoning in children follows ingestion of only 2.5 g ferrous sulfate (0.5 g Fe), causing vomiting, ulceration of the GI mucosa, shock, metabolic acidosis, liver and kidney damage and coagulation defects. Inhalation of large amounts of iron oxide causes siderosis. Hematite miners exposed to a combination of iron oxide and silica dust may develop siderosilicosis. Iron oxide alone does not cause pulmonary fibrosis in humans or in experimental animals. Hematite miners that develop pulmonary fibrosis do so because of the silica present in the mine dust. Hematite miners also are predisposed to tuberculosis.

The high incidence of lung cancer observed in hematite miners and iron smelter workers can not be completely accounted for by smoking habits. Studies in humans and in Syrian hamsters given BaP with hematite dust[29] indicate that iron oxide by itself is not a pulmonary carcinogen but that it serves a cocarcinogenic role, possibly by serving as a carrier of carcinogenic substances, resulting in prolonged BaP retention in the lung.

Lead

Lead, one of the oldest known poisons, was first smelted about 6,000 years ago; Hippocrates described the symptoms of lead poisoning in 370 B.C.

Lead poisoning occurs from ingestion of old paint chips by children, eating food from improperly lead-glazed earthenware, drinking water from lead pipes (thought by some to the one of the causes for the fall of the Roman Empire), and exposures in the mining, smelting or processing of lead. The particulate lead compounds include the monoxide (PbO) and tetroxide (PB_3O_4, or red lead) forms; the monoxide is a major by-product of the leaded gasoline burned in internal combustion engines. About 160,000 tons of lead are emitted into the atmosphere of the U.S. in 1975; 142,000 tons originated from the combustion of gasoline containing lead additives.[30]

Since the introduction of lead akyls, such as tetraethyl lead (TEL), into gasoline there has been a sharp rise in worldwide air lead levels. Typical urban air lead levels range from 0.1 to 10 µg/m³, which results in significantly higher blood lead levels in residents of urban areas than are seen in residents of rural areas;[31] measurements of tissue lead levels in wild rats demonstrates a significantly higher lead level in urban-dwelling rats than in rural-dwelling rats (Table 6.2). A depression in delta-amino levulinic acid dehydratase in kidney and red blood cells and the presence of renal intranuclear lead inclusion bodies confirms the lead-poisoning of urban rats.

TEL is added to gasoline in order to eliminate "knock" in internal combustion engines. One part TEL is added to 1300 parts gasoline, along with a halogen-like ethylene dibromide to assist in the removal of lead from the engine. TEL readily penetrates the skin and crosses the blood-brain barrier, concentrating in brain tissues. Lead from TEL in brain exceeds that

TABLE 6.2. Lead concentrations in tissues of wild rats (D. Mouw et al. 1975. Lead. Possible toxicity in urban and rural rats. Arch. Environ. Health 30: 276–280). **Values are means ± SE (number of samples)**

Tissue Sample	Concentration of Lead (µg/g net weight) Urban	Rural
Blood	0.55 ± 0.04(39)	0.17 ± 0.02(28)
Liver	3.34 ± 0.45(21)	0.44 ± 0.09(19)
Kidney	22.7 ± 2.80(25)	1.14 ± 0.28(20)
Lung	1.24 ± 0.20(23)	0.24 ± 0.03(19)
Brain	1.11 ± 0.17(22)	0.21 ± 0.06(19)
Bone	200. ± 19.0 (24)	10.3 ± 1.90(18)

found in bone. Gasoline sniffing or accidental exposure during siphoning can lead to serious lead encephalopathy. Ingested TEL transported to liver is converted enzymatically to inorganic lead. Renal damage is also prominent in TEL intoxication, as much or more than from inorganic lead.[61,62]

Healthy adults absorb about 10% of ingested inorganic lead, while young children may absorb as much as 50%.[32] Inhaled lead oxides are solubilized in phagosomes following their phagocytosis by macrophages and type I alveolar epithelial cells (Figure 6.8).[67] They are transported mostly to the liver and kidney; the body burden of lead is then redistributed to bone, or it is excreted. The blood-brain barrier to lead uptake is not as well developed in young children as in adults, making children more susceptible to brain damage from lead poisoning. The whole-body retention half-time for lead in humans is about one month.

Blood lead levels in children and adults vary from 10–40 µg/100 ml. The EPA considers the highest safe blood lead level for children of less than 5 years of age as 30 µg/100 ml. OSHA concluded that a safe blood lead value for the occupationally exposed is <40 µg/100 ml. The normal daily intake of lead in the diet is about 300 µg. Lifetime exposures to 600 µg lead per day have not resulted in toxicity, but a 4-year, continuous daily exposure to 2,500 µg lead caused damage in humans; toxicity developed in a few months at a daily dietary intake level of 3,500 µg. Mild toxic effects have been seen in children at blood levels of 60–80 µg/100 ml, and clearcut CNS effects were seen at blood lead levels of 120 µg/100 ml.[48] Health effects associated with elevated blood lead values in adults and children are listed in Table 6.3.

The occupational standard for lead exposure in air over an 8-hour period is 50 µg/m^3. The National Ambient Air Quality standard for environmental exposure to lead is only 1.5 µg/m^3, based upon levels needed to protect children under the age of 5.

Characteristic discrete, dense-staining, intranuclear lead inclusion bodies are found in renal tubular epithelium of heavily lead-exposed individuals. Intranuclear inclusions have also been seen in hepatocytes and astrocytes of rats given lead. The formation of lead inclusions is probably a protective mechanism for storing excess lead in nondiffusible complexes. Intranuclear lead inclusion body formation in the kidney precedes the development of clinical and pathological effects. The pathological effects of lead exposure are most evident in the CNS, the bone marrow and the kidneys. The most serious manifestation of high lead exposure is lead encephalopathy in children, characterized by cerebral edema, endothelial cell damage, gliosis, focal necrosis, neuronal degeneration, convulsions, mental retardation and, in 25% of cases, death. Over half the children who survive lead encephalopathy experience mental retardation or seizures.[33] Low to moderate blood lead levels in children have been associated with visual, motor, perceptual and learning deficits and with hyperactivity.[53,54]

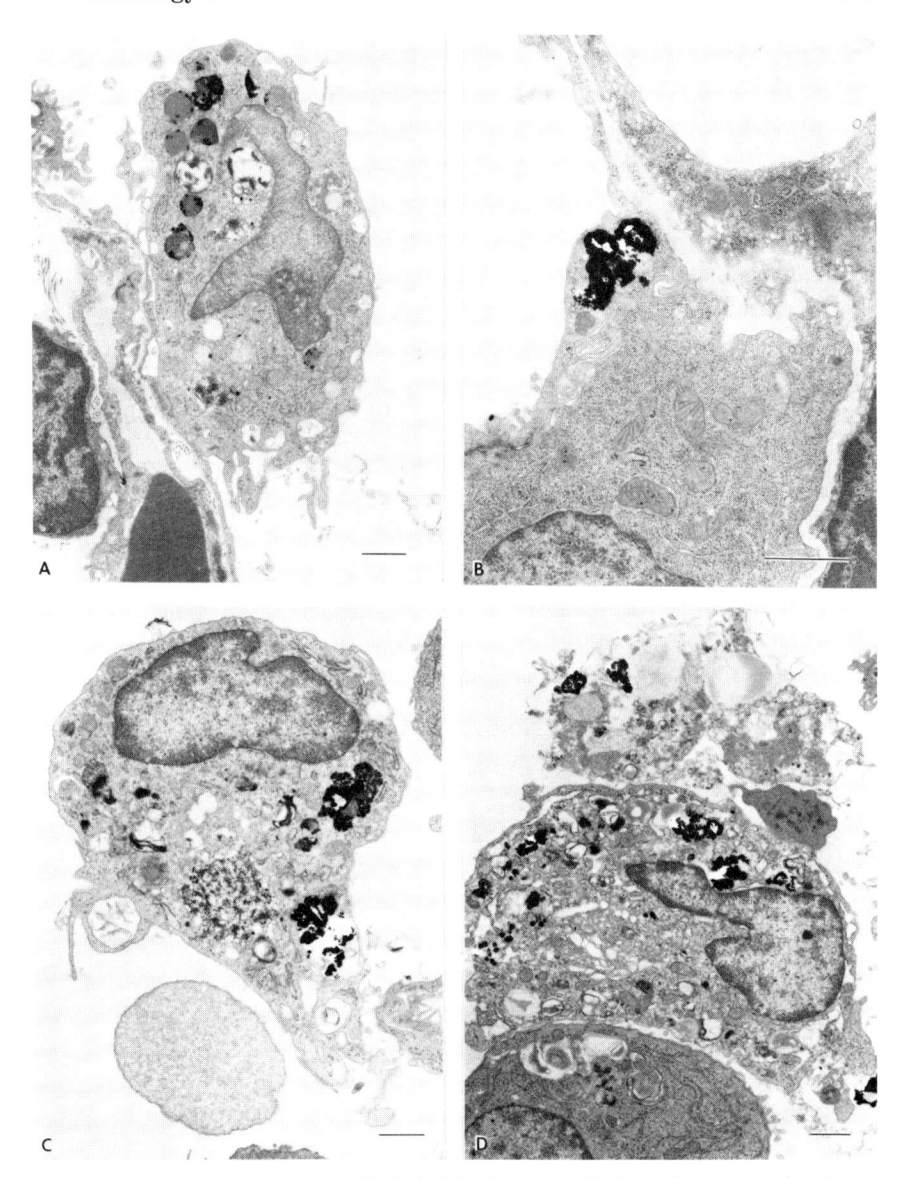

FIGURE 6.8 — Phagocytosis of inhaled lead monoxide by pulmonary, alveolar cells. All photomicrographs taken at 3 days after inhalation. (A) localization within phagolysosomes of a macrophage; (B) localization within attenuated cytoplasm of a type I cell; (C) solubilization of lead oxide and the recrystallization of lead ions within a macrophage; (D) cytotoxicity to a macrophage from phagocytized lead oxide. Bars in all cases = 1 micron (C. L. Sanders and R. R. Adee. 1977. Early fate of inhaled lean monoxide. Pacific Northwest Laboratory Annual Report for 1976, Part 1. Biomedical Sciences, BNWL-2100 PT1, pp. 180–183).

TABLE 6.3. Summary of EPA and OSHA guidelines for lead exposure in adults and children as related to the location and nature of resulting health effects caused by over-exposure to lead (D. R. Hattis, R. Goble and N. Ashford. 1982. Airborne lead: A clearcut case of differential protection. Environment 24:14–42).

Effect	Range in Blood Lead, µg/100 ml Adults	Children
Inhibition of heme biosynthesis	>20	>10
Buildup of free protoporphyrin	>15	>15
Reduced blood hemoglobin	>50	>40
Cognitive deficits	–	>50
Fatigue, nervousness, sleep disturbance	>60	–
Peripheral neuropathies	>60	>50
Slowing of nerve conduction	>50	>50
Lead encephalopathy	>100	>80
Lead nephropathy (>66% loss in kidney function)	>40–80	–
Decreased quantity of sperm; decreased fertility	>40	–
Excess risk of fetal damage and abortion	>30	–

Lead produces a microcytic, hypochromic anemia because it binds to erythrocytes and decreases their survival time in the blood; it also inhibits heme biosynthesis. Basophilic stippling in erythrocytes due to RNA accumulation is seen more frequently in children than in adults with lead poisoning. Lead inhibits aminolevulinic acid synthetase activity associated with heme biosynthesis, causing accumulation of delta-aminolevulinic acid (ALA) in blood and urine. Increased ALA levels in urine are seen prior to the appearance of clinical symptoms from lead poisoning. An increased level of porphyrins is also seen in erythrocytes and urine. In addition, lead depresses the incorporation of iron into the heme molecule because it inhibits ferrochelatase activity.

Renal alterations following lead poisoning include reversible tubular dysfunction, with resorption anomalies. Chronic nephropathy is uncommon, resulting only from high-level, long-term lead exposure. Lead is carcinogenic in the kidneys of rodents but not in humans.

Calcium disodium ethylenediaminetetraacetic acid (EDTA) enhances the excretion of lead from the body.[55,56] However, EDTA should not be given orally while lead is in the GI tract since it will also enhance lead uptake.

Manganese

Manganese is used in steel production, ceramics, electronics, welding rods, matches, glass, dyes and in medicine as antiseptics and germicides. Manganese is an essential element involved in fatty acid and cholesterol biosynthesis. The liver, kidney, intestine and pancreas concentrate manganese following oral administration; normal body burden in humans is 20 mg.

Inhaled manganese dioxide causes a chemical pneumonitis at high exposure levels. Chronic exposure to inhaled manganese causes CNS injury resulting in a psychiatric disorder characterized by irritability, motor disturbances, compulsive behavior and Parkinson-type symptoms; selective damage occurs to the subthalamic nucleus. Recovery is slow following removal from further manganese exposure and may be accompanied by liver cirrhosis.[68] Current workplace standard set by OSHA is 5000 µg Mn/m³, as a ceiling exposure. WHO recommends an occupational limit in air of 300 µg Mn/m³. Methylcyclopentadienyl manganese tricarbonyl (MMT) is effective in raising the octane level of gasoline; MMT is currently being used in Canada as a gasoline additive (0.07 g Mn/U.S. gal) for this purpose. Experimental animals that inhaled the combustion products of MMT up to 1000 µg Mn/m³ for 9 months did not show toxic effects; rhesus monkeys, susceptible to toxic neurologic effects of manganese, did not show symptoms, even at exposures of up to 5000 µg Mn/m³ for 66 weeks. The small amounts of manganese added to the environment by the use of MMT as a fuel additive in the U.S. would not be expected to create environmental health hazards.[69]

Mercury

Three forms of mercury are of toxicological importance: inorganic mercury vapor and ions, aryl organic mercury compounds and alkyl mercury compounds (Figure 6.9). From 30,000 to 150,000 tons of mercury are released annually into the atmopshere by degassing from the earth's crust and oceans. In addition, about 20,000 tons of mercury are released into the air and water by industry and the combustion of fossil fuels.[34] Mercury is used in chlorine production, electrical apparatus production, mildewproofing, drugs, as a catalyst in the chemical industry and in agriculture as protection of seeds from fungi.

```
                                                           Toxicological Aspects of Energy Production
```

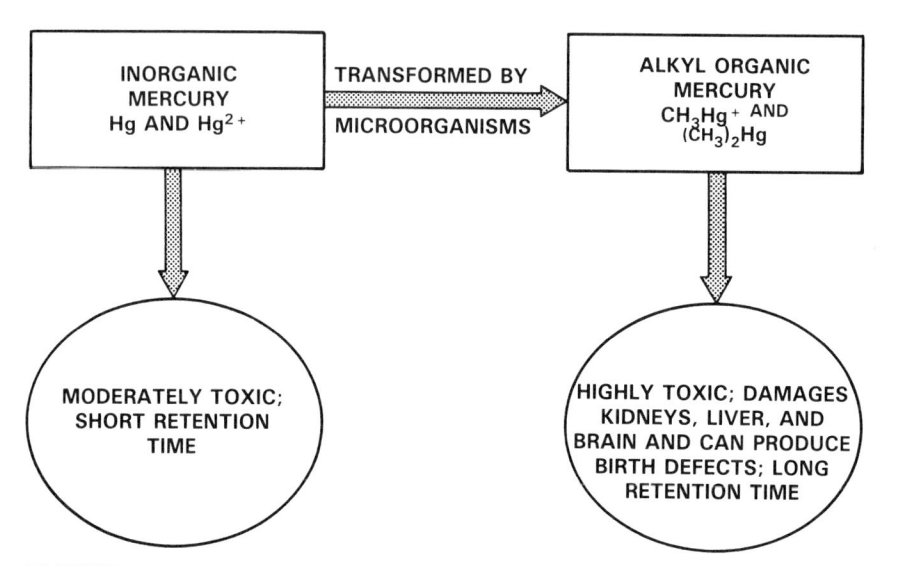

```
┌─────────────┐  TRANSFORMED BY   ┌─────────────────┐
│ INORGANIC   │ ════════════════▶ │ ALKYL ORGANIC   │
│ MERCURY     │                   │ MERCURY         │
│ Hg AND Hg²⁺ │  MICROORGANISMS   │ CH₃Hg⁺ AND      │
│             │                   │ (CH₃)₂Hg        │
└──────┬──────┘                   └────────┬────────┘
       ║                                   ║
       ▼                                   ▼
   ╱─────────╲                       ╱───────────╲
  │MODERATELY│                      │HIGHLY TOXIC;│
  │  TOXIC;  │                      │   DAMAGES   │
  │  SHORT   │                      │KIDNEYS, LIVER,│
  │ RETENTION│                      │   AND BRAIN │
  │   TIME   │                      │ AND CAN PRODUCE│
   ╲─────────╱                      │BIRTH DEFECTS;│
                                    │    LONG     │
                                    │ RETENTION TIME│
                                     ╲───────────╱
```

FIGURE 6.9—The relationships of different forms of mercury and their biomedical effects (G. T. Miller. 1975. Living in the environment. Concepts, Problems, and Alternatives. Wadsworth publishing Co., Inc., Belmont, CA, p. E99).

Organic mercury compounds are produced by the action of microbes on mercury vapor and salts. A large portion of the mercury in the environment is methylmercury, an alkyl compound that is highly toxic. The methylation of mercury in aquatic systems emphasizes the importance of elemental transformations in trace element transport. Microbial activity is thought to be mostly responsible for the production of highly toxic mono- and dimethylmercury compounds:

$$Hg^{++} + 2R-CH_3 \rightarrow CH_3HgCH_3 \rightarrow CH_3Hg$$

$$Hg^{++} + R-CH_3 \rightarrow CH_3Hg^+ \rightarrow CH_3HgCH_3$$

Dimethylmercury is more volatile than monomethylmercury, causing a greater release into the air and decreasing the extent of bioaccumulation. In acidic water, more of Hg^{++} will be converted to monomethylmercury and remain in aquatic ecosystems.[47]

Mercury is concentrated as much as 100,000-fold in aquatic invertebrates. Fish which have consumed contaminated invertebrates can cause toxic symptoms in humans when ingested. In 1953 an epidemic of methylmercury poisoning occurred in residents of Minimata Bay, Japan following industrial pollution of the bay with mercury; the clinical syndrome was called Minimata disease. The fetus was shown to be especially sensitive to methylmercury poisoning. Mercury vapor is readily absorbed from the lung and rapidly oxidized by erythrocytes to divalent mercuric ion. Uptake of oral metallic mercury by the GI tract is negligible. Mercury vapor, in contrast to mercury salts, causes fetal damage; fortunately, nearly all mercury vapor is oxidized before it reaches these critical tissues. Less than 10% of inorganic mercury salts are absorbed from the GI tract, concentrating mostly in the kidneys. In contrast alkyly mercury compounds are readily absorbed from the GI tract (e.g., 95% of ingested methylmercury) and lung and concentrated in the CNS and fetal tissues because of their high lipid solubility. The half-life for mercury compounds in the body averages about 70 days. Mercury is concentrated in kidney and liver. Mercury compounds translocated to the brain are cleared slowly from this organ. Levels of mercury in the urine and hair are good indicators of exposure and deposition levels.

The toxicity of aryl mercury compounds is relatively low; the LD_{50} is about 10 times more than inorganic mercury. The enzymatic conversion of a small amount of aryl organic mercury in the body to inorganic mercury can increase overall toxicity of mercury compounds.

Chronic poisoning from mercury vapor is characterized by nonspecific neuropsychiatric disorders, weakness, fatigue, weight loss, fine tremors of fingers and eyelids, inflammation of the gums and excess salivation. Paresthesia, visual changes, hearing defects, dysarthria and ataxia are associated with alkyl mercury intoxication. The kidney is the target organ for inorganic mercury exposure; significant damage occurs at concentrations of 10–70 mg/kg in kidney tissues. Intakes of methylmercury that result in body burdens of less than 0.5 mg/kg body weight are not likely to produce neurological symptoms.[34] The clinical symptoms of toxicity from various mercury compound classes are list in Table 6.4. Several agents, such as dimercaptopropanol (BAL) and penicillamines have been tried in humans with limited success to enhance the excretion of mercury compounds.[57,58]

Nickel

Nickle found in sulfide and oxide ores is used in steel production, the manufacture of coins, electronic components and nickel-cadmium batteries. Normal daily intake of nickel in the U.S. ranges from 300–500 µg; nickel air levels range from about 6 ng/m³ in rural areas to 20 ng/m³ in cities.[35]

Nickel is an essential element at oral daily ingestions of <0.5 mg; greater deposition levels can be rapidly toxic. In experiments, nickel oxide cleared slowly from the lung of hamsters (Figure 6.7). Gaseous nickel compounds as, Ni carbonyl [$Ni(CO)_4$] are rapidly cleared from the lung and excreted in urine; nickel carbonyl is rapidly catabolized in tissues to nickel

TABLE 6.4. Toxicity of common mercury chemical forms (M. W. Neathery and W. J. Miller. 1975. Metabolism and toxicity of cadmium, mercury, and lead in animals: A review. J. Dairy Sci. 58:1767-1780).

Mercury Vapor	Mercury Salts	Methylmercury
Bronchitis	Anorexia	Paraesthesia
Interstitial pneumonia	Weight loss	Hearing loss
Circulatory collapse	Memory loss	Ataxia
Renal failure	Metallic taste	Peripheral vision loss
	Gingivitis	Intellectual deterioration
	Increased saliva	Cerebral palsy neonatal syndrome

and carbon monoxide. About 10% of ingested nickel is absorbed from the GI tract.

The lung is the target organ for both inhaled and parenterally administered nickel carbonyl. Hypersensitive dermatitis is seen in some individuals following percutaneous exposure to nickel compounds. Nickel dust and nickel oxide produce a simple pneumoconiosis in exposed workers; in addition, alveolar lipoproteinosis is seen in nickel-exposed animals. Nickel carbonyl, in animals, produced squamous cell metaplasia and carcinoma; however, evidence for carcinogenicity in humans is lacking. Nickel dust, nickel subsulfide, nickel oxide and nickel biscyclopentadiene are carcinogenic in lungs and nasal cavity of animals. An overall relationship between nickel compounds and human cancer risk has not been established.[37]

Palladium and Platinum

Palladium is the lightest metal in the platinum group. Palladium and platinum are used in electronics, dental alloys, as catalysts in chemical industries and especially in automotive exhaust converters to reduce emissions of NO_x. Industrial toxicity of both metals has been very limited due to their limited uses. Platinum is more toxic than palladium. Palladium causes lung and skin hypersensitivities. Platinum compounds have caused asthma, rhinitis, conjunctivitis and dermatitis, indicative of its potential allergenic reactions in sensitizied individuals. High doses of platinum aerosols cause pulmonary fibrosis and emphysema in animals.[70,71] The current TLV for platinum and palladium is 2 µg/m^3. A highly toxic organoplatinum compound, cis-platinum, is used in chemotherapy of cancer.

Selenium

Selenium is a nonmetallic elemental by-product of the production of lead, copper, mercury and nickel from the treatment of sulfide ores.[38] It is also released into the environment from the combustion of fossil fuels. World production of selenium is about 1,500 tons annually. Selenium is used in electronics, glass, pigments, rubber, and as a feed additive for cattle in areas where selenium concentrations in the soil are low. Daily human dietary intake of selenium varies from less than 50 µg to 300 µg, depending upon geographical location. Selenium is an essential dietary constituent in the range of 0.1–0.3 ppm in food or water. Atmospheric concentrations of selenium are usually in the range of a few nanograms/m^3.

More than 90% of selenium present in food or water is absorbed by the GI tract; both selenite and selenate compounds are readily absorbed. Selenium is metabolized in the body to methyl derivatives, some of which may be exhaled, giving a "garlic" odor. The "therapeutic window" for beneficial effects from selenium is narrow; selenium deficiency occurs at dietary Se levels of <0.1 ppm and overt Se toxicity at levels >5 ppm.[40] Excess selenium exposure leads to skin, cardiovascular and pulmonary damage. There is a weak relationship between the blood levels of selenium in humans and the total incidence of cancer deaths throughout the U.S. Blood selenium levels are usually depressed in patients with cancer.[56] In cities with populations having low blood levels of selenium, the cancer death rate is twice as great as places where blood selenium levels are high. Selenium is believed to be teratogenic in humans.[60]

Uranium

Uranium occurs naturally as ^{238}U (99.3%), ^{235}U (0.7%) and ^{234}U (0.05%); it is used almost exclusively as a fuel in nuclear energy. Minor applications include its use in pigments and chemical processes. Uranium exists in biological fluids in soluble form in the hexavalent state as uranyl ions, (UO$^{++}_2$), which form complexes with carbonate ions and proteins.

Uranium exhibits both chemical and radiological health hazards. Because of its very long physical half-life, the chemical toxicity of ^{238}U is the limiting factor in exposure, whereas the radiological toxicity of ^{235}U is the limiting factor for this isotope.[45] Only microgram quantities of uranium are ingested in food, water and air.[43] According to the ICRP (International Commission on Radiological Protection) only 1% of uranium passing through the GI tract is absorbed into the blood. However, insoluble oxides are retained if inhaled. Inhaled high-fired UO_2 was found to be very insoluble in the lungs of monkeys and dogs; pulmonary retention times varied from 500 to 2,000 days. More soluble compounds of uranium, such as UF_6, $UO_2(NO_3)_2$ and uranium sulfates and carbonates, exhibit retention half-times in the lungs of 50 to 100 days. Soluble uranium leaves the lung as a cation complexed with bicarbonate. In the blood plasma, the bicarbonate complex rapidly equilibrates with the extracellular fluid and serum proteins. About 70% of the uranium in the blood will pass into the urine through the kidneys during the first 24 hours after exposure. Clearance from the blood is mainly into the skeleton and kidneys; bone is not a critical tissue for uranium but acts as a long-term storage depot.[44]

Accumulation of uranium in the kidney is the crucial factor in determining the chemical toxicity of uranium. A slightly acid pH of the glomerular filtrate causes the uranium-bicarbonate complex to dissociate. Bicarbonate is resorbed by the kidney tubules, leaving unbound uranyl ions, which bind to phosphate and organic acid groups of membranes of proximal tubule epithelium. The resulting renal disease at higher doses is due to alterations in membrane permeability and transport and a progressive, slowly developing, tubular necrosis with little evidence of regeneration. Catalase in urine is a sensitive early indicator of renal uranium toxicity. Later, more severe renal damage causes an increase in albumin and other substances in the urine as the resorption capabilities of the tubule epithelium are increasingly impaired. The nephrotoxic limit for uranium in blood is 3 µg U/hour excreted from the kidney, equivalent to a single dose of 2.7 mg U in the blood. The TLV for soluble chemical forms of uranium is 200 µg/m³.

Vanadium

More than 20,000 tons of vanadium are produced in the world annually, principally for use in special steels and nonferrous alloys. The largest source of vanadium air pollution is combustion of fossil fuels; 0.2 to 2.0 kg V is released per 1,000 tons of coal consumed. Vanadium is found over a wide range of concentrations in coal and crude oil, ranging from <1 ppm to >1,000 ppm.[41]

Vanadium levels in most water and food samples range from 0.1 to 5 µg/g. Air levels of vanadium range from <1 ng/m³ in western states to about 50 ng/m³ in eastern states; urban areas may have vanadium air levels ranging from 100 to 1,000 ng/m³. Vanadium pentoxide (V_2O_5) is the most common chemical form of vanadium. It readily dissolves in water or tissue fluids to give a mixture of oxy-acids such as metavanadate. Chemical reaction with vanadium compounds is strongly influenced by pH.

About 2% of vanadium in the diet is taken up by the GI tract. Clearance of vanadium pentoxide from the deep lung is rapid, with the skeleton as the primary site for retained vanadium.[58] Excretion of vanadium in humans is mostly in the urine.

The lethal dose of injected vanadium is >6 mg/kg. Toxicity is low when vanadium was given orally. The LC_{50} of intratracheally instilled vanadate ion is 12 mg/kg in the rat; vanadate ions are potent inhibitors of ATPases, phosphatases, kinases and RNase.[59]

Acute or chronic exposure of animals or humans to vanadium pentoxide causes mostly reversible respiratory effects, such as tracheitis, pulmonary edema, respiratory irritation, cough and bronchopneumonia. Conjuctivitis, hypersensitization with repeated chronic exposure and a green coloration of the tongue are seen in occupationally exposed workers, mining or smelting vanadium ores. Vanadium compounds are not mutagenic, teratogenic or carcinogenic.[42] Ascorbic acid is an effective antidote for intoxication from vanadate or vanadyl ions because of its reducing action.[60]

References

1. L. Friberg, G. F. Nordberg and V. B. Vouk. 1979. Handbook on the Toxicology of Metals. Elsevier/North-Holland Biomedical Press, Amsterdam.
2. J. L. Huisingh, J. A. Campbell and M. D. Waters. 1977. Evaluation of trace-element interactions using cultured alveolar macrophages. In, "Pulmonary Macrophage and Epithelial Cells," CONF-760927, TIS, Springfield, VA, pp. 346-357.
3. M. D. Waters, D. R. Abernethy, H. R. Garland and D. L. Coffin. 1974. Toxic effects of selected metallic salts on strain WI-38 human lung fibroblasts. In Vitro 10:342.
4. M. E. Frazier and T. K. Andrews. 1979. In vitro clonal growth assay for evaluating toxicity of metal salts. In, "Trace Metals in Health and Disease," Raven Press, NY, pp. 71-81.
5. F. W. Sunderman. 1979. Carcinogenicity and anticarcinogenicity of metal compounds. In, "Environmental Carcinogenesis," Elsevier/North-Holland Biomedical Press, Amsterdam, pp. 165-188.
6. T. Norseth. 1979. Aluminum. In, "Handbook on the Toxicology of Metals," Elsevier/North-Holland Biomedical Press, Amsterdam, pp. 275-281.
7. N. C. Bowdler et al. 1979. Behavioral effects of aluminum ingestion on animal and human subjects. Pharmacol. Biochem. Behav. 10:505-512.

8. C. -G. Elinder and L. Friberg. 1979. Antimony. In, "Handbook on the Toxicology of Metals," Elsevier/North-Holland Biomedical Press, Amsterdam, pp. 283–292.
9. T. Dutkiewicz. 1977. Experimental studies on arsenic absorption routes in rats. Environ. Health Perspect. 19:173–177.
10. E. A. Crecelius. 1977. Changes in the chemical speciation of arsenic following ingestion by man. Environ. Health Perspect. 19:147–150.
11. J. S. Woods and B. A. Fowler. 1977. Effects of chronic arsenic exposure on hematopoietic function in adult mammalian liver. Environ. Health Perspect. 19:209–213.
12. Wen-Ping Tseng. 1977. Effects and dose-response relationships of skin cancer and blackfoot disease with arsenic. Environ. Health Perspect. 19:109–119.
13. M. G. Ott, B. B. Holden and H. L. Gordon. 1974. Respiratory cancer and occupational exposure to arsenicals. Arch. Environ. Health 29:250–255.
14. I. R. Tabershaw. 1972. The toxicology of beryllium. Public Health Service Publication No. 2173, U.S. Public Health Service, Washington, D.C., 50 p.
15. C. L. Sanders, W. C. Cannon, G. J. Powers, R. R. Adee and D. M. Meier. 975. Toxicology of high-fired beryllium oxide inhaled by rodents. Arch. Environ. Health 30:546–551.
16. N. L. Sprince and H. Kazemi. 1980. U.S. Beryllium Case Registry through 1977. Environ. Res. 21:44–47.
17. J. K. Wagoner, P. F. Infante and D. L. Batliss. 1980. Beryllium: An etiologic agent in the induction of lung cancer, non-neoplastic respiratory disease, and heart disease among industrially exposed workers. Environ. Res. 21:15–34.
18. L. Friberg, M. Piscator, G. F. Nordberg and T. Kjellstrom. 1974. Cadmium in the environment. CRC Press, Inc., Cleveland, OH, 248 p. 1.
19. J. G. Hadley, A. W. Conklin and C. L. Sanders. 1980. Rapid solubilization and translocation of 109-Cd following pulmonary deposition. Toxicol. Appl. Pharmacol. 54:156–160.
20. R. G. Thomas, J. S. Wilson and J. E. London. 1980. Multispecies retention parameters of cadmium. Environ. Res. 23:191–207.
21. L. Friberg, T. Kjellstrom, G. Nordberg and M. Piscator. 1979. Cadmium. In, "Handbook on the Toxicology of Metals," Elsevier/North-Holland Biomedical Press, Amsterdam, pp. 355–381.
22. K. -T. Wong and C. D. Klaasen. 1980. Age difference in the susceptibility to cadmium-induced testicular damage in rats. Toxicol. Appl. Pharmacol. 55:456–466.
23. S. Langard and T. Norseth. 1979. Chromium. In, "Handbook on the Toxicology of Metals," Elsevier/North-Holland Biomedical Press, Amsterdam, pp. 383–397.
24. S. Abe et al. 1982. Chromate lung cancer with special reference to

its cell type and relation to the manufacturing process. Cancer 49:783–787.
25. W. C. Hueper and W. W. Payne. 1959. Experimental cancers in rats produced by chromium compounds and their significance in industry and public health. Am. Indust. Hyg. Assoc. J. 20:272–280.
26. C. -G. Elinder and L. Friberg. 1979. Cobalt. In, "Handbook on the Toxicology of Metals," Elsevier/North-Holland Biomedical Press, Amsterdam, pp. 399–410.
27. M. Piscator. 1979. Copper. In, "Handbook on the Toxicology of Metals," Elsevier/North-Holland Biomedical Press, Amsterdam, pp. 411–420.
28. C. -G. Elinder and M. Piscator. 1979. Iron. In, "Handbook on the Toxicology of Metals," Elsevier/North-Holland Biomedical Press, Amsterdam, pp. 435–450.
29. U. Saffiotti, F. Cefis and L. H. Kolb. 1968. A method for the experimental induction of bronchogenic carcinoma. Cancer Res. 28:104–124.
30. Environmental Protection Agency. Air quality crieteria for lead, EPA 600/8–77–017, U.S. Government Printing Office, Washington, D.C. (1977).
31. K. Tsuchiya. 1979. Lead. In, "Handbook on the Toxicology of Metals," Elsevier/North-Holland Biomedical Press, Amsterdam, pp. 451–484.
32. R. A. Kehoe. 1961. The metabolism of lead in man in health and disease. J. R. Inst. Pub. Health 24:101–120.
33. E. B. McCabe. 1979. Age and sensitivity to lead toxicity: A review. Environ. Health Perspect. 29:29–33.
34. M. Berlin. 1979. Mercury. In, "Handbook on the Toxicology of Metals," Elsevier/ North-Holland Biomedical Press, Amsterdam, pp. 503–530.
35. T. Norseth and M. Piscator. 1979. Nickel. In, "Handbook on the Toxicology of Metals," Elsevier/North-Holland Biomedical Press, Amsterdam, pp. 541–553.
36. F. W. Sunderman. 1973. Current status of nickel carcinogenesis. Ann. Clin. Lab. Sci. 3:156–180.
37. J. P. W. Gilman. 1962. Metal carcinogenesis. II. A study on the carcinogenic activity of cobalt, copper, iron and nickel compounds. Cancer Res. 22:158–165.
38. J. Glover, O. Levander, J. Parizek and V. Vouk. 1979. Selenium. In, "Handbook on the Toxicology of Metals," Elsevier/North-Holland Biomedical Press, Amsterdam, pp. 555–577.
39. C. L. Sanders and A. W. Conklin. 1981. Influence of cadmium, selenium and vanadium on acute mortality and lung clearance. In, "Pacific Northwest Laboratory Annual Report for 1980," PNL-3700, Part 1. Biomedical Sciences, NTIS, Springfield, VA, pp. 22–23.
40. R. J. Shamberger. 1970. Relationship of selenium to cancer. I. Inhibiting effect of selenium in carcinogenesis. J. Natl. Cancer Inst. 44:931–936.
41. National Academy of Science-National Research Council. 1978. Medical

and biological effects of environmental pollutants: Vanadium. Washington, D.C.
42. V. Vouk. 1979. Vanadium. In, "Handbook on the Toxicology of Metals," Elsevier/North-Holland Biomedical Press, Amsterdam, pp. 659–674.
43. M. Berlin and B. Rudell. 1979. Uranium. In, "Handbook on the Toxicology of Metals," Elsevier/North-Holland Biomedical Press, Amsterdam, pp. 645–658.
44. L. J. Leach, C. L. Yuile, H. C. Hodge, G. E. Sylvester and W. B. Wilson. 1973. A five year inhalation study with natural uranium dioxide (UO_2) dust. II. Post-exposure retention and biologic effects in the monkey, dog and rat. Health Physics 25:239–258.
45. J. A. Orcutt. 1949. The toxicology of compounds of uranium following application to the skin. In, "Pharmacology and Toxicology of Uranium Compounds," McGraw-Hill, New York, pp. 377–422.
46. T. C. Hutchinson and J. Fitchko. 1974. Heavy metal concentrations and distributions in river mouth sediments around the Great Lakes. In, "Proceedings of the Internationl Conference on Transport of Persistent Chemicals in Aquatic Ecosystems," National Research Council of Canada, Ottawa, pp. I-69-I-77.
47. A. Finkel. 1983. Hamilton and Hardy's Industrial Toxicology. Mercury. pp. 93–104, John Wright. PSG Inc., Boston, MA.
48. D. R. Hattis, R. Goble and N. Ashford. 1982. Airborne lead: A clearcut case of differential protection. Environment 24:14–42.
49. A. J. Finkel. 1983. Hamilton and Hardy's Industrial Toxicology. Arsenic., pp. 17–24. John Wright. PSG Inc, Boston, MA.
50. T. Kjellstrom, C. -G. Elinder and L. Friberg. 1984. Conceptual problems in establishing the critical concentration of cadmium in human kidney cortex. Environ. Res. 33:284–295.
51. C. L. Sanders and J. A. Mahaffey. 1984. Carcinogenicity of single and multiple intratracheal instillations of cadmium oxide in the rat. Environ. Res. 33:227–233.
52. S. Takenaka, H. Oldiges, H. Konig, D. Hochrainer and G. Oberdorster. 1983. Carcinogenicity of cadmium chloride aerosols in W rats. J. Natl. Can. Inst. 70:367–373.
53. J. Marecek, I. M. Shapiro, A. Burke, S. H. Katz and M. L. Hediger. 1983. Low-level lead exposure in childhood influences neuropsychological performance. Arch. Environ. Health 38:355–359.
54. H. Haenninen et al. 1978. Psychological performance of subjects with low exposure to lead. J. Occup. Med. 20:683–689.
55. J. J. Chisolm. 1970. Treatment of acute lead intoxication-choice of chelating agents and supportive therapeutic measures. Clin. Toxicol. 3:527–540.
56. P. B. Hammond. 1971. The effect of chelating agents on the tissue distribution and excretion of lead. Toxicol. Appl. Pharmacol. 18:296–310.

57. J. Aaseth. 1976. Mobilization of methyl mercury in vivo and in vitro using N-acetyl-DL-penicillamine and other complexing agents. Acta Pharmacol. Toxicol. 39:289–301.
58. F. Bakir, A. Al-Khalidi, T. W. Clarkson and R. Greenwood. 1976. Clinical observations on treatment of alkyl-mercury poisoning in hospital patients. Bull. W.H.O. (Suppl.) 53:87–92.
59. E. Sabbioni, L. Clerici and A. Brazzelli. 1983. Different effects of vanadium ions on some DNA-metabolizing enzymes. J. Toxicol. Environ. Health 12:737–748.
60. D. S. E. Robertson. 1970. Selenium—A possible teratogen. Lancet 1:518–519.
61. L. W. Change, P. R. Wade, K. R. Reuhl and M. J. Olson. 1980. Ultrastructural changes in renal proximal tubules after tetraethyllead intoxication. Environ. Res. 23:208–223.
62. R. A. Goyer and B. C. Rhyne. 1973. Pathological effects of lead. Int. Rev. Exp. Pathol. 12:1–77.
63. S. P. Andreoli, J. M. Bergstein and D. J. Sherrard. 1984. Aluminum intoxication from aluminum-containing phosphate binders in children with azotemia not undergoing dialysis. New England J. Med. 310:1079–1084.
64. G. Martinez, M. Cebrain, M. Chamorro and P. Jauge. 1983. Urinary porphyrins as an indicator of arsenic exposure in rats. Proc. West. Pharmacol. Soc 26:171–174.
65. M. Webb. 1972. Biochemical effects of Cd^{2+}-injury in the rat and mouse testis. J. Reprod. Fertil. 30:83–98.
66. L. A. Shinobu, M. M. Jones, M. A. Basinger, W. M. Mitchell. D. Wendel and A. Razzuk. 1983. In vivo screening of potential antidotes for chronic cadmium intoxication. J. Toxicol. Environ. Health 12:757–765.
67. C. R. deVries et al. 1983. Acute toxicity of lead particulates on pulmonary alveolar macrophages. Ultrastructural and microanalytical studies. Lab. Invest. 48:35–42
68. World Health Organization. 1981. Environmental Health Criteria 17. Manganese. International Program on Chemical Safety, World Health Organization, Geneva.
69. W. C. Cooper. 1984. The health implications of increased manganese in the environment resulting from the combustion of fuel additives: A review of the literature. J. Toxicol. Environ. Health 14:23–46.
70. D. J. Holbrook et al. 1975. Studies on the evaluation of the toxicity of various salts of lead, manganese, platinum and palladium. Environ. Health Perspect. 10:95–101.
71. S. O. Freeman and J. Krupey. 1968. Respiratory allergy caused by platinum salts. J. Allergy 42:233–237.

CHAPTER 7

COAL AND OIL

Introduction

Coal utilization in the U.S. was 15 quad or 650 million tons in 1975 and is expected to grow to 28 quads by 1985. By 1995, it is estimated that 1,400 million tons of coal will be used in the U.S. for the production of electric power and space heating; by comparison, world consumption of coal in 1977 was 4,000 million tons, of which 2,400 million tons was used in electrical power production. Worldwide consumption of coal for electricity production and space heating is projected to be 7,500 million tons by 1995. There are about 1,000 fossil-fueled, steam generating, power plants in the U.S.; of these, 550 are coal-fired. A typical 1000 MW_e coal-fueled power plant consumes 6,400 tons of coal per day and releases, daily, 100 tons sulfur dioxide, 50 tons NO_x, 450 tons particulates, 650 µg BaP and 1300 µg benzo(a)anthracene per 1000 m³ from the stack, and a variety of metals into the air.[2] Bituminous coal contains traces (<200 ppm) of a wide variety of toxic metals. Coal dust and coal combustion products are clearly harmful to human health, particularly in the respiratory tract.

Coal Workers Pneumoconiosis

Coal itself is not a uniform substance and is classified according to increasing content of carbon and decreasing content of oxygen; that is, into lignite, sub-bituminous coal, bituminous coal and anthracite. The airborne dust coal mines is a complex mixture, mainly derived from the coal seam and associated geological strata. The mineral content of respirable dust in coal mines is derived from the non-coal strata, comprising kaolin, mica, small quantities of iron, calcium, magnesium, large amounts of silicates and varying amounts of quartz, usually at levels of <5%.

Inhaled coal dust deposited in the pulmonary region of the lung has a long retention time. In dogs, the mean half-life for inhaled coal dust is

1.9 years; only 4% of pulmonary deposited coal dust was translocated to tracheobronchial lymph nodes.[58]

An area of considerable health concern in the increased use of coal for energy production is the health of coal miners and processors. Respirable coal dust can cause mild to severe pulmonary disease, where the pathological damage is manifested in nonreversible fibrotic lesions causing progressive loss of pulmonary function with increasing lung dust burden. Disease progression can continue after removal of the coal miner from further occupational exposure.

Coal workers' pneumoconiosis (CWP) affects over 70% of all underground coal miners, although it is the cause of death in only about 5% of miners.[3] CWP constitutes a much larger health problem than other types of pneumoconioses, such as silicosis. The Federal Coal Mine Health and Safety Act (1969) was designed to reduce occupational health and safety risks. The current U.S. respirable dust standard for coal mines is 2 mg/m³. In CWP, alveoli opening into respiratory bronchioles become consolidated by tightly packed concentrations of coal-dust-laden macrophages, eliciting a moderate proliferation of reticulum fibers with few fibroblasts (grade l pneumoconiosis). Prolonged alveolar dilatation leads to proximal acinar emphysema. In contrast to silicotic nodules, the simple dust lesions of CWP are comprised mostly of dust accumulations and not connective tissues in response to irritating stimuli (Figure 7.1). The cytotoxicity of coal dust may, in part, be explained by the quartz content of the dust, which averages < 5% in most mine atmospheres. However, quartz alone can not explain the toxicity of coal dust with < 5% quartz. Even some natural dusts from coal mines containing up to 15% quartz are no more fibrogenic than dusts containing no quartz. This may be due to an intimate coating of quartz particles with material in coal dust. The quartz content of coal dust in the lungs of CWP patients averages 2–3%; the mean amount of quartz in the lungs of miners with simple CWP is only 200 mg. The mean clearance half-time of quartz-free coal dusts from the lungs of exposed dogs was about 2 years, as compared to 4–5 year half-time clearance in humans.[34]

No differences have been found in biological responses to low and high BTU or rank coals given by inhalation. Thus, the coal particles are not a likely explanation for the high epidemiological association of CWP with workers in high-rank coals. It is more likely that the mineral contents, particularly quartz, is higher in dusts from high-rank coals (especially anthracites) than in low-rank coals; however, coal miners may also develop simple CWP following prolonged exposure to coal mine dusts containing little quartz.

The extent of CWP in coal miners, determined radiographically, is a function of total dust exposure rather than of quartz content (Figure 7.2). The simple dust lesions of coal workers and others exposed to similar dust may best be interpreted as a nonspecific reaction to large accumulations

7 − Coal and Oil 183

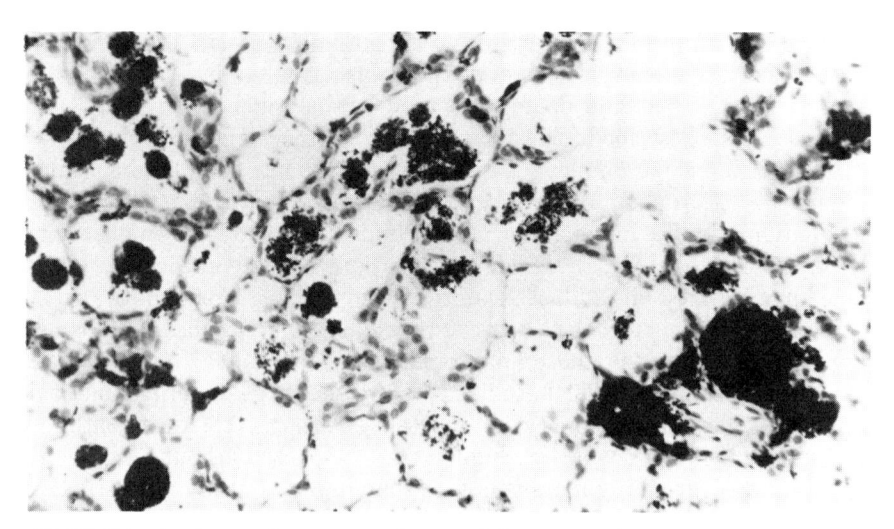

FIGURE 7.1−Lung of rat exposed to coal dust aerosol continuously for 20 months (Courtesy of Dr. R. H. Busch, PNL, Richland, WA).

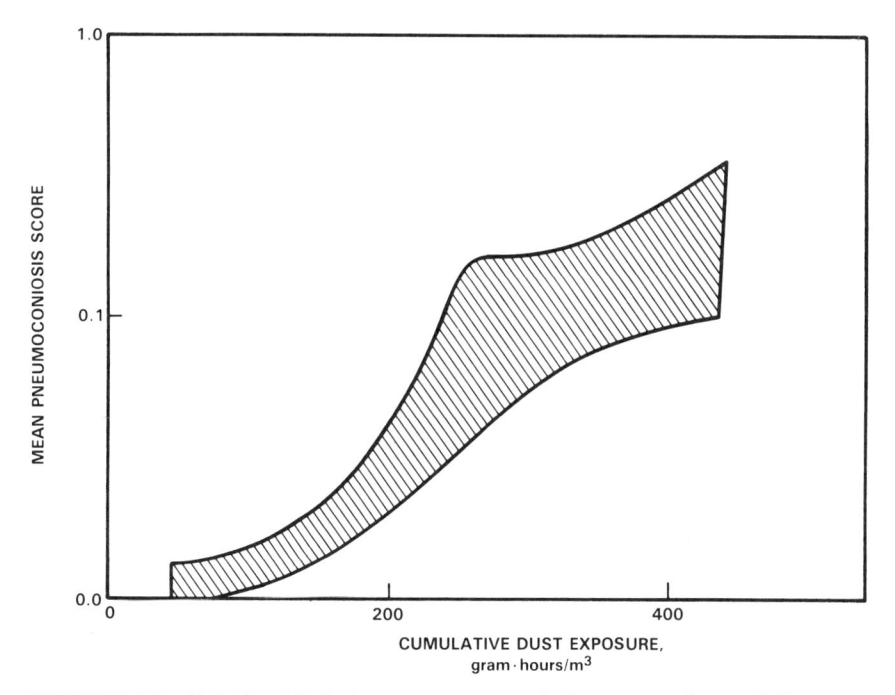

FIGURE 7.2−Relationship between pneumoconiosis score and cumulative working-life exposures to respirable dust in 2600 miners from 10 coal mines (M. Jacobsen and J. Dodgson. 1981. Long-term experience in collecting and using occupational health data in the coal industry. Ann. Occup. Hyg. 24:391–398).

of dust which become immobilized, with minimal fibrosis. Coal-dust exposure also leads to alveolar lipoproteinosis (as a result of irritative stimulation of type II cells and damage to type I cells) much like the alveolar lipoproteinosis described following exposure to other mixed-silicate dusts.[5]

The incidence of emphysema is high in coal miner populations, ranging up to 80%, or about double the incidence for a comparable non-mining population.[56] In coal miners, coal dust accumulations and focal emphysemic lesions occur together; these combined lesions have been termed "black hole lung."[55] In regions of dust deposits, the supporting tissues of the bronchiolar walls (e.g., fibroelastic layer and smooth muscle) disappear causing bronchiolostenosis and emphysema.

Overall, epidemiological studies of the diseases of underground coal miners involve both morbidity and mortality. Although the death rate for miners with simple CWP is no higher than for those without CWP, mortality in underground miners is higher than in above-ground miners. For example, age-adjusted mean annual death rates for cancer of the stomach, in a 1949–1953 study in England, ranged from 62% to 226% higher in miners than in non-miners.[35] Similar stomach tumor incidences were found in American coal miners.[36] Inhaled coal dust is cleared by the muco-ciliary eschalator from the respiratory tract and swallowed, going into the stomach. Nitrosation of ingested coal dust by nitrites in processed foods increases the genotoxicity of coal dust extracts and may be responsible for the elevated incidence of stomach cancer.[57] No evidence was found, in recent studies, for an association between exposure to coal dust and lung cancer, although the risk for lung cancer in miners who were cigarette smokers was increased. No evidence was seen for any interaction between coal mine dust and cigarette smoking in lung cancer risk in coal miners.[37]

Overall, epidemiological data clearly show a sharply reduced life span for coal miners with complicated CWP. Mortality was also high for miners who smoked cigarettes. Thus, many of the health problems associated with coal mining were due, in part, or were exacerbated by, cigarette smoking. Mortality rates ranging from 13% to 90% were noted for the following diseases: influenza, emphysema, asthma, tuberculosis, stomach cancer and hypertension.[36] The total number of coal miners suffering from lung diseases is expected to rise markedly during the 1980s because of their increasing age and number of years worked in coal mines. Coal miners working at dust levels < 2 mg/m^3 would not be expected to experience coal dust-related disease progression.[19]

Consequences of Carbon Dioxide Production

The earth's carbon is distributed among the crustal sediments, the oceans, a thin layer of living organisms, a small amount of atmospheric carbon

7 – Coal and Oil

dioxide (CO_2) and accumulations of geological organic material in the forms of natural gas, oil and coal. By burning fossil fuels and converting carbon-rich forest to thinly vegetated agricultural land, we translocate more carbon into the air as CO_2.

The atmosphere contains about 700 billion tons of carbon, mostly as carbon dioxide. Carbon stored in living and dead biomass on the earth is about 1,800 billion tons; 32,000 billion tons of carbon is stored in the oceans. The movement of carbon between the land, ocean and atmosphere occurs as the result of photosynthesis and absorption by the small amount of carbonate present in ocean waters. Carbon dioxide is emitted into the air by respiration and the combustion/oxidation of carbon-containing compounds (Figure 7.3). Globally we are adding >2 billion tons carbon annually into the atmosphere, about 50% of which remains in the atmosphere.

Direct release of heat, reflectivity from increasing paved areas on the earth's surface and increased particulate levels in the atmosphere from anthropogenic emissions, causing backscatter of solar radiation, all influence the earth's heat-cool cycles as well as the weather. Carbon dioxide is a good absorber of infrared radiation; increasing atmospheric CO_2 increases the effectiveness of infrared absorption and reduces radiant energy reaching the surface of the earth.[53]

Increasing consumption of fossil fuels for energy production has steadily increased the carbon dioxide levels in the air; this steady increase has been documented for more than 100 years. The anthropogenic contribution to CO_2 in the air is expected to double in the next 50 years (Figure 7.4). The current mean value for atmospheric carbon dioxide is 340 ppm as compared to a pre-industrial value of about 250 ppm[6]; this value is increasing by about 1 ppm per year. In 1976, 1.3 billion tons, or 27% of the

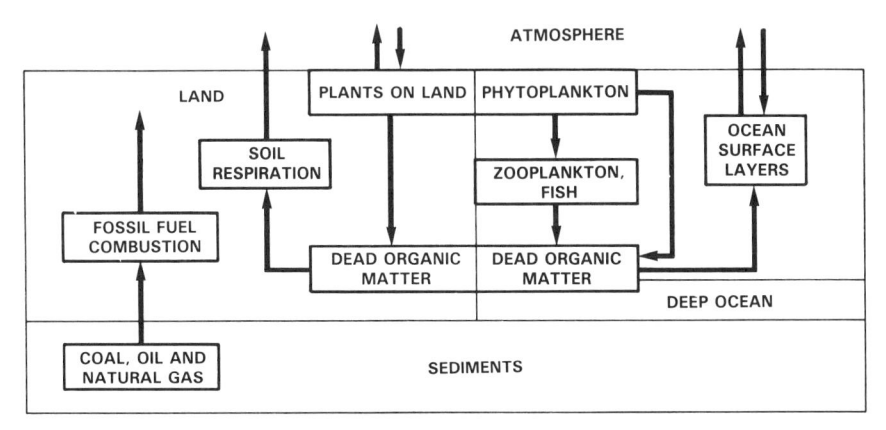

FIGURE 7.3 – The global carbon cycle (R. Allen. 1980. The impact of carbon dioxide on world climate. Environment 22:6–13).

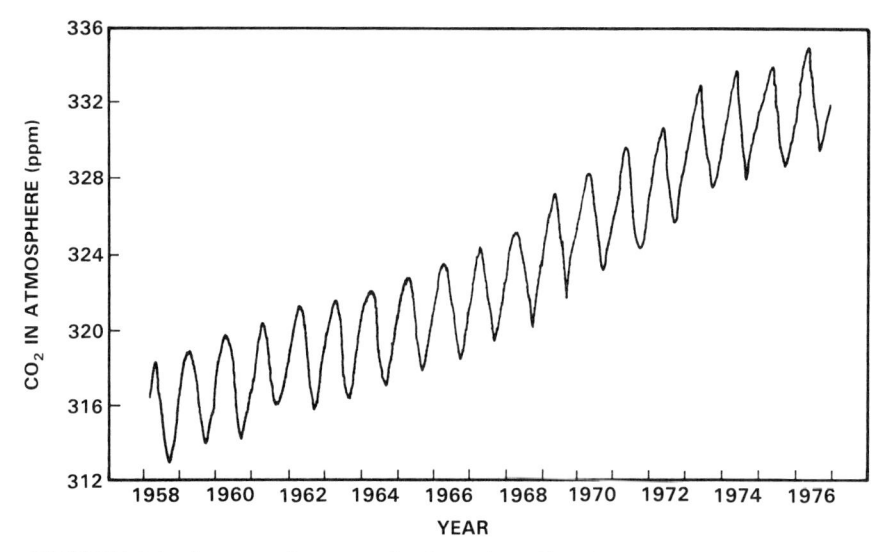

FIGURE 7.4—Increase in atmospheric carbon dioxide as recorded at Mauna Loa observatory in Hawaii; seasonal variations are clearly apparent (R. Allen. 1980. The impact of carbon dioxide on world climate. Environment 22:6–13).

world's total release of carbon into the air, was released as CO_2 by the U.S. alone. Increases in carbon dioxide release after 1985 are expected to be mostly from coal combustion as available supplies of oil and natural gas continue to dwindle and full utilization of nuclear reactors is precluded for political reasons.

The impact of increasing carbon dioxide on climate has been studied using several meteorological models. Some researchers believe that continuing increases in atmospheric CO_2 will cause a global warming of 2–4°C.[6] Major global climatic shifts from glacial to interglacial periods have occurred, with temperature shifts of 5°C, over many centuries. Other potential causes of potential atmospheric related "greenhouse warming" are increases in atmospheric methane, chlorofluorocarbons, carbon monoxide and trace gases.[54] Increased CO may deplete tropospheric hydroxyl radicals, slowing down the removal of dozens of man-made trace gases, influencing the ozone layer and future climate.[59] The rapid rise in global temperatures may have drastic effects on some ecosystems. There is, however, great uncertainty as to the climatic results from carbon dioxide.

The buildup in CO_2 could dramatically transform the balance of nature by its direct stimulation of photosynthesis and plant growth. Some plants experience a doubling in growth rates with a doubling of air CO_2 levels; this phenomena has long been known by greenhouse growers. As the atmospheric CO_2 levels increase, some plant and animals species could become extinct, while others will thrive and become dominant.

Acid Precipitation

The constituents of the gas-aerosol complex produced by the combustion of coal are divided into gases (principally SO_2 and NO_x) and particles. Acid rain, snow and other forms of precipitation are produced in the atmosphere from emissions of coal-fired power plants, smelters and other sources of fossil-fuel combustion. Emissions of acid precursors are significantly reduced by using fluidized-bed combustors for burning coal, rather than conventional combustors.[49] Emitted SO_x, NO_x, and, to a much lesser extent, HCl react with the air, eventually forming sulfuric, nitric and hydrochloric acids that can make rainwater as acid as vinegar. From 50–75% of the acid rain in the east is derived from coal-fired power plants in Ohio and Pennsylvania. Total SO_2 emissions from the eastern U.S. are more than five times as great as those from eastern Canada; >50% of Canadian emissions are from six copper-nickel smelters and one iron-processing operation.[62]

Sulfur dioxide emissions are directly related to the sulfur content of coal; typically emissions contain 1,000–2,000 ppm SO_2 in flue gas. Emission control equipment achieves 85–90% removal of SO_2. About 1–2% of the emitted SO_x is in the form of the trioxide, which rapidly reacts with water vapor to form sulfuric acid mist. A small amount of SO_2 is absorbed onto the surface of fly ash particles, where it reacts with metals for form metal sulfates (large amounts of calcium sulfate and iron alkali trisulfates and trace levels of many metal sulfates are formed).

The second most important source of acid rain is NO_x, which is produced by oxidation of nitrogen in the air at high temperatures; typical emission rates in flue gas are 1,000 ppm NO_x, mostly as NO_2 and NO. NO_x emission control is much more difficult to achieve than control of SO_x. Very tall smokestacks, such as the 1,250-foot stack of International Nickel in Sudbury, Ontario, Canada, are used to disperse emissions over a large area, resulting in acid precipitation as far as 700 miles from the source of production. NO_x also exhibits a longer lifetime in the atmopshere than does SO_x.

The pH of unpolluted rain water is about 5.6, the slight acidity being due to the production of dilute carbonic acid following reaction with carbon dioxide in the air. In the middle 1950s, the pH of most rainfall in the eastern half of the U.S. was below 5.6, particularly in the industrialized northeastern states. By 1973, the regions of low pH were considerably extended, and the degree of acidification had intensified in the northeast. Currently, the mean pH of rain in much of the northeastern states ranges from 4.0 to 4.2, with values between 2.1 and 3.6 observed at local areas (Figure 7.5); about 60–70% of the acidity is due to sulfuric acid. In the Los Angeles basin, pH values for rain have fallen from about 6.0 in the 1950s to about 4.5–5.0 in the 1980s.

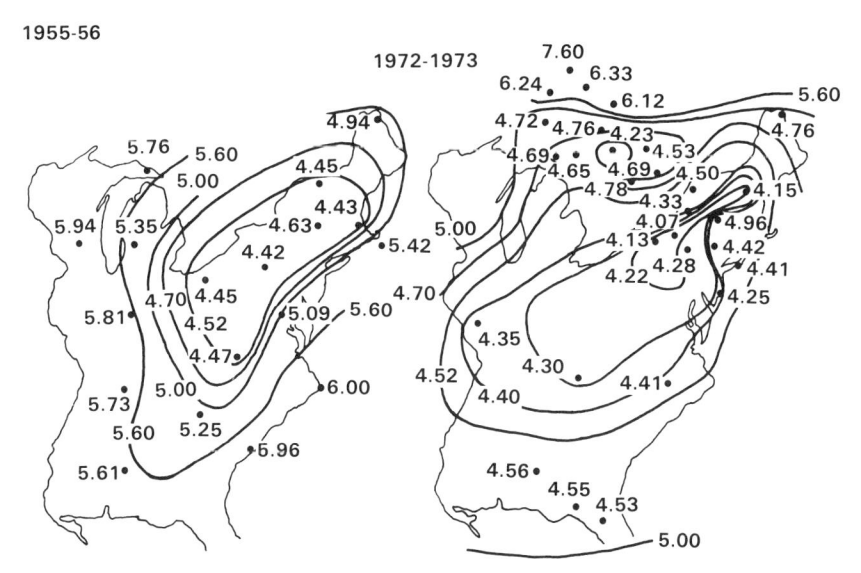

FIGURE 7.5 — Changes in pH levels in precipitation in the eastern half of the U.S. from 1955–1956 to 1972–1973 (N. R. Glass. 1979. Environmental effects of increased coal utilization: Ecological effects of gaseous emissions from coal combustion. Environ. Health Perspect. 32:249–272).

The relative contribution of NO_x to snow acidity in some areas of the eastern U.S. and Canada may be as high as 75%; the average is 60%. Although NO_x is not a problem for lakes and streams in the summer, when it is absorbed by vegetation before it reaches surface waters, it does constitute substantially to acidity during snowmelt and spring runoff; this "acid-shock" is thought to be responsible for damage in early spring to fish, amphibians and other aquatic life.[62]

Marked ecological effects of air pollutants have been noted in remote areas of Norway, Sweden, Canada and the eastern U.S. that are far removed from producers of air pollution. Both SO_2 and ozone increase the incidence and severity of plant diseases due to plant pathogens. Sulfur dioxide penetrates the plants through the stomata, and is converted to bisulfate and sulfite ions within the plant. Conifers and some fruits appear to be the plants most sensitive to sulfur dioxide damage. Air pollution may also increase insect infestation; an example is the increased susceptibility of ponderosa pine to invasion by western and mountain pine beetles. NO_x, although less toxic to plants than ozone, may exhibit greater ecological effects because of its wider dispersion in the environment. At threshold doses below toxicity, plants may serve as metabolic sinks for the removal of atmospheric nitrogen dioxide; ozone and peroxyacyl nitrates are the most phytotoxic air pollutants. Slight to heavy damage has been observed in nearly 200,000 acres of ponderosa and Jeffrey pine of the San Bernardino and Angeles National Forests. Conifer damage is also associated with ozone exposure in the eastern U.S. Severe forest damage has been observed at higher elevations in local areas of New England states, possibly due to acid precipitation. One theory is that the acid precipitation increases the solubility and availablitiy of metals like aluminum for plant uptake; it is this increased uptake in metals by plants that produces the damage.

Acid precipitation adversely affects freshwater bodies in the northeastern U.S. more than in other parts of the U.S. One of the reasons for enhanced freshwater toxicity, other than the greater acid levels in precipitation, is the relative low buffering capacity of the soils; buffering capacity occurs mostly through bicarbonate-rich rock formations. Nearly half the lakes in the Adirondack Mountains of upstate New York have pH levels below 5.0; most are devoid of fish. At a pH of <5.5–6.0 the reproduction capability of fish begins to drop. Fish eggs and fry are very sensitive to acidification; bacteria in the water die as pH drops, while fungi thrive. Organic matter accumulates in the bottom of affected lakes; acid-loving spaghnum moss grows on the bottom, and the water becomes very clear. The high acidity solubilizes more pollutant metals, elevating water levels of aluminum, manganese, cadmium, nickel, mercury, copper and lead. Levels of zinc, copper, aluminum and lead in domestic water supplies are also increased. Levels of toxic metals in fish may increase to such an extent that they are no longer fit for human consumption. In addition, acid rain attacks calcium in mortar, as well as corrodes metals on buildings and equipment, causing billions of dollars of loss annually.

Coal Fly Ash

U.S. coal reserves are estimated at 3.6 trillion metric tons, of which about 400 billion metric tons can be presently, economically mined. In 1974, the electric power industry consumed 400 million tons of coal, or about 60% of all coal mined in the U.S. Fly ash produced by coal-burning power plants amounts to about 75% of all coal fly ash produced in the U.S., or about 100 million tons fly ash annually; of this amount, between 1 and 5% will be released into the environment.[7] The amount of fine fly ash particles emitted into the air is expected to continue to increase despite the use of electrostatic precipitators (Figure 7.6). Of the fly ash currently generated by power plants, less than 10% is recycled for use in manufacturing cement, road construction and anti-skid devices, soil conditioner, and filler materials for a variety of uses. The remaining ash is converted to an aqueous slurry and transported to disposal sites, where it is lagooned. Long-term impacts on surface and ground waters in the vicinity of such lagoons have not been evaluated.

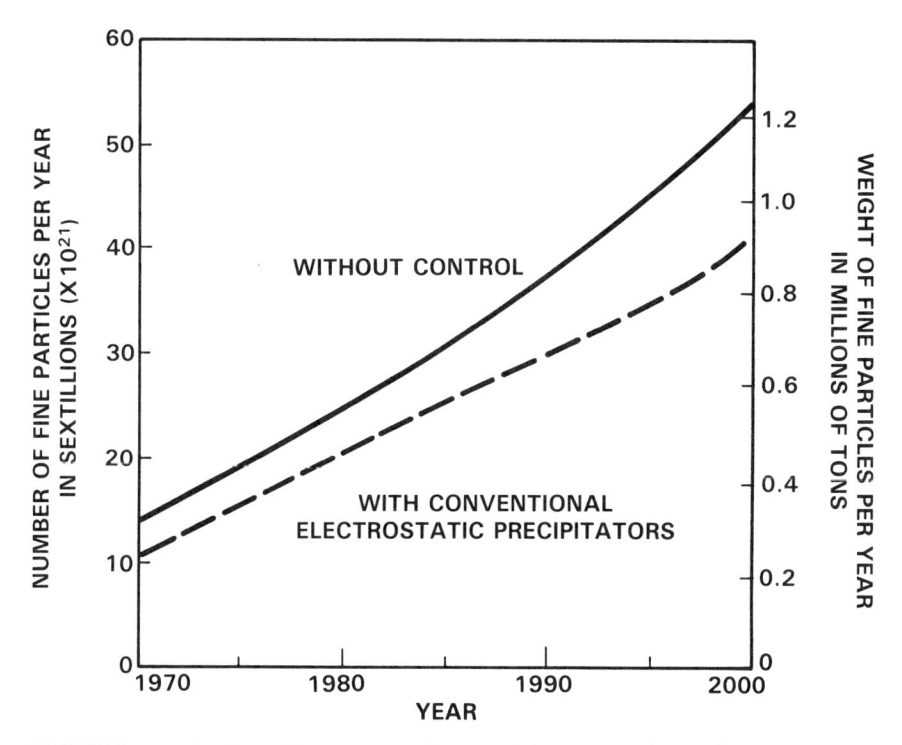

FIGURE 7.6 – Projected increase in fine particle (<1.0 micron diameter) emissions from coal burning sources in the U.S. with and without electrostatic precipitation (G. T. Miller. 1975. Living in the environment. Concepts, problems and alternatives. Wadsworth Publishing Co., Inc., Belmont, CA, p. E131).

Coal fly ash is characterized by low permeability, low bulk density, high specific surface area, often alkaline pH and particle shapes that are mostly spherical, with particles within hollow particles that have a strong tendency to aggregate. The spherical shape of the particles and their small size in stack samples that pass through electrostatic precipitators (count median diameter of <2.0 microns) provide a large surface area for trace-metal bonding, since total surface area is inversely related to particle size.

A typical matrix element composition for U.S. coal fly ash consists of high levels of Al, Si, K, Ca and Fe. Smaller amounts of Li, Na, Mg and Ti are also found in fly ashes. Also present commonly in fly ash are silicates, mullite, hematite, magnetite, ferrite, quartz and gypsum.

The larger sized fly ash particles of a fossil-fuel power plant are captured in the stack by an electrostatic precipitator. Most respirable, submicron particles pass through the precipitators into the atmosphere. Elements which are volatilized at the high temperatures of combustion, about 1500°, recondense as the flue temperature decreases up the stack, mostly on the surface of submicron particles.[9] Highly volatile elements, such as Hg, Se, Sb, As, Cl, F and I, are enriched by condensation on small fly ash particles, prior to their release into the atmosphere. Because of this enrichment, some elements may be passed into the environment at over 20 times the amounts predicted based upon their levels in coal (Table 7.1). The amount adsorbed on ash particles is an inverse function of particle size, with most adsorption occuring on a 1000 Å thick surface on ash particles. Natusch[9] found that for large particles (>75 µm diameter) only a small fraction of an element was present on the particle surface whereas smaller particles (<1 µm diameter contained as much as 80% of the mass of some elements. The concentration of lead and vanadium increases with decreasing particle size in coal fly ash.[17] There is a fivefold increase in cadmium, copper and nickel, and a tenfold increase in mercury, molybdenum, and tungsten for coal combustion emission rates.[16]

TABLE 7.1. Concentration of metals in 101 coals and coal fly ash (A. L. Page, A. A. Elsewi and I. R. Straughan. 1979. Physical and chemical properties of fly ash from coal-fired power plants with reference to environmental impacts pp.84–120; K. K. Bertine and E. D. Goldberg. 1971. Fossil fuel combustion and the major sedimentary cycle. Science 173:233; D. F. S. Natusch. 1978. Potential carcinogenic species emitted to the atmosphere by fossil-fueled power plants. Environ. Health Perspect. 22:79–90).

Element	Amount in coal, ppm Mean	Range	Atmospheric Release ×10⁹g/y	Fly Ash Concentration, µg/g for size diameter 1–3 micron	>3 micron
Al	10,000	4,300–30,000	1,400	–	–
As	14	0.5–93	0.7	–	17
Be	1.6	0.2–4.0	0.4	–	–
Cd	2.5	0.1–65	–	2.2	1.5
Co	9.6	1.0–43	0.7	–	15
Cr	14	4.0–54	1.4	–	65
Cu	15	5.0–61	2.1	23	70
Hg	0.2	0.02–1.6	0.002	–	–
Mn	49	6.0–180	7.0	100	120
Ni	21	3.0–80	2.1	63	46
Pb	35	4.0–220	3.5	90	52
Sb	1.3	0.2–8.9	–	–	4.4
Se	2.1	0.5–7.7	0.4	–	22
V	33	11–78	3.5	–	–

Elements are mobilized into the atmosphere by weathering and by combustion of fossil fuels. Some elements are mobilized more by weathering than fossil fuel combustion; beryllium is mobilized by fossil fuel combustion at an annual rate of 0.41×10^9 g compared to 5.6×10^9 per year from weathering, and lead is mobilized from fossil fuel at 3.6×10^9 g per year compared with 110×10^9 g per year from weathering.[18]

The matrix elements, Fe, Si, Ba, Ca and Mg exhibit low extractability in aqueous environments while trace elements absorbed to particle surfaces are often highly solublized in water. The estimated amounts of various elements released into the atmosphere from the combustion of coal are given in Table 7.1. An estimate of the trace element emission rate from a 10,000 ton/day coal power plant was determined by Vaughn et al[16] for 30 elements. About half the elements would adsorb onto the surface of ash particles.

Soils and plants in the vicinity of coal-fired power plants become enriched in some of these elements. Combustion of coal with a high level of a toxic element, such as "arsenic" coals, can have serious health implications for those living in the vicinity or down-wind from such power plants.

Experiments with animals have shown that coal fly ash is cleared from the lungs as a double exponential; in Syrian hamsters, the lung burden decreased to about 10% of initial value by 99 days after exposure (Figure 7.7). Alveolar macrophages rapidly phagocytize inhaled ash particles, greatly increasing the elemental concentration of toxic metals in individual cells; for example, selenium concentration increased 75-fold; arsenic, 80-fold; lead, 25-fold; vanadium, 1000-fold; and barium, 40-fold in one study.[11] In addition, coal fly ash may impair the bactericidal capacity of macrophages.[12] Low sulfur fly ash from coal causes damage to macrophage membranes while oil fly ash from oil with a high vanadium content was not toxic to macrophages.

Trace elements in the atmopshere from coal-fired emissions do not appear to be a significant ecological hazard, except for those from certain coals that contain very high levels of toxic elements, such as arsenic.[10] Contamination of soil and water may occur from leaching of trace elements during mining, cleaning and storage of coal and from storage areas for solid-waste fly ash. The acidic nature of the drainage from eastern mines tends to solubilize trace elements to a greater extent and promotes transport to surface and ground waters. Trace metals are concentrated in aquatic and terrestrial organisms; the degree is a function of the element and the type of organism(s). Concentration of some elements by aquatic invertebrates and fish may be of sufficient magnitude to make them hazardous for humans to eat.

Few studies have been carried out to examine the long-term effects of chronic exposure to ash. No pulmonary lesions have been found in monkeys or guinea pigs exposed to a fly-ash aerosol for up to 78 weeks.

7 – Coal and Oil

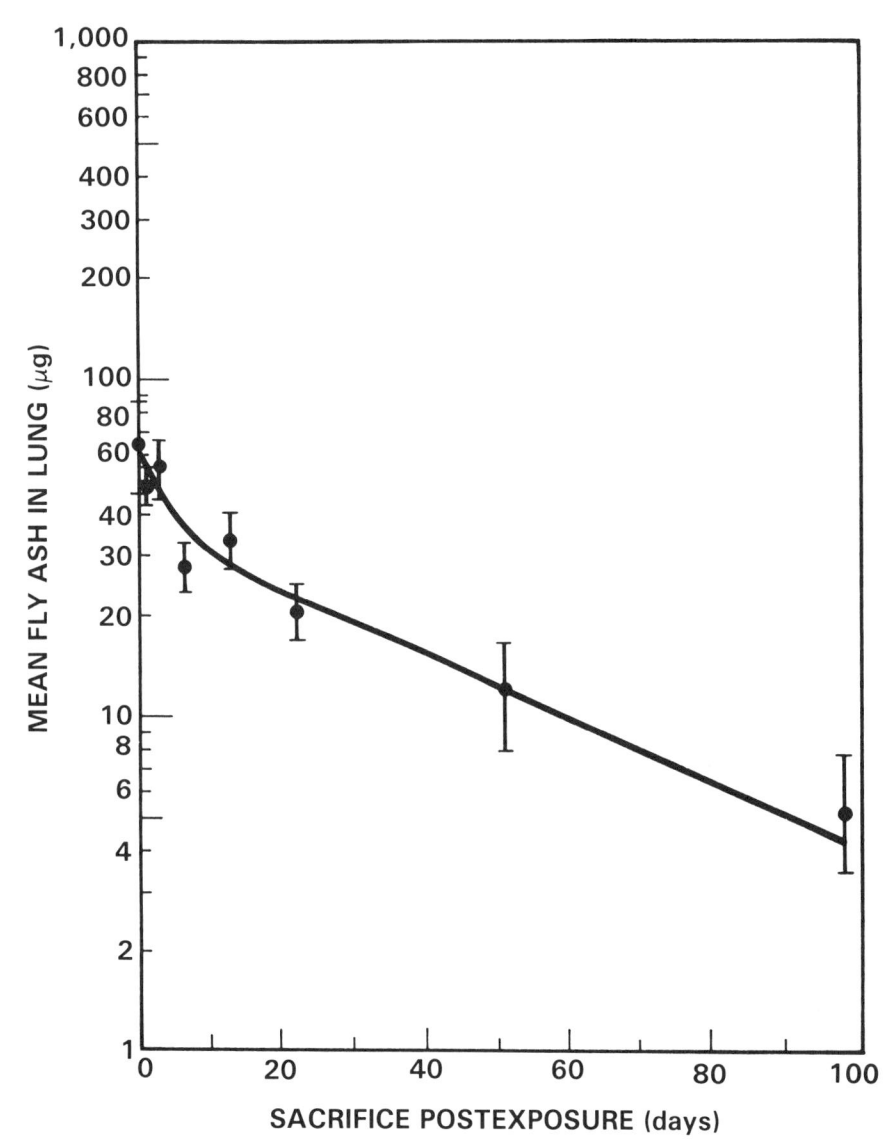

FIGURE 7.7 – Pulmonary clearance of inhaled neutron activated coal fly ash in Syrian hamsters based upon 46-Sc activation (A. P. Wehner, C. L. Wilkerson, J. A. Mahaffey and E. M. Milliman. 1980. Fate of inhaled fly ash in hamsters. Environ. Res. 22:485–498).

Histologically, the pulmonary reaction to fly ash was similar to that for "nuisance" dust and caused no permanent pathological reaction in the lung. [41,42] Coal fly ash was found to be a potent inhibitor of prolyl hydroxylase in vitro with the extent of inhibition inversely proportional to fly-ash particle size. Fly ash in lung organ culture increased collagen synthesis and deposition, indicative of fibrogenic potential in humans.[63] A mild inflammatory response was noted in lungs exposed to nickel-enriched fly ash. At high doses, this would probably lead to development of simple pneumoconiosis, similar to that seen for long-term exposure to soil or volcanic ash particles.[13] The degree of mutagenicity of coal fly ash, if any, has not been adequately defined.

Carcinogens From Fossil-Fuel Sources

Fossil fuels may comprise various chemical structures; a proposed structure for coal is given in Figure 7.8. Of particular interest with respect to carcinogenicity are the presence of polycyclic aromatic and heterocyclic

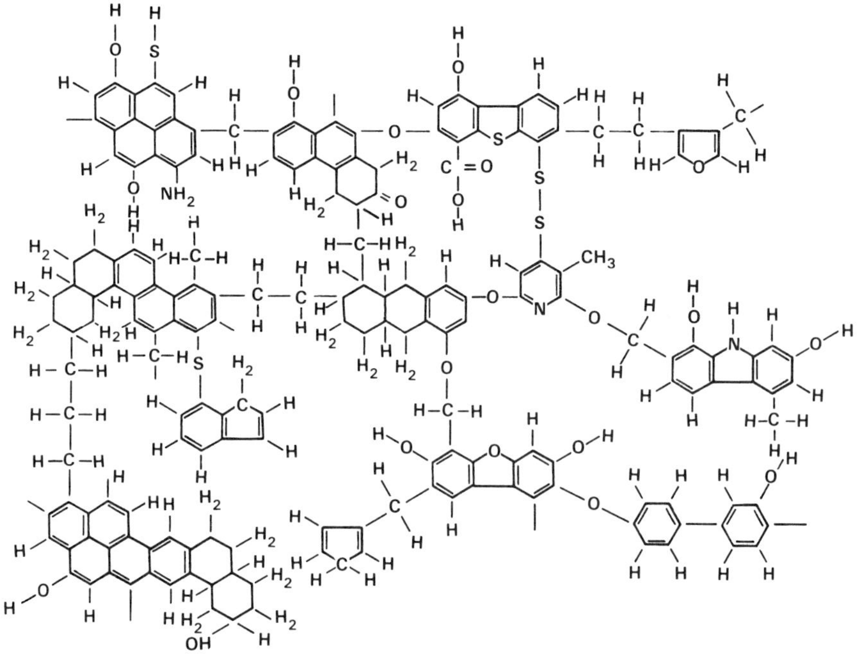

FIGURE 7.8 – Proposed structure for bituminous coal (R. A. Wadden. 1976. Coal hydrogenation and environmental health. Environ. Health Perspect. 14:201).

hydrocarbons in fossil fuels and their combustion products. Chemical bonds are broken and free radicals formed during the combustion of fossil fuels. During the process, the polycyclic aromatics are broken down or may lose their alkyl sidechain. Following incomplete combustion, soot particles are formed along with various adsorbed hydrocarbons on particle surfaces. Some fossil fuels, such as bituminous coal, contain very low levels of PAH; distillate fractions such as coal tar and pitch contain very high PAH levels (Table 7.2).

Average concentrations of BaP in the atmopshere are higher in winter than in summer due to fossil–fuel combustion for heating. The amount of PAH released from fossil fuels is determined by the type of fuel utilized and the type of burner; the range of BaP is from <50 µg. metric ton to >1 mg/metric ton burned, released in the flue gas of coal-fired power plants. Residential and open burning of coal constitutes nearly 90% of BaP atmospheric emissions in the U.S.; coal-fired steam, power plants account

TABLE 7.2. Benzo(a)pyrene content of energy-related substances (M. R. Guerin. 1978. Energy sources of polycyclic aromatic hydrocarbons. In, "Polycyclic Hydrocarbons and Cancer," Vol. 1, Academic Press, New York, pp. 3–42).

Substance	ppm BaP
Crude Oils	
Petroleum	1.
Shale-derived	3.
Coal-derived	3.
Petroleum products	
Gasoline	0.4
Motor oil, new	<1.
Motor oil, used	4.
Diesel fuel	0.05
Petroleum pitch	2,000.
Miscellaneous	
Creosote	200.
Coal tar	2,000.
Coal tar pitch	10,000.
Raw shale	0.015
Processed shale	0.03
Bituminous coal	<0.001

for very little of the total BaP or PAH emissions (Table 7.3). N-nitrosamines, although not present in coal, may be produced during the combustion of coal following reaction of NO_x and amines. Little is known at present about the health significance of chemical reactions involving nitrosamine formation.[14]

Aliphatic Hydrocarbons

The saturated aliphatic series is comprised of gases (natural gas, methane, ethane, and bottled gases, propane and butane), liquids of 6 to 16 carbon chains and longer-chain solids. Methane and ethane are not toxic except as asphyxiants, due to their replacement of oxygen. Propane and butane depress the CNS without causing any morphological damage. The C_5-C_8 aliphatic hydrocarbons cause CNS depression and polyneuropathy due to demyelination and axonal degeneration. In general, C_3 to C_8 saturated hydrocarbons show increasingly strong narcotic actions. Heavier compounds have decreasing volatility, making for less concentration in the air.

Exposure to gasoline vapors produces narcosis, with CNS depression; death can result from fatal cardiac arrhythmia due to the release of epinephrine, which acts on the myocardium, and to respiratory failure.[39] There is no convincing evidence that prolonged, chronic exposure to gasoline vapors, as in the case of gasoline station attendants, causes toxicity. Subtle neuropsychological changes have been observed, however, in workers exposed to jet fuel vapors.[38]

Gasoline and less volatile, low-viscosity liquids, such as kerosene, can enter the lungs by aspiration following ingestion and spontaneous or induced vomiting. In the absence of vomiting, ingestion is not especially harmful. NIOSH-recommended limits for the C_5 to C_8 aliphatics range from 75 to 120 ppm.[40]

Crude Oil

Crude oil from natural sources accounts for about 45% of all energy production in the U.S. (for 1978); approximately 6 billion barrels of oil are required to produce this amount of energy. Each barrel contains 140 kg of oil, composed, on the average, of 1 wt% of PAH and 1 ppm BaP. Thus, the petroleum industry handles about 84 million metric tons of PAH and 840 metric tons of BaP annually.[14] PAH content of crude oil influences the potential toxicity of refinery waste streams and products as well as environmental pollution from their combustion. Values of 0.04, 1.3 and 1.6 ppm BaP, respectively, have been reported for Persian Gulf, Libyian and Venezuelan crude oils.

TABLE 7.3. Estimated annual atmospheric emission of benzo(a)pyrene in the U.S.; units are in metric tons per year (Scientific and Technical Assessment Report on Particulate Polycyclic Organic Matter (PPOM). 1975. EPA-600/6-74-001. U.S. Environmental Protection Agency, Washington, D.C.).

Source	BaP Emissions in 1972, Metric Tons
Coal	
Hand-stoked and underfeed-stoked residential furnaces	270
Intermediate sized furnaces	6
Open burning, coal refuse	281
Coke production	153
Coal-fueled steam power plants	<1
Total, Coal	710
Petroleum and Natural Gas	
Oil-fueled steam power plants	2
Gas-fueled steam power plants	2
Petroleum catalytic cracking	6
Asphalt air-blowing	<1
Asphalt hot-road-mix plant	<1
Gasoline-powered motor vehicles	<1
Diesel-powered motor vehicles	<1
Rubber tire degradation	10
All others	8
Total, Petroleum and Natural Gas	32
Other Materials	
Wood, home fireplaces and wood stoves	23
Enclosed incinerators	3
Open burning, vehicle disposal	5
Open burning, forest and agriculture	10
Open burning, other	50
Total, Other	91
Grand Total	833

Crude oil is refined into eight fractions or distillates (Figure 7.9.), which are further upgraded in petrochemical plants by purification, catalytic cracking, hydrotreating and other processes to a variety of feedstockes, which are used to make final products. PAH is concentrated in higher-boiling-point distillates and solid residues. Air concentrations of up to 250 µg BaP/100 m³ are seen in process stages dealing with high-boiling-point products;[14] air concentrations an order of magnitude higher are commonly seen in coking operations. The BaP levels of diesel fuel, fuel oil and unused motor oils are normally <0.05 ppm, increasing with use or combustion. The BaP content of motor oil increases 5 to 200-fold after use in vehicles.

Environmental pollution of aquatic ecosystems with petroleum waste products has been associated with proliferative lesions in fish. The Duwamish River, a polluted estuary near Seattle, WA, contains high levels of PAH; McAllister Creek, 80 miles to the south of Seattle, is a "clean" estuary. The incidence of gill hyperplasia in English sole in McAllister Creek is nil, while in the Duwamish River it is 15%. Likewise, the incidence of liver tumors in the Duwamish River sole is 32%, but it is nil in McAllister Creek sole.[15]

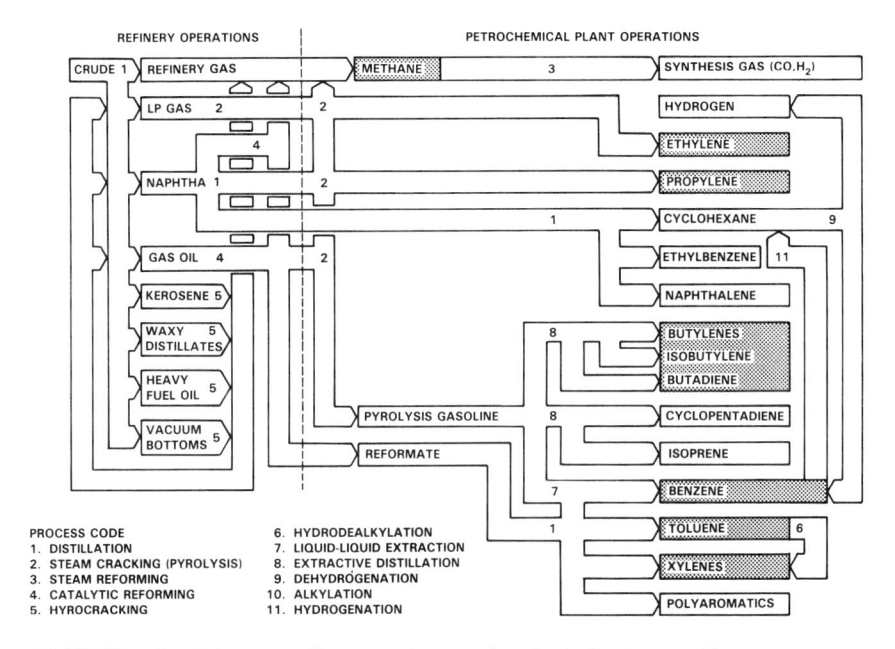

FIGURE 7.9 — Primary refinery and petrochemical plant operations associated with processing crude oil (H. E. Wise and P. D. Farenthold. 1981. Predicting priority pollutants from petrochemical processes. Environ. Sci. Technol. 15:1292–1304).

Oil Fly Ash

The ash produced during the combustion of oil is considerably finer and smaller in particle size than the ash produced during the combustion of coal. Oil fly ash contains a greater inventory and quantity of PAH; however, the quantity of ash released is considerably less than ash released by coal combustion. Thus, the quantity of trace metals and PAH found in oil fly ash is considerably less than in coal fly ash (Table 7.4). Exceptions are nickel, which is often found in comparable concentrations as in coal and vanadium, molybdenum and mercury, which are found in many oils at higher concentrations than in coal. Vanadium-rich oil fly ash damages

TABLE 7.4. Concentration of trace elements in oil fly ash emissions (data taken from: G. Anderson and P. Grenufelt. 1973. Determination of heavy metals in fuel oil and an estimation of emission of heavy metals from oil combustion. In, "Handbook on the Toxicology of Metals," Elsevier/North Holland Biomedical Press, Amsterdam, p. 50; D. F. S. Natusch. 1978. Potentially carcinogenic species emitted to the atmopshere by fossil-fueled power plants. Environ. Health Perspect. 22:79–90; K. K. Bertine and E. D. Goldberg. 1971. Fossil fuel combustion and the major sedimentary cycle. Science 173:233).

Element	Concentration in Oil, ppm	U.S. Amount Released into the Atmosphere by Combustion, $\times 10^9$/yr	Specific Concentration in Fly Ash, µg/g
Al	0.5	0.08	100–5,000
As	0.01	0.002	30
Ba	0.1	–	500–10,000
Co	0.2	0.03	90
Cr	0.3	0.05	66
Cu	0.1	0.02	50–2,000
Hg	10.	1.6	–
Mn	0.1	0.02	1–100
Ni	10.	1.6	–
Pb	0.3	0.05	200–2,000
Se	0.2	0.03	5
V	50.	8.2	100–200,000

tracheal epithelium, inhibiting normal mucociliary clearance; the mechanism of vanadium cytotoxicity may be similar to cadmium.[60]

Little is known about the toxicity of oil fly ash in humans, in comparing its toxicity with that of coal fly ash. One such study[64] revealed more organic substances and more S, Ni and V in oil fly ash than in coal fly ash; the latter contained a larger amount of Al, Si, Cl, K, Ca, Ti, Mn, Fe, Se, Rb, Y, Zr, Ba, and Pb. Biological testing of the two ashes by intratracheal administration in hamsters, culturing with alveolar macrophages and studies on lung clearance indicated a greater acute and subacute toxicicty, a greater toxicicty to macrophages and a greater retention of BaP coated on the ash particles from oil combustion than from coal combustion.

Gasoline

Particulate and NO_x emissions from gasoline-powered vehicles are considerably less than those from diesel engines, but PAH contents tends to be comparable with that from diesel engines. In 1968, the annual consumption of gasoline in the U.S. was 81×10^9 gallons, increasing to a peak level of consumption in 1973-1974, followed by a small but steady decline into the early 1980s. The odor threshold for gasoline vapor is 0.15-0.7 ppm, the TLD is 300 ppm gasoline and the explosive threshold for gasoline is 10,000 ppm. The BaP content of gasoline typically ranges from 0.2 to 0.5 µg/g for both regular and premium grades. BaP emissions in vehicles without pollution control equipment range from 45-170 µg/gallon consumed; emissions are only 20-30 µg/gallon in vehicles with controls. BaP emissions in gasoline engine exhaust increase 30-fold when the air:fuel ratio is decreased from 14:1 to 10:1; they increase 5-fold in engines used for over 50,000 miles. Increases are also significant in the exhausts of vehicles that use a large amount of motor oil, or in those in which an oil-gasoline fuel mixture is used. Thus, exhaust emissions from 2-cycle engines, such as those in outboard motors, motorcycles and lawn mowers, which operate on oil:gasoline mixtures of 1:30, release nearly 10,000 ug BaP/gallon. Other fossil fuels that are burned in confined spaces, such as kerosene in space heaters or natural gas for cooking, may significantly increase indoor levels of toxic gases, such as CO, NO_x and SO_2, and PAH, resulting in pulmonary disease.[43]

Diesel Fuel

According to the EPA, from 10-25% of all U.S. automobiles will use diesel engines by 1990, primarily because of improved mileage (25-30%), lower fuel costs, and ease of maintenance. The annual consumption of diesel fuel in 1968 was 5×10^9 gallons; it has increased yearly since then.

While diesel-powered vehicles provide about 25% higher fuel economy than comparable vehicles with gasoline engines, and reduce emissions of Co, they produce more particulate matter. These particulates consist mostly of carbonaceous soot, on which are absorbed solvent-extractable organic compounds. Several factors are known to influence particulate emission rates in diesel engines: fuel density, aromaticity and distillation properties, fuel additives, and the car make, model and age. Reduction in aromatic content of diesel fuel decreases the particulate emissions in exhaust.[50]

Although the BaP content of diesel fuel is similar to levels found in fuel oil and unused motor oil (Table 7.2), diesel-exhaust emissions may be of significant health concern. Aromatic hydrocarbons in diesel fuels increase the smoking tendencies of fossil fuels (naphthalenes > benzens > olefins > paraffins); reduction of the aromatic content of fuels reduces soot emissions.[62] Particularly important are emissions of NO_x and particulates; the latter are associated with adsorbed PAH. The amount of particles released in the exhaust of diesel engines is 30–100 times greater than that of a catalytic-converter-equipped gasoline engine. The submicron size of diesel exhaust particles increases their deep lung deposition. PAH emissions in diesel exhaust, mostly concentrated on soot particles, are estimated at 60–100 µg/gallon.[20]

Currently, most diesel engines in cars do not meet the EPA standards of 0.6 g/mile for particles and 1 g/mile for NO_x. It is difficult to control both particles and NO_x simultaneously, since control of one pollutant increases the level of the other. Diesel odor by itself does not appear to be a health hazard, but it is often the factor that the public finds most objectionable when diesel fuel is used.

Toxic Effects of Fossil-Fuel Combustion

Combustion products derived from the burning of fossil fuels, particularly coal, are clearly damaging to human health. The relationship between inhalation of coal emissions and pulmonary disease has been known for centuries. Several studies have demonstrated a potential carcinogenic effect in humans from exposure to fossil-related pollutants. Cancer incidence is higher in urban regions than in rural regions of the U.S., as it is higher in smokers and in certain occupational groups. Increased incidences of cancer of the lung, GI tract and skin have been found in coal miners and for certain coal-processing technologies, such as coke production industry. In 1950, the standard mortality ratio (SMR) for coal miners aged 20–64 was 195 for all causes, 179 for all cancers, 192 for lung cancer and 491 for all respiratory diseases.[21]

The association of cancer and environmental air pollution due to coal combustion is difficult to establish. Several studies have indicated a positive correlation between PAH air levels and lung-cancer incidence; other studies have indicated no such correlation. One study concluded that as much as 25% of the mortality from lung cancer and 15% from all cancers was attributable to air pollution.[22] The most convincing data for a relationship between air pollution and lung cancer stems from studies of migrants. Individuals migrating from a polluted environment to a less polluted environment had a lower risk of lung cancer than those in their original location.[21]

High-boiling-point distillates of fossil fuels are enriched in naphthylamines, which have known carcinogenicity. Gas retorts and coal tar and pitch associated with coke production are rich in naphthylamines. Occupational exposure in these industries is associated with a high risk of urinary bladder and kidney tumors. In experimental studies, the incidence of liver tumors in mice and urinary bladder tumors in dogs increased after administration of 1-naphthylamine.

Petroleum refinery workers may experience a nonsignificant increase in mortality from cancer of the lung and from lymphomas and genital cancer.[23] However, other studies have failed to demonstrate any association between cancer in oil refinery workers and surrounding populations. One study showed a high incidence of brain tumors in refinery and petrochemical workers.[24] Overall, an annual increase of 1–2 deaths from lung cancer/100,000 may be associated with occupational exposure in the petroleum industry.[23] States along the east and gulf coasts, where petroleum industries are concentrated, appear to have significantly increased levels of cancer in the general population as compared to Rocky Mountain states, with few such industries. For example, the cancer rate is highest in New Jersey (205 cases/100,000 per year) and lowest in Utah (133 cases/100,000/year). Lifestyle plays an important role in cancer incidence.

Tetraethyl-lead [$Pb(C_2H_5)_4$], a gasoline antiknock additive that poses a potential occupational and environmental health hazard, is readily absorbed through the skin, lungs and GI tract. High lead concentrations in air and soil are found near heavily used highways, and lead blood levels are significantly higher in city dwellers than in rural town inhabitants. About 98% of all environmental lead pollution in the U.S. comes from the combustion of gasoline-containing lead additives, with the average vehicle using regular gasoline, releasing 2–4 pounds of lead into the air each year.[28] Lead levels in the environs have declined during the last few years due to controls on lead use in gasoline and the increasing use of lead-free fuels. Still, lead contamination is world-wide, being found even in ice of the North and South Poles.

Automobile exhaust constitutes a major source of air pollution in densely populated urban centers. The public health impact of urban air pollution in the past has been significant. The average automobile, in the 1960s, emitted exhaust containing 3.5% CO. Today, as much as half of all manmade CO comes from gasoline engines.[44] High levels of blood CO and lead are found in tunnel and turnpike workers exposed daily to motor vehicle exhausts.[45]

The genotoxicity of automobile exhaust particles using the Ames bioassay are greatest for spark-ignition vehicles, and somewhat less for light-duty diesel vehicles; similar results were obtained with mutation and sister chromatid exchanges bioassays using Chinese ovary cells. Environmental particles obtained from tunnels exhibited substantially less genotoxicity.[61] Gasoline exhaust condensates have produced skin cancer when painted on the back of mice; neutral and benzene soluble fractions containing the highest levels of PAH were the most carcinogenic.[25] Benign pulmonary adenomas were produced in hamsters following intratracheal instillation of gasoline exhaust condensate containing BaP levels as low as 340 µg/g.[26]

Unused diesel fuel is not mutagenic in the Ames test, although extracts of diesel-fuel combustion products do show mutagenicity at a level higher than that for extracts of gasoline-engine or home-heating-oil combustion products (Figure 7.10). The toxicity of exhaust emissions to humans is a function of the amount of carcinogenic PAH on particles, the particulate emission rate, the residence time of particles in the atmosphere and deep lung deposition of particles. Mutagenicity for combustion-product extracts has been demonstrated with organic solvents but not with biological fluids used to extract organic compounds adsorbed onto diesel-exhaust particles.[29] Fractionation studies of these extracts indicate that 1-nitropyrene and various dinitropyrenes account for a significant fraction of the mutagenic activity. Nitroaromatics also form in the atmosphere from reaction of aromatic hydrocarbons with nitrogen dioxide in the presence of smog. Nitropyrene induces sarcomas when injected subcutaneously in rats.[33] The mutagenicity of diesel-exhaust particles is not very different from that of gasoline-exhaust particles. However, a much greater mutagenic risk is seen with diesel-exhaust particles because they produce a much greater amount of particles.[51]

Mice exposed to diesel exhaust show increased pulmonary susceptibility to streptococcus infection than those exposed to gasoline exhaust, probably because of the higher NO_x emissions from diesels. Only mild inflammatory lesions are seen in the lung after deposition of diesel-exhaust particles. The progressive accumulation of diesel exhaust particles in lungs of animals has been associated with progressive inflammatory and fibrotic lung disease.[60] Skin papillomas were seen in mice after topical application of diesel-exhaust extracts. Skin tumor incidence is also higher for diesel-exhaust products than for gasoline-exhaust products. In some cases, it was comparable to that in mice given topical applications of roofing tar

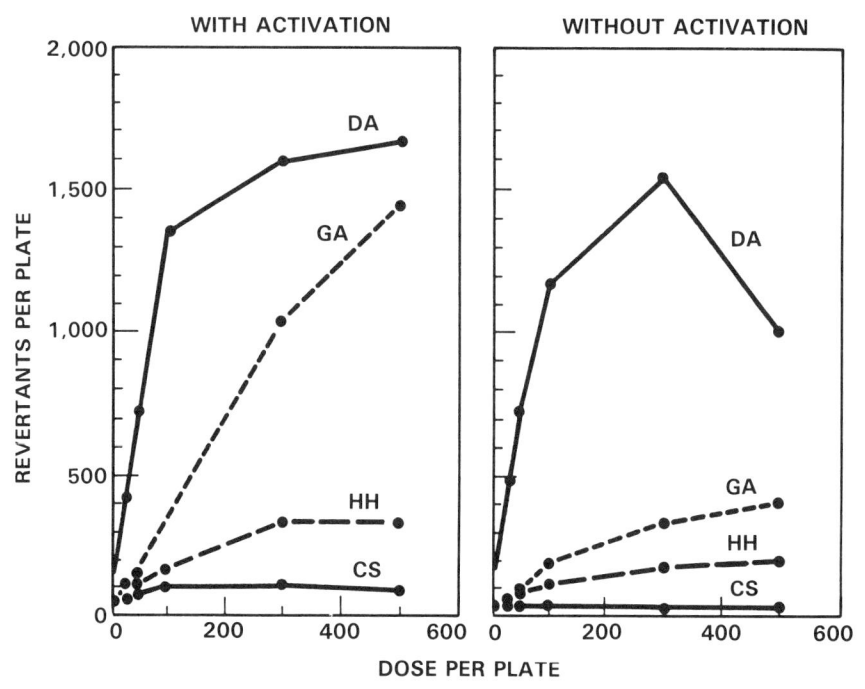

FIGURE 7.10—Mutagenicity of extracted combustion organics from diesel engine (DA), gasoline engine (GA) and home heating oil (HH) in strain TA-98 Salmonella typhimurium (L. Claxton and J. L. Huisingh. 1980. Comparative mutagenic activity of organics from combustion sources. In, "Pulmonary Toxicology of Respirable Particles," CONF-791002, NTIS, Springfield, VA, pp. 453–465).

and products for coking operations. Malignant lung tumors have not been produced in animals by exposure to high levels of inhaled diesel exhaust.[31]

Diesel exhaust exposure has been proposed as a risk factor in human cancers, particularly of lung and bladder, even though animal studies with inhaled whole diesel exhaust have not led to the induction of tumors.[52] Nine major epidemiological studies examined the health of workers exposed to diesel exhausts, including bus and railroad mechanics and other transportation workers, coal miners, and salt, iron and potash miners. No significant increase in lung-cancer mortality or in mortality due to other causes, nor increased in other respiratory diseases, were noted in any of these studies. Reduced pulmonary function and increased complaints have been reported from diesel-exhaust exposure, due probably to NO_x. Data are insufficient to demonstrate any relationship between diesel-exhaust exposure and bronchitis or pulmonary infections in humans.[32]

Overall, studies of emissions from gasoline, or diesel-fueled vehicles or those with artificial smog have shown either negative or only a small significant increase in the incidence of lung tumors in experimental animals.[46-48]

References

1. E. Marshall. 1980. Planning for an oil cutoff. Science 209:246–247.
2. E. K. Diehl, F. du Brevil and R. A. Glenn. 1967. Polynuclear hydrocarbon emission from coal-fired installations. J. Energy Power 89:276.
3. V. S. Snodgrass. 1982. "Black lung" takes toll in suffering, not death. J. Amer. Med. Assoc. 247:2332.
4. K. Bridbord et al. 1979. Occupational safety and health implications of increased coal utilization. Environ. Health Perspect. 33:285–302.
5. A. G. Heppleston. 1984. Pulmonary toxicology of silica, coal and asbestos. Environ. Health Perspect. 55:111–127.
6. R. M. Rotty. 1979. Atmospheric CO_2 consequences of heavy dependence on coal. Environ. Health Perspect. 33:273–283.
7. G. L. Fischer, C. E. Chrisp and O. G. Raabe. 1979. Physical factors affecting the mutagenicity of fly ash from a coal-fired power plant. Science 204:879–881.
8. A. L. Page, A. A. Elseewi and I. R. Straughan. 1979. Physical and chemical properties of fly ash from coal-fired power plants with reference to environmental impacts. Environ. Health Perspect. 33:84–120.
9. D. F. S. Natusch et al. 1975. Characterization of trace elements in fly ash. In, "Proc. International Conference on Heavy Metals in the Environment," Vol. II, Part 2, pp. 553–575.
10. R. I. Van Hook. 1979. Potential health and environmental effects of trace elements and radionuclides from increased coal utilization. Environ. Health Perspect. 33:227–247.
11. T. L. Hayes, J. B. Pawley, G. L. Fisher and M. Goldman. 1980. A model for the exposure of individual lung cells to foreign elements contained in fly ash. Environ. Res. 22:499–509.
12. P. C. Brenna, F. R. Kirchner and W. P. Norris. 1980. The effect of reaerosolized fly ash from an atmospheric pressure fluidized-bed combustor on murine alveolar macrophage function. In, "Pulmonary Toxicology of Respirable Particles," CONF-791001, NTIS, Springfield, VA, pp. 279–288.
13. A. P. Wehner, O. R. Moss, E. M. Milliman, G. E. Dagle and R. E. Schirmer. 1979. Acute and subchronic inhalation exposures of hamsters to nickel-enriched fly ash. Environ. Res. 19:355–370.
14. M. R. Guerin. 1978. Energy sources of polycyclic aromatic hydrocar-

bons. In, "Polycyclic Hydrocarbons and Cancer," Vol. 1, pp. 3-42. Academic Press, New York.
15. B. B. McCain and D. C. Malins. 1979. Uptake, fate, and effects of aromatic hydrocarbons on selected fish and crustaceans from the northeast Pacific Ocean. Marine Ecosystems Analysis (MESA)-New York Bight meeting held in New York, June, 1979.
16. B. E. Vaughan et al. 1975. Review of potential impact on health and environmental quality from metals entering the environment as a result of coal utilization. Battelle Energy Program Report, Pacific Northwest Laboratory, Richland, WA.
17. R. E. Lee and D. J. von Lehmden. 1973. Trace metal pollution in the environment. J. Air Pollut. Control Assoc. 6:1025-1030.
18. R. H. Ross. 1977. Environmental Interactions. In, "Environmental. Health, and Control Aspects of Coal Conversion: An Information Overview," Vol. 2, ORNL/EIS-95, Energy Research and Development Administration, Washington, D.C., pp. 6.1-6.131.
19. M. Attfield, R. Reger and R. Glenn. 1984. The incidence and progression of pneumoconiosis over nine years in U.S. coal miners. II. Relationship with dust exposure and other potential causative factors. Am. J. Ind. Med. 6:417-425.
20. D. S. Barth and S. M. Blacker. 1978. The EPA program to assess the public health significance of diesel emissions. J. Air Pollut. Control Assoc. 28:769-771.
21. P. E. Enterline. 1972. A review of mortality data for American coal miners. Ann. N.Y. Acad. Sci. 200:260-272.
22. L. B. Lave and E. P. Seskin. 1970. Air pollution and human health. Science 169:723.
23. M. A. Schneiderman. 1976. Carcinogenesis as an end point in health impact assessments. In, "Impact of Energy Production on Human Health," CONF-761022, NTIS, Springfield, VA, pp. 65-71.
24. V. Alexander, S. S. Leffingwell, J. W. Lloyd, R. J. Waxweiler and A. L. Miller. 1981. Brain cancer in petrochemical workers: A case series report. Am. J. Ind. Med. 1:115-123.
25. D. Hoffman, E. Theisz and M. Thomas. 1965. Studies on the carcinogenicity of gasoline exhaust. J. Air Pollut. Control Assoc. 15:162-165.
26. H. Reznik-Schuller and U. Mohr. Pulmonary tumorigenesis in Syrian golden hamsters after intratracheal instillations with automobile exhaust condensate. Cancer 40:203-210.
27. R. L. Boeckx, B. Postl and F. J. Coodin. 1977. Gasoline sniffing and tetraethyllead poisoning in children. Pediatrics 60:140-145.
28. T. J. Chow and J. L. Earl. 1979. Lead aerosols in the atmosphere: Increasing concentrations. Science 169:577-580.
29. T. L. Chan, P. S. Lee and June-Sang Siak. 1981. Diesel-particulate col-

lection for biological testing. Comparison of electrostatic precipitation and filtration. Environ. Sci. Technol. 15:89–93.
30. K. I. Campbell, E. L. George and I. S. Washington. 1980. Enhanced susceptibility to infection in mice after exposure to dilute exhaust from light duty diesel engines. In, "Health Effects of Diesel Engine Emissions," Vol. 2, EPA-600/9-80-057b, Washington, D.C., p. 772.
31. M. T. Karagianes, R. F. Palmer and R. H. Busch. 1981. Effects of inhaled diesel emissions and coal dust in rats. Am. Ind. Hyg. Assoc. J. 42:382–391.
32. E. J. Calabrese, G. S. Moore, R. A. Guisti, C. A. Rowan and E. N. Schulz. 1980. A review of the literature: Human health effects associated with exposure to diersel fuel exhaust. In, "Health Effects of Diesel Engine Emissions," Vol. 2, EPA-600/9-80-057b, Washington, D.C.
33. H. Ohgaki et al. 1982. Carcinogenicity in rats of the mutagenic compounds 1-nitropyrene and 3-nitrofluoranthene. Cancer Lett. 15:1–7.
34. P. E. Morrow and C. L. Yuile. 1982. The disposition of coal dust in the lungs and tracheobronchial lymph nodes of dogs. Fund. Appl. Toxicol. 2:300–305.
35. P. Stocks. 1962. On the death rates from cancer of the stomach and respiratory diseases in 1949–53 among coal miners and other male residents in counties of England and Wales. Br. J. Cancer 16:592.
36. H. Rockett. 1977. Mortality among coal miners covered by the UMWA Health and Retirement Funds. NIOSH Research Report, Morgantown, WV, M37. R. G. Ames, H. Amandus, M. Attfield, F. Y. Green and V. Vallyathan. 1983. Does coal mine dust present a risk for lung cancer? A Case-control study of U.S. coal miners. Arch. Environ. Health 38:331–333.
38. B. Knave et al. 1978. Long-term exposure to jet fuel. A cross-sectional epidemiologic investigation on occupationally exposed industrial workers with special reference to the nervous system. Scand. J. Work Environ. Health 4:19–45.
39. C. F. Reinhardt et al. 1971. Cardiac arrhythmias and aerosol sniffing. Arch. Environ. Health 22:265–279.
40. NIOSH/OSHA. 1978. Pocket Guide to Chemical Hazards. NIOSH 78-210, National Institute for Occupoational Safety and Health, Washington, D.C.
41. Y. Alrie, R. I. Kantz, C. E. Ulrich, A. A. Krumm and W. H. Basey. 1973. Long-term continuous exposure to sulfur dioxide and fly ash mixtures in Cynomolgus monkeys and guinea pigs. Arch. Environ. Health 27:251–253.
42. H. N. MacFarland et al. 1968. Chronic exposure of Cynomolgus monkeys to fly ash. In, "Inhaled Particles," Vol. l, Unwin Bros., Ltd., Surrey, England, pp. 313–327.

43. J. H. Wave, D. W. Dockery, A. Spiro, F. E. Speizer and B. G. Ferris. 1984. Passive smoking, gas cooking, and respiratory health of children living in six cities. Am. Rev. Respir. Dis. 129:366–374.
44. R. D. Stewart. 1976. The effect of carbon monoxide on humans. J. Occup. Med. 18:304–309.
45. D. J. Tollerud, S. T. Weiss, E. Elting, F. E. Speizer and B. Ferris. 1983. The health effects of automobile exhuast. VI. Relationship of respiratory symptoms and pulmonary function in tunnel and turnpike workers. Arch. Environ. Health 38:334–340.
46. P. Kotin, H. L. Falk and C. J. McCammon. 1958. The experimental induction of pulmonary tumors and changes in the respiratory epithelium in C57B mice following their exposure to an atmosphere of ozonized gasoline. Cancer 11:473–481.
47. P. Nettesheim, D. A. Creasia and T. J. Mitchell. 1975. Carcinogenic effects of inhaled synthetic smog and ferric oxide particles. J. Natl. Cancer Inst. 55:159–169.
48. W. E. Pepelko. 1984. Experimental respiratory carcinogenesis in small laboratory animals. Environ. Res. 33:144–188.
49. F. A. Seiler, C. H. Hobbs and R. B. Cuddihy. 1983. Potential health and environmental effects of the fluidized bed combustion of coal. Final Report, LMF-107, ITRI, Albuquerque, NM, pp. 372–376.
50. A. Levy. 1983. Unresolved problems in SO_x, NO_x and soot control. In, "Combustion, 19th Symposium on Combustion," The Combustion Institute, Pittsburgh, PA., p. 1223.
51 C. R. Clark. 1982. Mutagenicity of diesel exhaust particle extracts. LMF-96, ITRI, Albuquerque, NM, pp. 1–29.
52. N. E. L. Hall and E. L. Wynder. 1984. Diesel exhaust exposure and lung cancer: A case-control study. Environ. Res. 34:77–86.
53. P. S. Liss and A. J. Crane. 1983. Man-made carbon dioxide and climatic change: A review of the scientific problems. Geo Books, Norwich, UK, 127 p.
54. R. A. Kerr. 1984. Doubling of atmospheric methane supported. Science 226:954–955.
55. W. Giese. 1960. Pathmorphologie der Ventilation. Die pulmonal bedingten Ventilationsstorungen. Verh. Dtsch. Ges. Path. 44. Tagung, S. 35–46. Stuttgart. Fischer.
56. G. Worth. 1984. Emphysema in coal worker. Am. J. Ind. Med. 6:401–403.
57. J. D. Tucker, W-Z.Whong, J. Xu and T-M. Ong. 1984. Genotoxic activity of nitrosated coal dust extract in mammalian systems. Environ. Res. 35:171–179.
58. P. E. Morrow and C. L. Yuile. 1982. The disposition of coal dusts in the lungs and tracheobronchial lymph nodes of dogs. Fundamental Appl. Toxicol. 2:300–305.

59. M. A. K. Khalil and R. A. Rasmussen. 1983. Carbon monoxide in the earth's atmosphere: Increasing trends. Science 224:54–56.
60. H. L. Kaplan, K. J. Springer and W. F. MacKenzie. 1983. Studies of potential health effects of long-term exposure to diesel exhaust emissions. Final Report, SwRI Project No. 01-0750-103, Southwest Research Institute, San Antonio, TX, June.
61. A. L. Brooks et al. 1984. A comparison of genotoxicity of automobile exhaust particles from laboratory and environmental sources. Environ. Mutagenesis 6:651–668.
62. B. Hileman. 1984. Acid rain perspectives: A tale of two countries. Environ. Sci. Technol. 18:342A–344A.
63. H. P. Misra and M. Tolentino. 1981. Coal fly ash induced fibrosis in rat-lung organ cultures. In, "Coal Conversion and the Environment," CONF-801039, NTIS, Springfield, VA, pp. 378–386.
64. M. Ahlberg et al. 1983. Chemical and biological characterization of emissions from coal- and oil-fired power plants. Environ. Health Perspect. 47:85–102.

CHAPTER 8

FOSSIL FUEL CONVERSION

Coal Liquefaction and Gasification

In 1970, 434 million tons of coal were used for electrical power production and space heating or 85% of total coal consumption in the U.S. that year; the remaining 15% was mostly converted into coke for the production of steel. By 1977, coal consumption had increased to 700 million tons, of which only 8% was used for coke production. By comparison, world consumption of coal in 1977 was 4,000 million tons. By 1995, U.S. use of coal for electrical power is projected to be 1,400 million tons; a considerably increased amount of coal will be converted to liquid or gas for energy utilization.[25-27]

The production of coal-derived liquids used for roofing, wood preservation and as solvents was first started in the coke industry in the mid-19th century. Feedstocks from coal liquids became the basis of the early chemical industries, and coal gasification, to produce CO and H_2, was developed in the 1920s by Fischer and Tropsch. Coal conversion consists of two basic steps: (1) cracking of heavy hydrocarbon molecules into lighter molecules and (2) enrichment of lighter molecules with hydrogen. The resultant products may be either liquid (liquefaction) or gas (gasification) or, more commonly, a mixture of both liquids and gases. Coal conversion is a generic term for about 100 different processes designed to produce liquids, gases or solids as energy substitutes for crude oil. Undesirable ash and sulfur are removed as the coal is converted into a form suitable for direct use as a low, medium or high BTU gas, as gasoline or fuel oil or as feedstock for further chemical processing.[1] The chemical objective of coal conversion is to increase the atomic hydrogen/carbon ratio from 0.8/1.0 to a ratio of 1.75/1.0 for liquefaction; ratios of about 4/1 are sought for gasification. Depolymerization (cracking) of coal, followed by hydrogenation, is required to alter the hydrogen/carbon ratio.[2] Conversion technologies are divided into four types:[2]

1. Carbonization, in which coal is heated at 1000°F in the absence of air, causing the release of tars and gases. Hydrogen is added at low

pressure (600 psi) to increase the yield of volatiles and tars. The remaining solid char is used to generate synthesis gas for the production of process hydrogen. An example of how carbonization temperature influences the fraction of coal tar into common fractions is shown in Figure 8.1.
2. Direct hydrogenation, in which a coal slurry is heated to 850°F in the presence of hydrogen and a catalyst at 3500 psi pressure.
3. Extraction, in which coal is dissolved in a solvent at 500°F and 300 psi pressure. The solubilized coal liquid is then hydrogenated to produce a crude oil and unused solvent is extracted for use to dissolve further coal.
4. Synthesis, in which coal is burned in a gasifier in the presence of oxygen and steam to produce mostly carbon monoxide and hydrogen. The gas mixture is passed over a catalyst to yield liquids, natural gas, liquid petroleum gas and a variety of other products.

An outstanding feature of a coal conversion facility is its size. A standard, 250 million cubic feet per day (SCFD) gasification plant or a 50,000 barrels per day (BPD) liquefaction plant will process 15,000 to 25,000 tons of coal per day, often at high temperatures and pressures. One plant, processing 25,000 tons of coal per day, might be expected to produce 205 SCFD of gas, 5,000 BPD of liquid petroleum gas, 40,000 BPD of motor fuels and 5,000 BPD of fuel oil, for an investment of 1.5 million dollars.[1] Twenty such plants could produce 1 million BPD of liquids and 4.1×10^9 ft^3 of gas each day, equivalent to about 6% of the nation's present energy needs. Large reaction vessels and containment buildings are required. Large amounts of waste materials must be treated, stored and disposed of, requiring amounts of water equivalent to a small river for each plant. A single plant may employ 4,000 people, take up a square mile of land, and will influence a population in an area of 200 square miles around the plant. The various processing routes for coal conversion are shown in Figure 8.2.

During gasification processes, sulfur and nitrogen are removed from the original coal in the form of hydrogen sulfide and ammonia, respectively. When the gasification process is accomplished at high temperature, with oxygen and water, the coal is completely destroyed, and synthesis gas (CO + H$_2$) is produced. The four basic chemical reactions occuring during coal gasification are:

carbon-oxygen: $C + O_2 \rightarrow CO_2 + $ heat exothermic

carbon-CO$_2$: $C + CO_2 + $ heat $\rightarrow CO$ endothermic

carbon-water: $C + H_2O + $ heat $\rightarrow CO + H_2$ endothermic

carbon-hydrogen: $C + 2H_2 \rightarrow CH_4 + $ heat exothermic

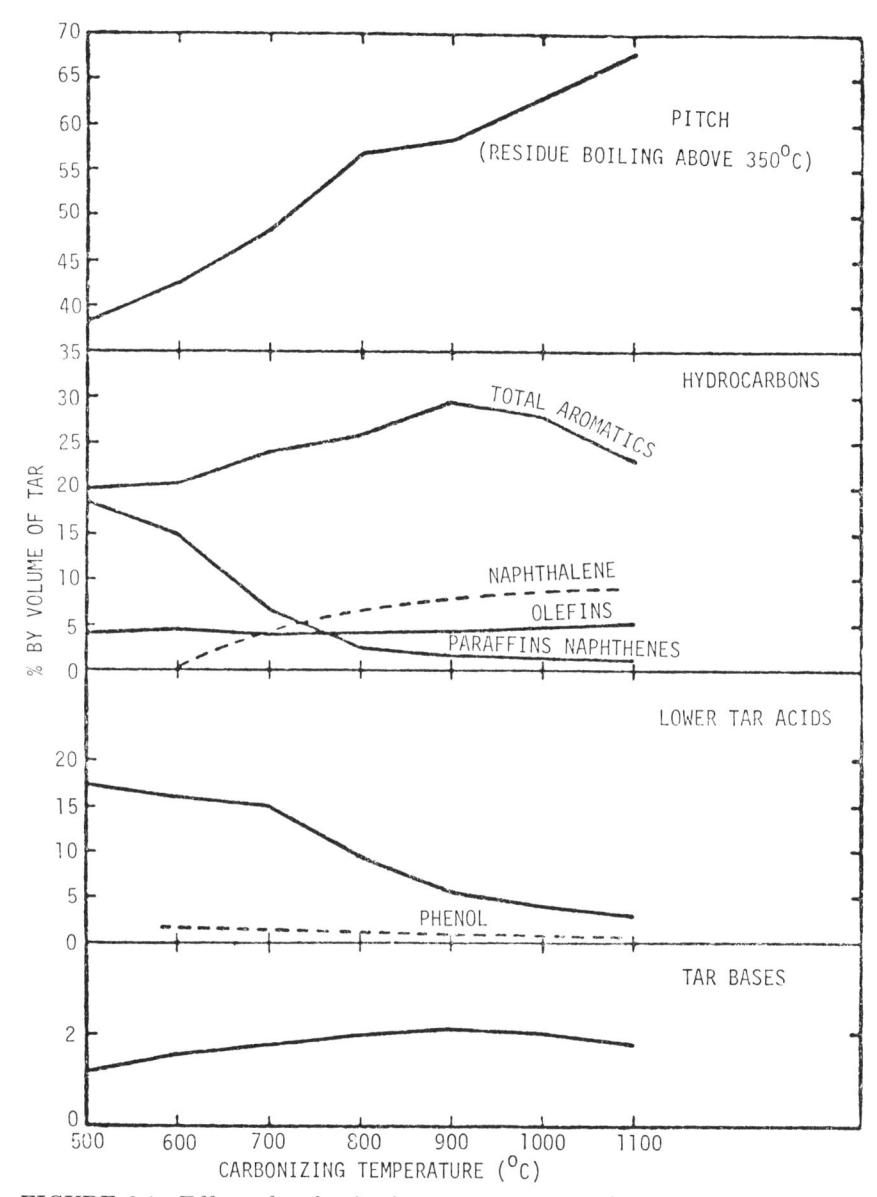

FIGURE 8.1—Effect of carbonization temperature on fractionation of coal tar (R. I. Freudenthal, G. A. Lutz and R. J. Michell. 1975. Carcinogenic potential of coal and coal conversions products. Battelle Columbus Laboratories, Columbus, OH.).

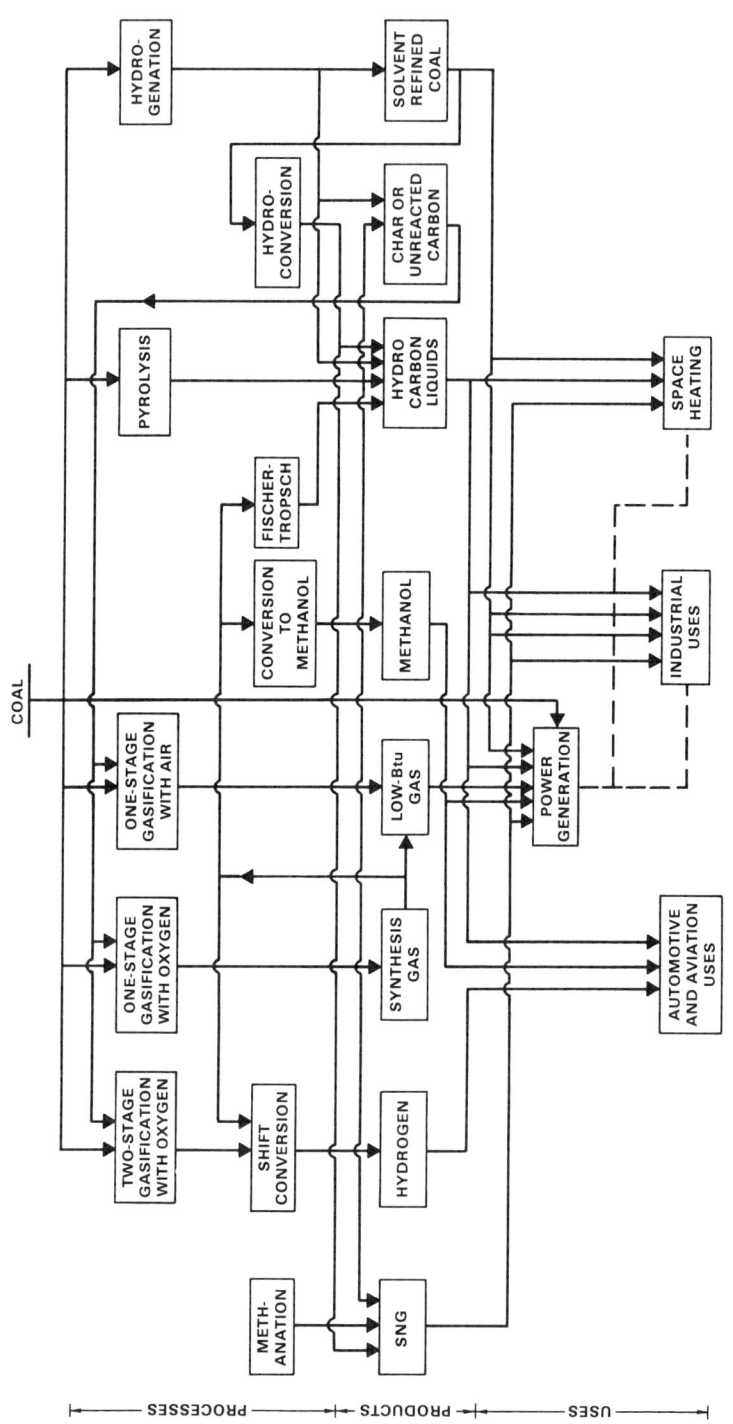

FIGURE 8.2.—Processing routes for coal conversion (M. R. Kornreich. 1976. Coal conversion processes; potential carcinogenic risk. The Mitre Corpration, Mitre Technical Report MTR-7155, Washington, DC).

8 – Fossil Fuel Conversion

The overall chemical reaction in coal gasification is endothermic; the carbon-oxygen reaction is sufficient to produce enough heat to sustain the chemical reactions required to produce synthesis gas.

A number of gasification processes have been developed; these include the carbon dioxide acceptor, Synthane, Bi-gas, Lurgi, Koppers-Totzek and Hygas processes. The most likely commercial gasification plants will treat a coal slurry at high temperatures and at pressure such that the coal is completely gasified, with very little residues remaining from the original coal. Once the gas is cleaned it can be burned in a gas turbine, and heat from the hot combustion gases can be used to make steam.[3] A simplified diagram of the coal gasification process is shown in Figure 8.3.

Coal gasification involves chemical reactions similar to those of coal carbonization. At any stage of coal gasification, high-boiling-point vapors or liquids may escape. These gasification reactor effluents are the most significant sources of potential carcinogenicity in emissions from coal gasification plants.[28] Liquid process streams likely to contain carcinogens include leachates from stored coal and ash and those that follow coal washing. The gaseous process stream leaving the gasifier is likely to contain carcinogenic chemicals. Trace elements and organic carcinogens will mostly be removed during the clean-up phase of gasification. The synthetic gas itself is not expected to exhibit a carcinogenic response.

Depending on the gasification process, potential toxic components of gasifier gas prior to cleaning include HCN, NH_3, M_2S, benzene, carbonyl sulfide (COS), toluene, thiophene, C_5 aromatics, methyl mercaptan, SO_2, CS_2, cresols, thiocyanides, anilines, trace metals (including carbonyls) and PAH. A Bi-gas plant will produce 180–300 tons of ammonia and 11–19 tons of HCN per day; the Lurgi process yields 9 tons SO_2 and 0.07 tons COS per day. Residual chars and tars formed in coal gasification are rich

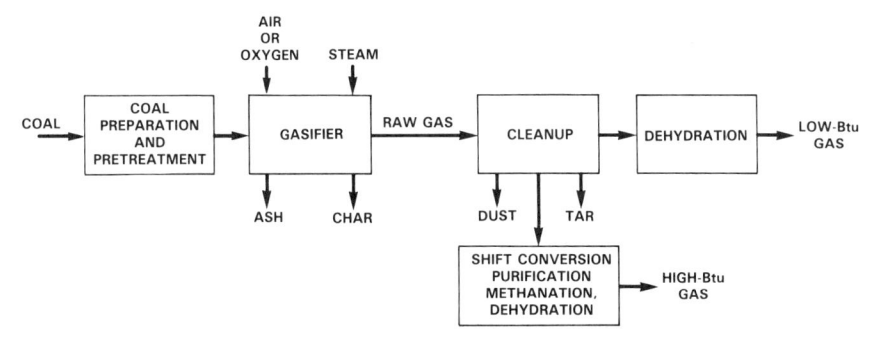

FIGURE 8.3 – A simplified diagram of the coal gasification process (M. R. Kornreich. 1976. Coal conversion processes; potential carcinogenic risk . MITRE Technical Report MTR-7155, The Mitre Corporation, Washington, D.C.).

in naphthalenes, fluorenes, 3-ring aromatics, phenols and N-heterocyclics. Waste tar, char and ash comprise about 20-30%, by weight, of raw coal. Primary aromatic amines (PAA) in gasifier tar process liquids and coal liquids are quantitatively correlated with mutagenicity in bacteria and carcinogenicity in small animal studies. A great deal of PAA mutagenicity can be eliminated in these fractions by acetylation or reaction with nitrous acid.[29] Little carcinogenic activity would be expected in finished gasifier gas. When the gasification effluent stream is cooled, a large amount of condensate is formed, mostly from the steam fed to the gasifier. This aqueous effluent condensate is mostly phenolic, containing 2,000-7,000 mg/l phenol in the Synthane process.

Concentrations of trace metals in gasifier gas varies according to their amounts in coal and their degree of volatility (Table 8.1). Nearly all mercury present in coal is found in the gas; little cobalt is lost due to vaporization from coal. Almost all methanation procedures use nickel catalysts, which result in the formation of nickel carbonyl, particularly at low temperatures used in some gasification reactions.

TABLE 8.1. Trace element emissions in coal gasification.

Element	lb/1000 Tons Coal in Gasifier Gas	Emission Rate from Gasifier Plant, x 10-1 g/sec	Percent Loss from Coal
As	28.	5.6	65
Be	0.1	0.42	20
Cd	–	0.52	60
Co	–	0.0087	<10
Cr	–	–	20
Cu	–	17.	20
Hg	0.4	0.86	99
Mn	6.8	0.87	<10
Mo	13.7	8.7	50
Ni	–	8.7	25
Pb	25.2	10.	60
Sb	1.9	0.52	30
Se	–	1.3	75
V	–	5.2	30

8 – Fossil Fuel Conversion

Coal can be processed to form liquid fuels by four methods, all of which utilize hydrogenation steps (Figure 8.4):

1. Direct hydrogenation, where coal is hydrogenated under high pressure and temperature (H-coal, Synthoil and Bergius).
2. Solvent extraction, where a solvent is used as the hydrogen donor to promote liquefaction under high pressure and temperature (solvent refined coal [SRC], and solvent extraction of lignite).
3. Pyrolysis, where thermal decomposition of coal occurs in the absence of oxygen to produce solids or char, liquids and gases (Char-Oil-Energy Development System [COED], Clean Coke System and Coalcon or Toscoal).
4. Indirect liquefaction where gasification of coal with steam and oxygen is used to produce synthesis gas, which is then catalytically converted to liquid products (Fischer-Tropsch System and Methanol System).

The SRC process converts high-sulfur and high-ash coals into cleaner-burning gaseous and liquid fuels by noncatalytic direct hydrogenation. In the SRC process, a highly aromatic solvent, which is a by-product of the process, dissolves coal at 850°F and 1000 psi. The solution is filtered to remove coal ash, and reacted carbon and solvent are removed by distillation. The product contains substantially less sulfur and ash than the feed coal, making it an attractive fuel for power plants and industry. The solid char from SRC may be further hydrotreated to yield liquid hydrocarbons.[4]

Two systems, SRC-I and SRC-II, have different products: SRC-I produces a solid coal-like, primary product containing <1% sulfur and 0.2% ash, while SRC-II produces a liquid fuel oil and naphtha that contains 0.2–0.5% sulfur. Both SRC systems produce significant quantities of gases, which are further processed to give synthetic natural gas (SNG) and liqui-

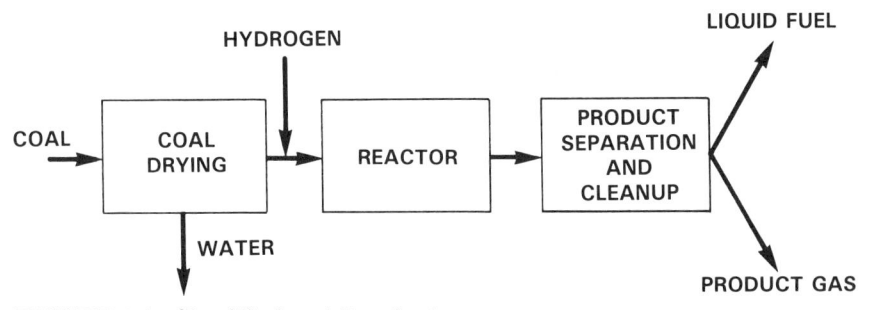

FIGURE 8.4—Simplified coal liquefaction process (M. R. Kornreich. 1976. Coal conversion processes: Potential carcinogenic risk. MITRE Technical Report MTR-7155, The Mitre Corporation, Washington, D.C.).

fied petroleum gas (LPG). In the Fort Lewis, Washington, SRC pilot plant, about 3,850 tons of bituminous coal are processed daily. The daily output from the plant includes 359 tons fly ash, 126 tons sulfur, 1.8 tons particulates, 438 tons carbon for plant energy, 1257 tons carbon for hydrogen production and 1037 tons carbon as fuels (490 tons as heavy fuel oil, 139 tons as light hydrocarbons, 79 tons as LPG and 332 tons as SNG).

In coal carbonization, bituminous coal is heated above its decomposition temperature in closed ovens or retorts. Solid coke is formed, and the volatile products are processed to form liquids or gases (Figure 8.5). In a typical coking facility about 75% of the coal, by weight, is converted into coke, with a yield of 31 gallon tar/ton coal, 2.3 gallon light oil/ton coal and 2000 ft^3 gas/ton coal.[6]

Coal conversion processes constitute significant sources of PAH emissions into the environment. A single coal-liquefaction plant that consumes 25,000 tons of coal per day, produces 210,000 kg PAH, including 7 kg BaP per day. An industry processing a million tons of coal a day would produce about 100 metric tons of BaP annually. However, most of the PAH, BaP and aromatic amines are retained in the process streams, causing little risk for occupational exposure.[5] Examples of carcinogens present in coal gasification process streams are seen in Table 8.2.

Coal and shale-derived liquid fuels contain 2–3 times more BaP than crude oils. Liquefaction processes deliver products that contain over 100 times more BaP than crude oils; tars and high-boiling-point liquid products may contain up to 40 wt% of PAH. In general, coal- and shale-derived liquid fuels contain larger quantities of alkylated PAH and heteroaromatics than does crude oil.[1] A much higher PAH content is found in SRC process water than in gasification process water or in water effluents in petroleum refineries. Large amounts of benzene and phenol are present in water process streams.

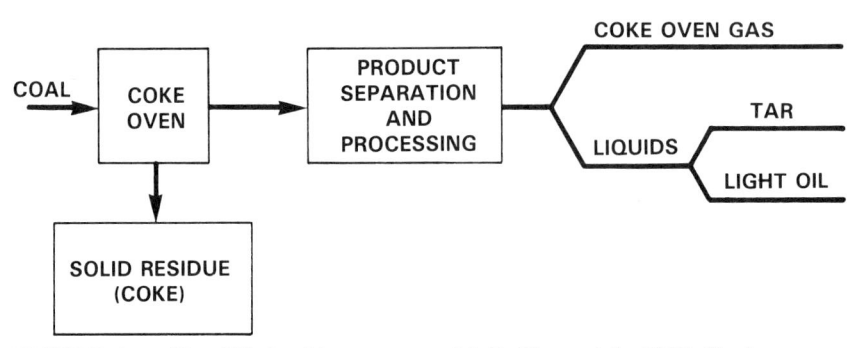

FIGURE 8.5 — Simplified coking process (M. R. Kornreich. 1976. Coal conversion processes: Potential carcinogenic risk. MITRE Technical Report MTR-7155, The Mitre Corporation, Washington, D.C.).

TABLE 8.2. Carcinogens in coal gasification process streams (S. S. Talmage. 1977. Humans: Metabolism and biological effects. In, "Environmental, Health, and Control Aspects of Coal Conversion: An Information Overview," Vol. 2, ORNL/EIS-95, Energy Research and Development Administration, Washington, D.C., pp. 10-1-10-99).

Chemical Class	Compound
Amines	Diethylamines
	Methylethylamines
Heterocyclics	Pyridines
	Pyrroles
Hydrocarbons	Benzene
Phenols	Cresols
	Alkyl cresols
Polynuclears	Anthracene
	Benzo(a)pyrene
	Chrysene
	Benzo(a)anthracene
	Benzo(a)anthrone
	Dibenzo(a,l)pyrene
	Dibenzo(a,n)pyrene
	Dibenzo(a,i)pyrene
	Indole(1,2,3-c,d)pyrene
	Benzoacridine
Trace metals	Nickel
	Beryllium
	Cadmium
Organo-metallics	Nickel carbonyl

Shale Oil

Oil shale is a sedimentary rock containing varying amounts of organic material (kerogen), which decomposes when heated above 900°F to form hydrocarbons and carbonaceous material. The hydrocarbons are condensed to form a heavy shale oil, which may either be burned directly for electrical power production or be refined into a variety of other fuels. Recoverable shale oil in the Green River formation of the Rocky Mountains is estimated at 600 billion barrels.[7]

Shale oil can be produced either by above-ground retorting or by the in situ retorting process (TIS) below ground. Of the surface retorting processes, three are of the indirect heating type (TOSCO II, Petrosix and Lurgi/Ruhrgas), and two are of the direct heating type (Gas Combustion and Union Oil Company). Indirect heating processes utilize a separate furnace for heating materials that are injected, while hot, into the shale in a retort. Direct heating processes rely on the internal combustion of recycle gas or residual carbon from spent shale within the bed of shale in the retort.

Commercial production of shale oil first began in Canada and the U.S. in the mid-19th century and in France in 1938. The shale oil industry in Scotland was not displaced by crude oil until 1962. At its height, the shale oil industry of the world produced 1.7 million barrels a day.[7]

Primary environmental problems associated with production of shale oil involve obtaining adequate supplies of water. A possible shale oil plant scheme is shown in Figure 8.6. Inputs for a 50,000 bbl/day plant using oil shale that contains 25 gal/ton of shale, would be 95,000 ton/day of shale and 2,000-6,000 acre-feet water/year. Outputs would be 82,000 ton/day of spent shale, 3,000 ton/day shale dust and coke, 14,000 ton/day carbon dioxide, 2-10 ton/day SO_x, 6-30 ton/day NO_x, 1-5 ton/day carbon monoxide and 50,000 bbl/day shale oil.

Raw shale oil differs from crude oil in having high concentrations of

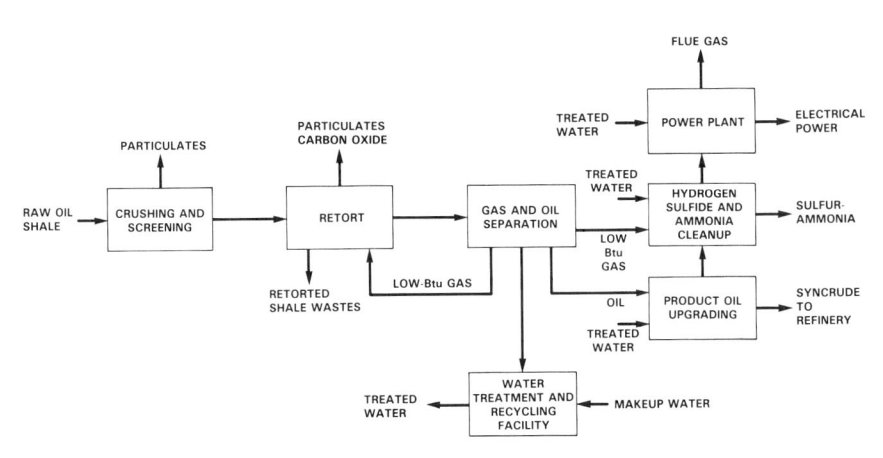

FIGURE 8.6—A possible process flow scheme for a shale oil retort plant producing 50,000 bbl/day of oil (The Aerospace Corporation. 1981. Energy technologies and the environment. Environmental Information Handbook. DOE/EP-0026, NTIS, Springfield, VA, p. 153).

nitrogen-containing and phenolic compounds. The basic nitrogen-rich fraction of shale oil is also rich in alkyl-substituted anilines, pyridines, pyrroles and quinolines. The PNA fraction is rich in naphthalenes, biphenyls, phenanthrenes and other PAH. BaP levels in raw shale oil range from 1.8 to 2.3 ppm; retorting enriches BaP levels to 3.3–22 ppm. Sulfur contents of shale oil average 0.6 wt%. Raw shale oil contains up to 20 ppm arsenic, 0.7 ppm beryllium, 0.7 ppm cadmium and 10 ppm lead.[8] Neither metal nor sulfur contents of shale oil are considered sufficient, in most cases, to constitute significant or unique environmental problems. However, the formation and exposure of workers to PAH during retorting of oil shale and other occupational diseases related to exposures have been of concern for several decades.

Tar Sand

Tar sand is a broad term for rock types, particularly sands, that contain 11–12 wt% bituminous material. Each sand particle consists of a silica nucleus, enveloped first by a layer of water, then by a layer of bitumen, which has a higher carbon/hydrogen ratio than crude oil. Therefore, processing bitumen from tar sand yields smaller amounts of light fuel derivatives, such as gasoline and kerosene, than does crude oil.

There are two commercial tar sand plants in North America, both located in the Athabasca region of northern Alberta, Canada. The tar sands are found in beds below the surface, 50–100 feet deep. Oil production from tar sand is currently about twice as expensive as oil obtained from wells because of the high cost of excavation (2.5 tons tar sand are needed to produce 1 barrel of oil) and the location of deposits in cold, muskeg-covered region of Canada.[9,10]

Sun Oil Company and Great Canadian Oil Sands produce 55,000 bbl/day of synthetic crude oil from tar sands. The overburden, consisting of a superficial growth of tamarack and black spruce, and the primary soil cover of muskeg (a semi-floating mass of decaying vegetation) are first removed to exposed the tar sands. Two years of drainage are required to remove all the muskeg. Excavation can be carried out only in winter, when the ground is frozen. The sand is separated from bitumen, using a mixture of hot water, steam and air.

Small amounts of vanadium and nickel are found in the bitumen, which also has a sulfur content of about 5 wt%. Most of the sulfur is removed during the early processing of bitumen and before the final coking process. Few data are available as to particulate or PAH emissions in processing of tar sands, but SO_2 emissions are of some concern in the region because of the high sensitivity of lichens to this air pollutant.[11]

Biological Effects

The mutagenicity of natural and synthetic oils from crude oil, coal and oil shale provides an estimate of potential carcinogenicity. The mutagenicity of crude oils is usually low but can vary over several orders of magnitude (Table 8.3). Shale oil tends to have more mutagenicity than most crude oils; in comparison, the mutagenicity of coal-derived liquids is often orders of magnitude higher than for crude oils.

Coal is subjected to severe conditions of temperature and pressure during the liquefaction process, which increases the formation of mutagenic organic compounds. A large fraction of SRC-II plant emissions are light hydrocarbons with low biological toxicity. SRC-II heavy distillate and SRC-I process solvent show high mutagenicity; the low-boiling-point distillates are inactive, while higher-boiling liquids (450–840°F) show high mutagenicity. The basic and tar fractions of SRC are responsible for most mutagenic activity; a high amount of 3- and 4-ring aromatic amines are

TABLE 8.3. **Mutagenicity of oils according to chemical class** (M. R. Guerin, C. H. Ho, T. K. Rao, B. R. Clark and J. L. Epler. 1980. Polycyclic aromatic primary amines as determinant chemical mutagens in petroleum substitutes. Environ. Res. 23:42–53). **Mutagenic activity is measured as number of revertants/mg material).**

Origin of Oil	Total Mutagenic Activity (rev/mg)	Neutral Fraction	Acid Fraction	Basic Fraction
Petroleum, crude	5–241*	95–100	0–2	0–3
LETC shale	178	54	2	42
Paraho shale	388	31	0	69
Coal-derived Synthoil	4100	9	3	83
SRC-II	1000	65	0	35
H-coal ASOH	4100	76	0	24
COED	500	87	0	13

*Mean of five crude oils

found in these fractions. Polyaromatic primary amines in SRC-I process solvent have their highest concentrations in subfractions that show the greatest mutagenicity in the Ames bioassay; even so, polyaromatic primary amines are relatively minor constituents of these complex liquid mixtures.[42] Hydrotreating SRC-II material substantially reduces both its mutagenicity and PAH levels.[12] Hydrotreatment also significantly reduces cell transforming activity and carcinogenicity in mouse skin-painting studies of coal conversion products and streams.[45] Likewise, chemical reaction of SRC materials with nitrous acid (nitrosation) is highly effective and apparently specific in reducing the mutagenicity of primary polycyclic aromatic amines using the Ames bioassay system.[43] This is in contrast to nitrosation of secondary amines that forms nitroso derivatives which are then converted to mutagenic and carcinogenic metabolites by biotransforming enzymes in tissues.[44]

Synfuel liquids from all processes with boiling points above 800°F are more biologically active in several "in vitro" systems than liquids with boiling points of <800°F; liquids boiling below 700°F have minimal biological activity.[33] The "in vitro" cytotoxicity of crude and synthetic liquids has been examined in VERO monkey kidney cells, using a clonal growth assay test (Table 8.4). Results indicate that SRC-I process solvent, shale oil and SRC-II heavy distillate are more cytotoxic than are crude oils. A significant correlation has been observed between the "in vivo" carcinogenicity of an organic compound and its "in vitro" malignant transformation ability in mammalian cells (Figure 8.7). Neutral aliphatic and neutral tar fractions of SRC-II show high cell transforming activity even though they are not highly mutagenic in the Ames test.[13] Crude oil, shale oil and high-boiling distillates of SRC cause transformation of mammalian cells; the polyaromatic-hydrocarbon-enriched fraction or SRC-II heavy distillate showed the greatest transforming activity.[48] Care must be taken in interpreting cell transforming data because of non-linear dose-responses; simple linear-regression analysis is not appropriate, and caution must be used in attempting to quantify carcinogenic potential of components in mixtures using this bioassay.

Fly ash from fluidized-bed combustion of coal was not cytotoxic to canine alveolar macrophages at concentrations of up to 100 $\mu g/10^6$ cells. There was no mutagenic activity using the Ames bioassay test in vapor-phase organics, although dichloromethane extracts of fly ash was often mutagenic. The mutagenic activity of fly ash from fluidized-beds generally increased with increasing distance from the combustor bed, increasing combustor temperature and decreasing particle size. However, overall, fly ash from fluidized beds was no more mutagenic or cytotoxic than fly ash obtained from conventional coal combustion.[40] Pulmonary deposition of fly ash from fluidized bed coal combustion in Syrian hamsters result in focal, grade I fibrotic lesions similar to those described for volcanic ash

TABLE 8.4. In vitro cytotoxicity of crude oils, shale oil and coal-derived liquids in VERO monkey kidney cell cultures. RPE$_{50}$ is the amount of material required to reduce plating efficiency by 50% (M. E. Frazier, M. R. Peterson, M. J. Hooper, T. K. Andrews and B. B. Thompson. 1979. Use of cell cultures in evaluation of solvent refined coal [SRC] and shale oil [SO] materials. Pacific Northwest Laboratory Annual Report for 1978, PNL-2850, Pt. 1, Biomedical Sciences, NTIS, Springfield, VA, pp. 1.43-1.44).

Test Material	RPE$_{50}$ Dose (μg/ml)	Test Material	RPE$_{50}$ Dose (μg/ml)
Fossil fuels		SRC-I light oil	500
Diesel oil	250	SRC-I wash solvent	200
Wilmington crude	190	SRC-II heavy distillate	30
Prudhoe Bay crude	350	SRC-II light distillate	200
Shale oil (S-11)	50	SRC-II middle distillate	180
PNA fraction	40		
Basic fraction	50	Reference compound	
Neutral fraction	200	Dihydroxybenzene	90
SRC-I process solvent	35		
PNA fraction	30		
Basic fraction	100		
Neutral fraction	75		

or mixed silicate dusts.[46] Thus, the fibrogenic potential of coal fly ash in the lung is low, whether obtained from conventional or unconventional combustion sources.

Highest boiling product of SRC, SRC-II heavy distillate, contains polynuclear aromatics (39%), tars (35%) and neutral compounds (22%); this fraction is highly mutagenic in the Ames bioassay, carcinogenic in skin-painting studies with mice, and teratogenic and embryotoxic when given by gavage.[47] The effects of SRC-I and II materials on fetal development has been investigated following oral administration to pregnant rats at 7-16 days gestation.[12] Following inhalation of coal liquids, fetal growth and survival were decreased only at doses that also produced maternal toxicity. The incidence of fetal abnormalities, particularly cleft palate, was increased by SRC materials.[34,37]

Synthetic fossil fuels are assumed to possess some human carcinogenic activity because of the chemical similarities of various mixtures to known

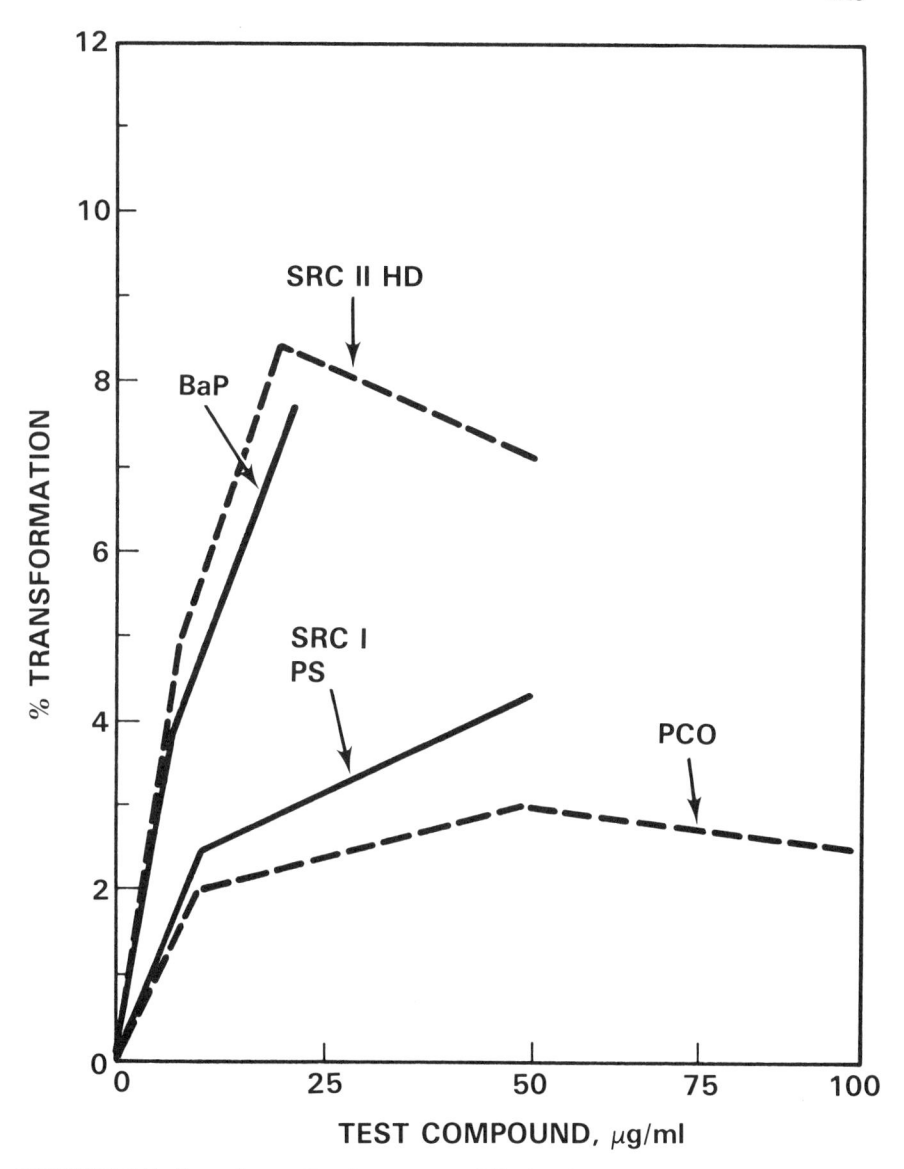

FIGURE 8.7—Transformation frequency of Syrian hamster embryo cells incubated with benzo(a)pyrene (BaP), petroleum crude oil (PCO), SRC-I process solvent (SRC-I PS) and SRC-II heavy distillate (SRC-II HD) (M. E. Frazier and T. K. Andrews. 1983. Transformation of Syrian embryo cells by synfuel mixtures. J. Toxicol. Environ. Health 11:591–606).

carcinogens in coal tar and because of their known mutagenicity, cell-transforming activities and carcinogenicity in mouse skin-painting studies. It is generally assumed that overall carcinogenicity of synthetic fossil fuels is due to their PAH and PAA contents. Exposures to coal carbonization emissions during coke formation and to coal-tar and shale-oil production process emissions have been associated with increased cancer incidence in both experimental animals and humans (Table 8.5).

Chemicals produced during the combustion of coal are carcinogenic in human skin.[14] PAH adsorbed onto soot particles is responsible for the high incidence of scrotal cancer in chimney sweeps of the last century. A high incidence of basocarcinoma and squamous cell carcinoma have been

TABLE 8.5. Epidemiological studies in humans demonstrating carcinogenic relationships between exposure to synfuels or their derivatives.

Reference	Exposure	Cancer Site
Pott, 1875	English chimney sweeps	Scrotum
Volkman, 1875	Carbonization of lignite, tar workers	Scrotum
Rehn, 1895	Dye and coal tar workers	Bladder
Lueke, 1907	Carbon workers	Skin
Henry, 1931	Coal tar workers	Bladder
Kuroda and Kawahala, 1936	Coal gas workers	Lung
Kennaway and Kennaway, 1947	Coal carbon workers, street pavers, asphalters	Lung, larnyx
Doll, 1952	Gas retort workers	Lung
Bruusgaard, 1959	Gas retort workers	Bladder
Sexton, 1960	Coal liquefaction workers	Skin
Doll, 1965	Gas workers	Lung, bladder
Kawai, 1967	Gas workers	Lung
Lloyd, 1969, 1974	Steel and coke workers	Lung
Doll, 1972	Gas workers	Bladder, scrotum
Mazumdar, 1975	Coke oven workers	Lung

seen in coal tar and pitch workers. Creosote exposure has also caused skin cancer in humans. The incidence of skin cancer in workers at a West Virginia coal liquefaction pilot plant was 16–37 times higher than for a nonexposed population; major risk to skin cancer was associated with coal hydrogenation.[15] The relationship between shale oil exposure and skin cancer was noted in 1924. In one study, a significant increase in skin carcinoma was seen in 2003 shale processing plant workers who had worked in the shale industry for longer than 10 years.[17]

Skin-painting studies in mice indicate a correspondence between mutagenicity and carcinogenicity of liquid fossil fuels, particularly for fractions with high PAH and PAA levels. Substantial agreement exists between results of the Ames bioassay test for mutagenicity and skin-painting carcinogenicity tests. Low-boiling-point naphtha fractions of crude coal oils show little or no mutagenic or tumorigenic activity. Higher-boiling-point fractions (e.g., gas oils and residues) and crude oils were both mutagenic and carcinogenic in skin. Coal-derived heavy fractions were significantly more active in both tests than equivalent boiling-point fractions from petroleum crude oil.[41] Most of the carcinogenic activity from crude oil, shale oil, and SRC-I and SRC-II liquids is found in neutral PAH and nitrogen polygenic aromatic compound fractions.

Synthoil PAH isolates had 1/119th the skin carcinogenicity of BaP; shale oil and shale oil PNA exhibited 1/1380th and 1/50th, respectively, the carcinogenicity in skin of BaP.[16] Light-oil hydrocarbons with boiling points of <260°C were not carcinogenic on the skin of mice. Products with higher boiling points are, however, often carcinogenic on mouse skin (Table 8.6). Hairless mice living in cages with spent shale bedding exhibited no greater incidence of skin cancer than those living on normal bedding materials. Likewise, crude shale oil appears to be no more carcinogenic in skin than used fuel oil. Hydrogenation of coal liquids and shale oil reduces its carcinogenicity. The neutral PAH fraction from SRC-II 750–800°F distillate is the most active initiator of skin carcinogenesis.[33,35,36] The latency period is shorter in materials with greater carcinogenic activity.

Interactions among various hydrocarbon solvents and mixtures may enhance or decrease the carcinogenic activity. For example, a 1000-fold increase in epidermal carcinogenicity was seen in mice given BaP or benz(a)anthracene when 50% of the neutral solvent was replaced with n-dodecane, a noncarcinogenic, long-chain, aliphatic hydrocarbon.[18]

Oil shale silicosis ("mohagany lung") has been found in shale workers. Oil-shale dust produced during drilling and processing of shale contains up to 8% quartz. Intratracheal instillation of large amounts of this dust in rats produced only a minimal grade fibrosis and a mild granulomatous and lipoproteinosis response. Thus, the fibrogenic potential of raw or spent shale dusts in the lung is less than would be expected based upon measurements of its free silica contents.[19]

TABLE 8.6. Incidence and length of time for microscopically confirmed epidermal tumors in mice following skin deposition of fossil fuel liquids and known pure chemical carcinogens (R. A. Renne. 1982. Health effects of synthetic fuels. In, "Pacific Northwest Laboratory Annual Report for 1981." PNL-4100, Pt.1, Biomedical Sciences, NTIS, Springfield, VA, pp. 11-14).

Material	Dose, mg	Mice per Group at Start of Study	Mice with Epidermal Tumors	Median Time to Observed Tumor, Days
Untreated	0	50	0	–
Acetone	vehicle control	50	0	–
Shale oil	0.21	50	0	–
	2.1	50	41	358
	21.	50	36	200
Crude petroleum	0.17	50	0	–
	1.7	50	10	501
	17.	50	39	442
SRC-II heavy distillate	0.23	50	22	542
	2.3	50	45	296
	23.	50	46	95
SRC-II light distillate	0.20	50	0	–
	2.0	50	0	–
	20.	50	0	–
2-Amino-anthracene	0.005	50	30	583
	0.05	50	22	386
Benzo(a)pyrene	0.0005	50	0	–
	0.005	50	46	358
	0.05	50	50	143

Inhalation exposure of rats to coal-derived liquids resulted in a suppurative necrotizing bronchiolitis and pneumonitis, followed by a histiocytic and fibroblastic proliferation, leading to granulomatous lesions in the lung. Exposure of guinea pigs to inhaled SRC process solvent (boiling range, 230–450°C) at 0.15 to 0.19 mg/l (40 times higher than the TLV for exposure to nuisance dusts established by the American Conference of Governmental Industrial Hygienists and 800 times higher than the TLV for exposure to coal tar products, including coal tar pitch) caused only a mild pneumonitis and decreased diffusion capacity during an exposure period of several months.[49] Epithelial metaplasia and dysplasia were later reactions in the lung.[22] Exposure of rodents to coal tar fumes has been reported to give negative and positive lung tumor responses.[30,31] Exposure of rats to a coal tar aerosol caused lung tumors; exposure to lamp black or carbon black did not induce lung tumors.[32]

In one study in rats, pellets of beeswax-tricaprylin mixed with crude shale oil, crude oil, or fractions thereof, were implanted in the lungs of rats, allowing for a slow release of carcinogenic hydrocarbons from the implant. A dose-related increase in lung-cancer incidence was observed for each substance; the highest lung tumor incidence was for fractions with the highest PNA contents (Table 8.7). Overall, crude shale oil was more carcinogenic than crude petroleum oil. Although skin cancer risk in oil shale workers has been well documented, no increased risk of lung cancer has been found in either the petroleum or oil shale processing industries.[20,21]

PAH and BaP levels have been measured in the air in coke plants at a mean concentration of about 3 µg BaP/m³. Above retorts in coking facilities the BaP level may be as high as 220 µg/m³ or 10,000 times the normal London air BaP levels.[24] Coke oven workers exhibit an increased incidence of cancer in the skin, lungs, urinary bladder and kidneys. Lung-cancer mortality rate is 7 times higher for men employed more than 5 years at the top of coke ovens; incidence increases with exposure to carbonization temperatures from 500°C to 1500°C. The rate of lung cancer mortality for all coke oven workers is 2–3 times that expected for the general population.[23] Humans exposed to coal tar fumes in generator gas plants experience a dose-related increase in lung cancer. The increased incidence of lung cancer following occupational exposure to soot, pitch and tars and coke oven emissions gives an indication of the potential carcinogenicity of coal and oil shale derivatives (Table 8.8).

A difficulty found in relating occupational cancers to specific etiologic or xenobioitic agents is their long latent period, which can be >20 years for skin and lung tumors after fossil fuel derivative exposure. The most comprehensive data on human lung cancer risk and exposure to effluents from fossil fuel conversion are for coke-oven workers. Risks for other tumors and for other diseases were also increased.[38] In one study of 11,449 British gasworks workers (coal carbonization), an increase in death from lung cancer and bronchitis was observed in the most heavily exposed groups (69% and 126%, respectively) which corresponded to continuous occupational exposures of >100 times as high as the normal PAH level in London.[24,39] These and other epidemiological studies of workers in fossil fuel conversion industries (e.g., coal gasification, liquefaction, coke) demonstrates an increased risk of lung, skin and other cancers and non-cancer pulmonary diseases.

TABLE 8.7. Induction of squamous cell metaplasia and primarily tumors in the lungs of rats given pulmonary implants of shale oil fractions, crude shale and petroleum oils in beeswax pellets comprised of beeswax, tricaprylin and test material (G. Dagle, personal communication).

Material	Dose (mg)	Total	Number of Rats With Lung Tumors	With Squamous Metaplasia
Shale oil				
Neutral fraction	0.6	30	0	4
	6.0	30	1	13
	60.	30	6	28
Basic fraction	0.6	30	2	13
	6.0	30	2	26
	60.	30	8	28
PNA fraction	0.6	30	1	17
	6.0	30	9	29
	60.	30	10	29
Crude oil	0.6	30	1	12
	6.0	30	3	24
	60.	30	9	28
Crude petroleum	0.6	30	1	8
	6.0	30	1	19
	60.	30	9	28
Methylcholanthrene	1.0	30	24	27
Vehicle control	–	30	0	1
Surgical control	–	30	0	0

References

1. M. R. Guerin. 1978. Energy sources of polycyclic aromatic hydrocarbons. In, "Polycyclic Hydrocarbons and Cancer," Vol. 1, Academic Press, New York, pp. 3–42.
2. N. P. Cochran. 1976. Oil and gas from coal. Scient. Am. 234:5.
3. P. H. Abelson. 1982. Clean fuels from coal. Science 215:351.
4. F. K. Schweighardt. 1981. Coal liquefaction: Process description and effluent characterization. In, "Coal Conversion and the Environment," CONF-801039, NTIS, Springfield, VA, pp. 1–15.

TABLE 8.8. Temperature of coal carbonization and risk of lung cancer (National Institute for Occupational Safety and Health [NIOSH]. 1973. Criteria for a recommended standard: Occupational exposure to coke oven emissions. Cincinnati, OH, NIOSH.).

Carbonizing Chamber	Temperature Range, °C	Percent Excess Lung Cancer
Vertical retorts	400–500	27
Horizontal retorts	900–1100	83
Coke ovens	1200–1400	255
Japanese gas generators	1500	800

5. L. R. Harris et al. 1980. Coal liquefaction: Recent findings in occupational safety and health. Am. Ind. Hyg. Assoc. 41:A50–A61.
6. J. W. Sheehy. 1980. Control technology for worker exposure to coke oven emission. NIOSH Technical Report Publication No. 80-114, Washington, D.C., 29 p.
7. The Aerospace Corporation. 1981. Energy technologies and the environment. Environment Information Handbook, DOE/EP-0026, MTIS, Springfield, VA, 512 p.
8. M. R. Petersen, J. Fruchter and J. C. Laul. 1976. Characterization of substances in products, effluents and wastes from synthetic fuel production tests. Quarterly Report for the U.S. Energy Research and Development Administration, Battelle, Pacific Northwest Laboratory, BNWL-2131, Richland, WA.
9. Energy Global Prospects 1985–2000. 1977. Report of the Workshop on Alternative Energy Strategies, McGraw-Hill, New York.
10. Extraction getting underway in Canada. 1977. Sci. Digest 82:
11. N. DeNevers, G. Bard and B. Clifford. 1979. Analysis of the environmental control technology for tar-sand development. University of Utah, COO-4043-2, Salt Lake City, UT.
12. R. H. Gray and H. Drucker. 1982. Assessing health and environmental effects of a developing fuel technology. In, "Beyond the Energy Crisis. Opportunity and Challenge," Pergamon Press, Oxford, pp. 499–507.
13. R. A. Pelroy, D. S. Sklarew and S. P. Downey. 1981. Comparison of the mutagenicity of fossil fuels. Mutat. Res. 90:232–245.
14. W. C. Heuper. 1953. Experimental studies on carcinogenesis of synthetic liquid fuels and petroleum substitutes. AMA Arch. Ind. Hyg. Occup. Med. 8:307–327.

15. R. J. Sexton. 1960. The hazards of health in the hydrogenation of coal. IV. The control program and the clinical effects. Arch. Environ. Health 1: 208–231.
16. J. M. Holland, D. A. Wolf and B. R. Clark. 1981. Relative potency estimation for synthetic petroleum skin carcinogens. Environ. Health Perspect. 38:149–155.
17. P. Bogovski. 1980. Historical perspective of occupational cancer. J. Toxicol. Environ. Health 6:921–939.
18. E. Bingham and H. L. Falk. 1969. Environmental carcinogens. The modifying effect of cocarcinogens on the threshold response. Arch. Environ. Health 19:779–783.
19. R. A. Renne, L. G. Smith, K. E. McDonald, C. A. Shields, A. J. Gandolfi and J. E. Lund. 1980. Morphologic and biochemical effects of intratracheally administered oil shale in rats. J. Environ. Pathol. Toxicol. 3:397–406.
20. M. Purde and S. Etlin. 1980. Cancer cases among workers in the Estonian oil shale processing industry. In, "Health Implications of New Energy Technologies," Ann Arbor Science, Ann Arbor, MI, pp. 527–528.
21. A. Scott. 1922. On the occupational cancer of the paraffin and oil workers of the Scottish oil shale industry. Br. Med. J. 2:1108.
22. W. M. Haschek, M. E. Bolin, M. R. Guerin and H. P. Witschi. 1980. Pulmonary toxicity of a coal liquefaction distillate product. In, "Pulmonary Toxicology of Respirable Particles," CONF-791002, NTIS, Springfield, VA, pp. 338–356.
23. J. W. Lloyd. 1971. Long-term mortality study of steel workers. V. Respiratory cancer in coke plant workers. J. Occup. Med. 13:53–67.
24. P. J. Lawther, B. T. Commins and R. E. Waller. 1965. A study of the concentration of polycyclic aromatic hydrocarbons in gas works retort houses. Br. J. Ind. Med. 22:13–20.
25. A. E. Vandergrift, L. J. Shannon, E. E. Saller, P. G. Gorman and W. R. Park. 1971. J.Air Pollut. Control Assoc. 21:321–328.
26. M. W. McElroy, R. C. Carr, D. S. Ensor and G. R. Markowski. 1982. Science 215:13–19.
27. Energy Information Administration. 1977. Statistics and Trends of Energy Supply, Demand and Prices. Vol. III, Annual Report to Congress. DOE/EIA 0036/3. Washington, D.C.
28. R. E. Royer, C. E. Mitchell and R. L. Hanson. 1979. Fractionation and chemical analysis of a low BTU coal gasifier effluent. LF-69, ITRI, Albuquerque, NM, pp. 3212–3314.
29. R. A. Pelroy et al. 1984. Comparison of in vitro carcinogenesis and in vitro genotoxicity of complex hydrocarbon mixtures. In, "Synthetic Fossil Fuel Technologies: Results of Health and Environmental Studies."

30. W. C. Hueper and W. W. Payne. 1960. Carcinogenic studies on petroleum asphalt, cooling oil and coal tar. AMA Arch. Pathol. 70:372-384.
31. M. Mestitzova and P. Kossey. 1961. Experimenteller Beitras Zum Problem der Genese der Lungenkrebses. Neoplasma 8:27-39.
32. W. E. Pepelko. 1984. Experimental respiratory carcinogenesis in small laboratory animals. Environ. Res. 33:144-188.
33. D. D. Mahlum et al. 1984. Synfuels biostudies. In, "Pacific Northwest Laboratory Report," PNL-5000, Biomedical Sciences, NTIS, Springfield, VA, pp. 3-12.
34. D. L. Springer, K. A. Poston, D. D. Mahlum and M. R. Sikov. 1982. Teratogenicity following inhalation exposure of rats to a high-boiling coal liquid. J. Appl. Toxicol. 2:260-264.
35. D. D. Mahlum. 1983. Skin-tumor initiation activity of coal liquids with different boiling point ranges. J. Appl. Toxicol. 3:254-258.
36. D. D. Mahlum. 1983. Initiation/promotion studies with coal-derived liquids. J. Appl. Toxicol. 3:31-34.
37. P. L. Hackett, D. N. Rommereim and M. R. Sikov. 1984. Developmental toxicity following oral administration of a high-boiling coal liquid to pregnant rats. J. Appl. Toxicol. 4:57-62.
38. C. K. Redmond, A. Ciocci, J. W. Lloyd and H. W. Rush. 1972. Long-term mortality study of steelworkers. VI. Mortality from malignant neoplasms among coke oven workers. J. Occup. Med. 14:621-629.
39. R. Doll et al. 1972. Mortality of gasworkers – final report of a prospective study. Br. J. Ind. Med. 29:394-406.
40. J. O. Hill et al. 1981. In vitro cytotoxicity and mutagenicity of potential effluents from fluidized-bed combustion of coal. In, "Coal Conversion and the Environment," CONF-801039, NTIS, Springfield, VA, pp. 338-351.
41. W. H. Calkins, J. F. Deye, R. W. Hartgrove, C. F. King and D. F. Krahn. 1981. Synthetic crude oils from coal: Mutagenicity and tumor-initiation screening tests. In, "Coal Conversion and the Environment," CONF-801039, NTIS, Springfield, VA, pp. 462-470.
42. R. A. Pelroy and B. W. Wilson. 1981. Relative concentrations of polyaromatic primary amines and azaarenes in mutagenically active nitrogen fractions from a coal liquid. Mut. Res. 90:321-335.
43. R. A. Pelry and D. L. Stewart. 1981. The effects of nitrous acid on the mutagenicity of two coal liquids and their genetically active chemical fractions. Mut. Res. 90:297-308.
44. W. Lijinski. 1976. Carcinogenic and mutagenic N-nitroso compounds. In, "Chemical Mutagens," Vol. 4, Plenum, NY, pp. 193-217.
45. M. E. Frazier and D. D. Mahlum. 1984. Mutagenic and carcinogenic activity of a hydrotreated coal liquid. J. Toxicol. Environ. Health 13:531-543.

46. R. Lantz and D. E. Hinton. 1984. Pulmonary toxicity associated with fly ash from fluidized bed coal combustion. Toxicol. Appl. Pharmacol. 75:44–51.
47. Biomedical Studies on Solvent Refined Coal (SRC-II) Liquefaction Materials. A status report, PNL-3189, NTIS, Springfield, VA (1979).
48. M. E. Frazier and T. K. Andrews. 1983. Transformation of Syrian hamster embryo cells by synfuel mixtures. J. Toxicol. Environ. Health 11:591–606.
49. S. M. Loscutoff, B. W. Killand, R. A. Miller, R. L. Buschbom, D. L. Springer and D. D. Mahlum. 1983. Pulmonary toxicity of inhaled coal liquid aerosols (boiling range 230–450°C). Toxicol. Appl. Pharmacol. 67:346–356.

CHAPTER 9

BIOMASS, SOLAR AND GEOTHERMAL

Biomass Conversion

Biomass energy, in the form of heat, steam, electricity or fuels, is derived from living matter and waste products from living matter, including animal manures. Biomass materials include wood, grasses and other herbage, grain and sugar crops, crop residues, animal manure, food processing wastes, kelp from ocean farms, oil-bearing plants and other materials. Energy from biomass currently provides nearly 2% of all energy in the U.S., about 1.5 quad per year, mostly in the form of wood from the forest products industry for home heating. Biomass energy has the potential to supply 12–17 quad by the year 2000.[1] Of this total, up to 10 quad would come from wood, up to 5 quad from grasses and herbage, 1 quad from crop residues, 0.3 quad as biogas from manure and 0.2 quad as ethanol from grain and other plant source fermentation. Biomass energy is domestically available, renewable, technologically feasible, economical on a small scale and nearly pollutant-free. These attributes appeal to those who favor decentralization of the energy supply.

Biomass offers some advantages over fossil fuels with respect to environmental pollution; these include a lower ash and sulfur content than coal. However, the heating value of wood is considerably less than for coal or oil (Table 9.1).

Biomass conversion technologies include direct combustion, thermochemical conversion processes, biological conversion processes and biophotolysis (Table 9.2). Direct combustion involves the burning of biomass, usually to produce steam for electricity production or for home heating, with wood the most common fuel. During thermochemical liquefaction, biomass is converted, under high temperature and pressure, to an oxygenated oil with a low-sulfur content. Gasification of biomass produces a medium-BTU gas (mostly methane and ethylene) by degrading and saturating organic materials with hydrogen. Oxygen-blown gasification substitutes oxygen for air and produces a high-BTU gas, mostly carbon

TABLE 9.1. Some fuel characteristics of wood, oil, and coal.

Characteristics	Wood	Oil	Coal
Heating value	5,000 BTU/lb	145,000 BTU/gal	12,500 BTU/lb
Ash content, wt%	1.0	0.7	10.
Sulfur content, wt%	0.1	2.0	1.0

TABLE 9.2. Biomass conversion processes.

Type	Product
Combustion	Steam
Thermochemical	
Liquefaction	Oxygenated oil
Pyrolysis	Medium BTU gas and char
Gasification	Medium BTU gas
Methanol synthesis	Methanol
Bioconversion	
Anaerobic digestion	Biogas
Fermentation	Ethanol
Biophotolysis	Hydrogen

monoxide and hydrogen. In pyrolysis, an external heat source is used to gasify biomass, producing a gas like that produced in oxygen-blown gasification and a solid char, like charcoal. These gases can then be altered, in the presence of a catalyst, to form methanol.

In anaerobic digestion of biomass, the feedstock (e.g., manure, grass, algae, seaweed) is degraded by microorganisms in the absence of air to produce a mixture of carbon dioxide and methane, which is referred to as biogas. Biogas is usually used for space heating or cooking. Fermentation of biomass employs plant material as a feedstock to produce ethanol, which can then be used as a fuel. The feedstock is first hydrolyzed by acids or enzymes to simple sugars, which are then fermented to ethanol by the action of yeast cells. Biophotolysis involves the direct production of hydrogen by photosynthesis, using various types of hydrogen-producing algae that are also termed biofuel cells. The environmental impact of biomass energy production is generally considered much less than that of other conventional technologies (Table 9.3).

Wood Combustion

Extensive deforestation in Europe and England in the 1600s caused many local fuel shortages, eventually resulting in a decline in English iron production and a movement of this industry to the U.S., where abundant wood supplies were found. In 1709, Darby substituted coke for charcoal in iron

TABLE 9.3. Environmental effects of biomass conversion technologies (S. E. Plotkin. 1980. Energy from biomass. Environment 22:6–13, 37–40).

Biomass Conversion Source	Pollution Problems
Wood stove	Particulate and PAH air pollution; fire safety problems.
Biomass boiler	High particulate, CO and organic emissions that are easily controlled.
Biomass gasifier	Ammonia, hydrogen sulfide, cyanide, phenols in raw gas and PAH in tars.
Ethanol distillery	"Stillage" wastes with high oxygen demand to dispose; potential water pollution and danger of high pressure steam.
Methanol synthesis	Similar to gasifier.
Anaerobic digester	Hydrogen sulfide; sludge with high oxygen demand; potential water pollution.

production, leading to a massive shift from wood to coal as England's major fuel source.

The number of homes using wood for heat is increasing yearly as fossil fuel costs continue to increase. The total energy equivalent in U.S. standing forests as wood heat is estimated at 324 quad; the net annual forest growth provides the equivalent of 8.2 quad.

Combustion of wood occurs in three steps: (1) water evaporation, (2) evaporation and combustion of volatile organic matter, and (3) combustion of fixed carbon. On the average, wood contains only 0.013 wt% sulfur and has a mean ash content of about 1%. Although these values are 5–100 times lower than those for coal, the environmental impact of burning wood is, in some respects, similar to that of burning coal (Table 9.4). Trace elements are present in wood in minute quantities and are not thought to constitute an environmental pollution problem except, possibly, for manganese and cadmium in biomass growing on soils highly contaminated with these elements.[2]

The sale and use of millions of wood stoves poses serious problems for maintenance of local air quality. Emissions of carbon monoxide, particulates and PAH are greater than for residential combustion of natural gas or fuel oil but somewhat less than for residential combustion of coal. Emissions in some areas of heavy wood use, such as Vail, Colorado, have been so high that legal restrictions have been placed on woodburning to avert

TABLE 9.4. Air pollution emissions for uncontrolled combustion of wood, oil and coal (Energy Research and Development Administration. 1977. Solar program assessment: Environmental factors from biomass fuels. ERDA 77-477/7, Washington, D.C.).

Pollutant	Emissions, lb/MM BTU		
	Wood	Oil	Coal
Particulates	3.0	0.055	6.8
SO_2	0.15	2.2	1.5
CO	0.2–6.0	0.020	0.040
Hydrocarbons	0.2–7.0	0.014	0.012
NO_x	1.0	0.72	0.72

health hazards. The new, airtight wood stoves offer increased heating efficiency but also increased emissions of carbon monoxide, particulates and PAH because they decrease the temperature of combustion and do not completely burn the wood. Attempts at secondary combustion of carbon monoxide and hydrocarbons in wood stoves have only been partially effective in reducing pollutant emissions.[1] In many homes equipped with wood stoves, indoor air levels of carbon monoxide, particulates and PAH are up to 10 times greater than outdoor levels measured at the same time.

Adverse effects of smoke inhalation have been clearly recognized since the Cleveland clinic fire of 1929 and the Coconut Grove fire in Boston in 1942. The major cause of death in these and most fires is carbon monoxide and thermal decomposition product inhalation.[24] The thermal decomposition products of wood produce carboxyhemoglobinemia, necrosis of the cornea, asphyxiation, convulsion, and, if inhaled at a high concentration, death.[25] Effects include necrosis of mucosal epithelium of the trachea and bronchi, with ulceration.[26] The antibacterial properties of rabbit pulmonary macrophages are adversely affected by exposure to wood pyrolysis products of Douglas fir.[11] However, compared to combustion of nitrogen-containing polymers (which cause acute lethality from thermal decomposition by production of complex mixtures of gases, particularly HCN), wood smoke is about 10 times less potent than smoke from polyvinylchloride, and animals recovered much more rapidly than with smoke from polyvinylchloride.[27]

Oil of turpentine is produced by steam distillation of pine wood and is used in artists' paints and in high-quality ship and house paint, shoe polish and in the manufacture of synthetic camphor and terpin hydrate. Commercial preparations of wood turpentine also contain methanol, formaldehyde, phenols and pyridine; this preparation is irritating to the skin and has been associated with outbreaks of dermatitis.[12,29]

Organic extracts of emissions from wood combustion have been shown to be highly mutagenic in the Ames bioassay, to cause chromosomal damage as demonstrated by the induction of sister chromatid exchanges, and to induce cell transformation in Syrian hamster embryo cells.[28] The majority of genotoxic activity was found in the volatile portion of wood thermal decomposition; however, storage for one year resulted in an almost total loss of mutagenicity. Nitroderivatives of otherwise unsubstituted PAH compounds, such as nitroarenes, may be the most potent mutagens in wood thermal decomposition products.[28]

The high temperatures ($>750°C$) of carbonization chambers, such as those found in horizontal retorts and charcoal ovens, produce a highly carcinogenic tar. In contrast, a noncarcinogenic soot extract was obtained from wood that had been heated to 400–450°C, indicating the importance of pyrolysis temperature in the formation of carcinogens.[10] Although little is known of the long-term health effects of exposure to wood combustion particulates, many of the hydrocarbons identified in wood smoke are known carcinogens. Preliminary studies have also indicated that these carcinogenic PAH are absorbed onto the smoke particles. The EPA estimated that in 1972, 23 metric tons of BaP was discharged into the air of the U.S. from fireplaces and wood stoves; this amount is greater than the combined BaP emissions from residential coal furnaces, and coal-fired steam power plants.[3]

Wood and Other Biomass Conversion

All biomass thermochemical conversion processes can produce tars, oils, chars, sulfur-containing gases (CS_2, H_2S and SO_2) and nitrogen-containing gases (HCN, NO_x and NH_3). Biomass, including wood, is relatively easy to convert to synthetic fuels by pyrolysis, gasification and liquefaction. Hemicelluloses of wood decompose first at 200–260°C, followed by decomposition of cellulose at 240–350°C and, finally, by decomposition of lignin at 280–500°C.[4,5] The approximate stoichiometry for wood liquefaction is:

Wood (100g) + CO (0.2 mol) → CO_2 (0.7 mol)
+ H_2O (0.6 mol) + products (60g)

The products include wood oil, comprised of a neutral fraction and three phenolic fractions, water-soluble organics and residues of tar and char.[6] Wood tars have a 30 wt% oxygen content, are corrosive and contain carcinogenic chemicals. The volatile phenolic species of wood oil and tar include guaiacol, methylguaiacol, ethylguaiacol, eugenol and isoeugenol. The liquid wood fuel is further refined by hydrotreatment and fractionation to yield liquids in the boiling-point range of gasoline (40%) and diesel (50%), with a 10% residuum of heavier fractions. Despite its low sulfur content, wood combustion may violate local air pollution standards for hydrogen sulfide.

Other bioconversion energy systems appear to be associated with less severe environmental and health risks. Generally, the gas produced by anaerobic digestors (55% methane and 45% carbon dioxide) has <1% H_2S and NH_3. However, levels of H_2S and NH_3, in some manure digesters, can reach levels sufficient to cause severe toxicity and even death.

Biofuel Cells

Hydrogen, the most abundant element on earth, is a colorless, odorless, tasteless and simple asphixiant gas present as H_2 in the atmosphere at levels <1 ppm. Hydrogen forms chemical bonds either by donating or accepting electrons. It is attractive as a fuel because, when burned, it produces only water vapor, because it can be burned in conventional internal combustion engines, and because it has the highest density of energy per unit weight of all non-nuclear fuels.

Hydrogen is produced commercially by chemical reaction with methane and steam and during fossil fuel conversions. The biggest hydrogen plant in the U.S. (Donaldson, Louisiana) produces 3.7 million m^3 hydrogen/day, which is mostly liquefied for use as rocket fuel and for the production of ammonia-based fertilizers. Another method of producing hydrogen is the electrolysis of water, which requires about 53 kWh of electrical energy per kg hydrogen produced. Recently, the General Electric Company has developed a solid polymer electrolyte (sulfonated fluoropolymer) which, when completely hydrated, takes the place of water and electrolytes during electrolysis.[7]

Solar cell electrolysis uses sunlight to produce a current in silicon spheres which immediately dissociates water into hydrogen and oxygen gases.[21] Sunlight is directly converted to chemical energy with a free energy efficiency of solar-to-hydrogen conversion of 13%, exceeding the solar-to-fuel conversion efficiency of green plants.[30] Biophotolysis has also been demonstrated in numerous species of algae and bacteria. Two types of hydrogen production have been observed during biophotolysis: (1) hydrogenase-catalyzed reactions in green algae and (2) nitrogenase-catalyzed reactions in blue-green algae and photosynthetic bacteria. In all cases, the organisms act as biocatalysts to split water into hydrogen and oxygen under certain environmental conditions.

The conventional hydrogen-oxygen fuel cell (Figure 9.1) is beginning to make a significant contribution to power production in the U.S. because it converts the free energy of a chemical reaction into electricity with high efficiency. Many types of biofuel cells have been demonstrated, during this century, using various microorganisms and substrates. Bacteria-containing biofuel cells have been commercially marketed; e.g., one producing 40 mA at 6V, fueled by powdered rice husks.[8] Other types of biofuel cells use Clostridium welchii and glucose, Nocardia salmonicolor and hydrocarbons, or Micrococcus cerificans and hexadecane. Fuel cells can operate essentially without causing pollution because of the low temperatures of operation and the production of water when hydrogen is burned. The environmental impact of biofuel cells for power generation is very low.[9]

Solar

Man has employed the sun as an energy source for thousands of years. In the 15th century B.C., the Egyptians used mirrors to concentrate the sun for lighting fires, drying food and grains and distilling liquids. By the 17th century A.D., sunlight was being used to melt gold, silver, lead and

CELL REACTION: $2H_2 + O_2 \longrightarrow 2H_2O$

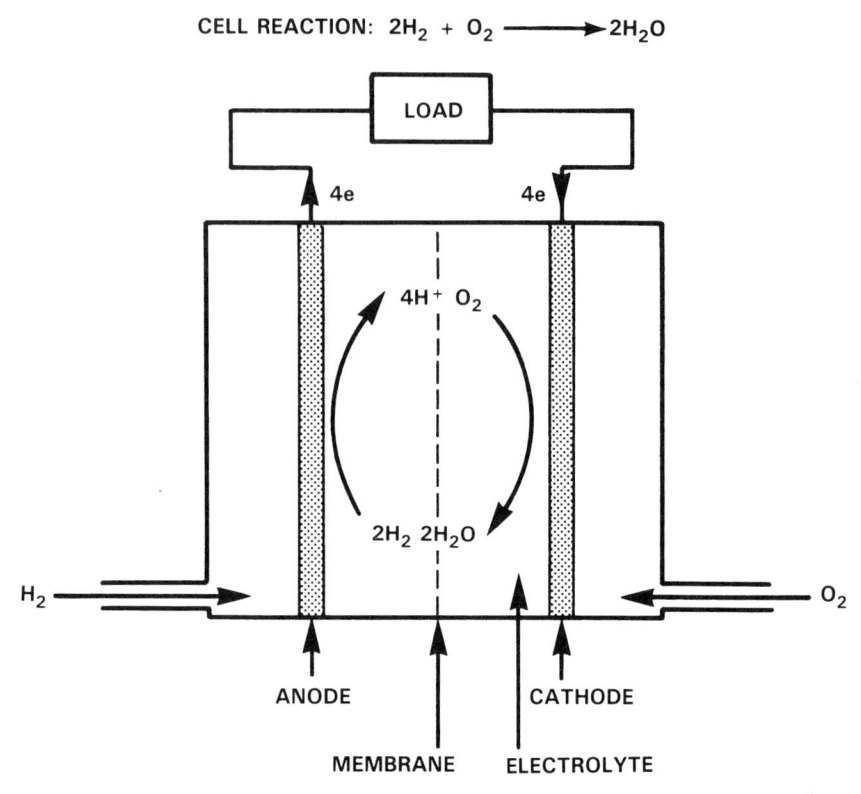

FIGURE 9.1 — Basic hydrogen-oxygen fuel cell (I. J. Higgins et al. 1980. Electroenzymology and biofuel cells. In, "Hydrocarbons in Biotechnology", The Institute of Petroleum, Cambridge University Press, London, pp. 181–193).

copper. By the 19th century, solar energy was used to produce steam in steam engines, and Mouchot developed the forerunner of the modern parabolic Molar collector.[13] Over 25,000 solar hot-water heaters were installed in southern Florida homes during the depression. The first solar-heated house in the U.S. was built by M.I.T. in 1939. It was not until the current increase in fossil fuel costs that widespread and renewed interest in solar energy was revived.[14]

At first glance, solar energy systems appear to present few health hazards. However, solar systems have their own peculiar toxicity problems. The most common thermal solar energy systems use flat-plate collectors, through which a heat transfer fluid circulates. This fluid may be a water/antifreeze mixture, antifreeze alone, a hydrocarbon oil, a silicone oil or a number of other commercial products (Table 9.5). The heat transfer fluid circulates between the collectors and the heat exchanger, inside a storage tank. The heat exchanger warms the potable water in the tank. The poten-

TABLE 9.5. Commercial solar heat transfer fluids (T. C. Marshall et al. 1979. Toxicological assessment of seven solar heat-transfer fluids. Inhalation Research Institute Annual Report for 1978–1979, LMF–69, ITRI, Albuquerque, NM).

Commercial Name	Manufacturer	Composition
DC Q2–1132	Dow Corning	Polydimethylsiloxanes
Mobiltherm	Mobil Oil	Paraffinic, high aromatic
Mobiltherm 603	Mobil Oil	Paraffinic, neutral aromatic
Suntemp	Resource Tech	Paraffinic, low aromatic
Caloria HT-43	Exxon	Paraffinic, low aromatic
Process Oil SX-0358	Exxon	Naphthenic mineral oil
Virco Pet 30	Mobil	Butoxyethyl acid phosphate diethylamine
Therminol 66	Monsanto	Terphenyls
Dowtherm A	Dow	Diphenyls
Ethylene glycol	Aldrich	Ethylene glycol
Dowtherm SR-I	Dow	Ethylene glycol + additives
Propylene glycol USP	Union Carbide	Propylene glycol
Dowfrost	Dow	Propylene glycol + inhibitors
Solar Winter Ban	Camco	"
Freeze Proof	Commonwealth	"
Ucon Food Freeze 35	Union Carbide	"
Nutek 835	Nuclear Tech	"
Solargard G	Daystar	Glycerine + water
Drewsol	Drew	Not available
Sunsol 60	Sunworks	Not available
UCON 50-HB-280X	Union Carbide	Ethylene & Propylene oxides
Nutek 876	Nuclear Tech	Concentrated inhibitors for aqueous system
UCAR 17	Union Carbide	Ethylene glycol + inhibitors
Norkool	Northern Petrochemical	"
Solargard P	Daystar	Propylene glycol
SF-96	General Electric	Polydimethylsiloxanes
Thermia C	Shell	Paraffin-hydrocarbon oil
Nutek 820	Nuclear Tech	Not available
H-30	Mark Enterprises	Synthetic hydrocarbon

tial for contaminating the potable water with heat transfer fluids constitutes a major potential toxicological problem.

Heat exchange fluids may contain additives such as corrosion inhibitors (chromates, nitrites, arsenates, triazoles, benzoate, silicate or organophosphates), or bactericides or fungicides (chlorine, pentachlorophenol, O-phenylphenol, trichlorophenol). A double-walled heat exchanger must be used as a safety measure to minimize leaks into the potable water supply. Epidermal exposure to heat transfer fluids may occur during filling of the solar apparatus or during periodic replacement of fluids. Primary skin irritation studies in rabbits indicated only mild inflammatory reactions for 24 commercial fluids. No significant increase in skin irritation was seen after heat-treating the fluids at temperatures expected in solar collectors (Table 9.6).

The oral toxicity of heat transfer fluids varies widely, although none are highly toxic (Table 9.7). Ethylene glycol is oxidized by the liver to

TABLE 9.6. Skin irritation of unaltered and heat-treated solar, heat-transfer fluids. Heat-treated fluids were heated at 232 °C for seven days in evacuated glass tubes. Irritancy scores were as follows: <2, mild or nil irritation; 2 to 5, moderate irritation; 6 to 8, severe irritation. (C. R. Clark et al. 1978. evaluation of primary skin irritation of solar heat transfer Fluids. Inhalation Toxicology Research Institute Annual Report, 1977–1978, LF–60, Albuquerque, NM).

| | Irritancy Score ||
Solar Fluid	Unaltered	Heat-Treated
Therminol 6b	2.0	1.7
Mobiltherm Light	1.2	1.3
Dowtherm A	0.4	1.1
Process Oil SX0358	0.6	0.7
Suntemp	1.1	0.5
Mobiltherm 603	0.2	0.2
Caloria HT-43	0.1	0.1
Ucon 50	0.0	0.1
SF 96	0.0	0.1

9 – Biomass, Solar and Geothermal

metabolites that cause CNS depression, pulmonary edema, cardiac collapse and irreversible kidney damage; these effects are seen only at high doses. The mean lethal dose for oral ethylene glycol in adult humans is about 100 ml.[16] Ethylene oxide causes skin burns; propylene oxide is a mild CNS depressant and a carcinogen of the nasal turbinates in rodents[32] and diphenyl is readily absorbed through the skin, causing CNS depression, pneumonia, myocarditis, nephritis and hepatitis at high doses.

Large-scale arrays of sun-tracking mirror reflectors, called heliostats, focus sunlight onto a receiver at the top of a high tower, in the middle

TABLE 9.7. Acute oral toxicity of solar heat transfer fluids in rats (T. C. Marshall et al. 1978. Acute oral toxicity of solar heat transfer fluids in rats. Inhalation toxicology heat transfer fluids in rats. Inhalation Toxicology Research Institute Annual Report, 1977–1978, LF-60, ITRI, Albuquerque, NM).

Solar Fluid	24 Hour Oral LD_{50} (g/kg)	Lowest Lethal Dose (g/kg) 24 Hours	48 Hours
Ucon 50	1.88	1.59	1.59
Ethylene Alycol	4.00	3.80	3.80
Dowtherm A	4.14	3.79	3.79
Virco Pet 30	5.90	5.64	5.04
Dowtherm SR-l	6.00	3.36	3.36
Propylene glycol ESP	22.75	20.95	19.78
Mobiltherm Light	24.00	16.99	4.27
Dowfrost	>24.00	>24.00	>24.00
UCON Food Freeze 35	>24.00	>24.00	>24.00
Nutek 876	>24.00	>24.00	>24.00
Solar Winter Ban	>24.00	>24.00	>24.00
Freeze Proof	>24.00	none	>24.00
DC Q2-1132	>24.00	none	none
Nutek 835	>24.00	none	none
Mobiltherm 603	>24.00	none	none
Suntemp	>24.00	none	none
Caloria HT-43	>24.00	none	none
Process Oil SX-0358	>24.00	none	none
Therminal 66	>24.00	none	none
Solargard G	>24.00	none	none
Drewsol	>24.00	none	none
Sunsol 60	>24.00	none	none

of the field where superheated steam is generated for electricity production. About 2,000 heliostats, involving 100 acres of land, are required for a 10-MW power plant. Ultraviolet reflections could pose serious eye hazards to maintenance personnel or to the surrounding population from unscheduled beam movements.[15]

Infrared and Ultraviolet Radiations

The infrared region of the electromagnetic spectrum is from 0.75 to 3,000 microns. All objects emit infrared radiation to other objects of lower temperature; the higher the temperature, the shorter the wavelength of the radiation. Thus, the initial effect of exposure to infrared radiation is heating of tissues. Prolonged exposure to intense infrared radiation (e.g., glass blowers or steel furnaces) can lead to cataract formation. Corneal burns rarely occur because of the accompanying pain associated with the exposure. Excessive exposure to infrared radiation may also lead to heat stress causing heat exhaustion, heat stroke and hyperthermia.[31]

The skin and the eyes are very susceptible to damage from ultraviolet radiation. The cornea and aqueous humor absorb essentially all of the ultraviolet radiation incident on the eye with little of the radiation penetrating as far as the lens. Solar exposure can lead to blepharitis, conjunctivitis, keratitis and keratoconjunctivitis.

The formation of epidermal carcinoma in humans following overexposure to sunlight is well documented. Ultraviolet light (UV) in sunlight at wavelengths of 250–310 nm is the most carcinogenic and damaging to skin. The greatest risk of epidermal damage occurs in light-skinned individuals exposed at noon on June 21 on a cloudless day at 30° latitude N; at this location, sunburn occurs within 5–10 minutes.

UV irradiation is maximally absorbed by purine and pyrimidines bases in the DNA of epidermal cells, forming several molecular lesions, such as thymine dimers. The ability of the skin to repair DNA damage determines, in large part, the tolerance of an individual to sunlight. The repair process consists of removal of the dimer by an endonuclease, restoration of the proper sequence of base pairs by a polymerase and restoration of the normal DNA chain integrity with a polynucleotide ligase.

Overexposure to sunlight leads to hyperkeratosis, senile keratosis and epidermal carcinoma, all indicators of aging in skin.[22,23] Following sunburn, some epidermal cells die, releasing lysosomal enzymes, followed by inflammation and erythema. The skin may become temporarily darkened due to the oxidation of melanin pigment and activation of tyrosinase. The incidence of both basal cell carcinoma and squamous cell carcinoma increases after exposure to large amounts of UV in sunlight, particularly in the head and neck region of the body, which usually receive the greatest exposure.

9 – Biomass, Solar and Geothermal

Genetic diseases associated with ineffective DNA repair mechanisms, such as xeroderma pigmentosa, Bloom syndrome or Hutchinson-Gilford syndrome, cause increased sensitivity to sunlight and a high incidence of senile keratosis and epidermal carcinoma.

Geothermal

Geothermal power production, often associated with continuing volcanic activity, is projected, for the next 30 years in the U.S., as 21,000 MW$_e$ for known resources and 72,000 MW$_e$ for undiscovered resources.[17] Overall, the use of geothermal energy in the U.S. is small compared to that in other countries (Table 9.8). Over 99% of geothermal resources in the U.S. are found in eight western states, and nearly one-third of this resource is found in the Imperial Valley of California (Figure 9.2), which has a potential generating capacity of 7,000 MW$_e$ for 30 years.

TABLE 9.8. World-wide use of geothermal energy (P. J. Lienau. 1980. Geothermal resource utilization. Geothermal Symposium, USDOE, WAOENG–80–16, NTIS, Springfield, VA, pp. 11–12).

Country	Future Predicted	Agriculture/ Aquaculture	Industrial Processes
Iceland	680	40	50
New Zealand	50	10	350
Japan	10	30	5
USSR	120	5100	–
Hungary	300	370	–
Italy	50	5	20
France	10	–	–
USA	75	5	5
Others	10	10	5
TOTAL	1245	5570	235

Energy Utilization, megawatts-thermal

Geothermal sources of energy are generally thought to be pollution-free. In fact, many geothermal sites discharge toxic trace metals to surface waters as well as discharging more sulfur (often in the form of hydrogen sulfide) per MW electric generating capacity than do coal-fired power plants. Geothermal fluids contain dissolved gases (H_2S, CO_2, radon, methane and NH_3) and other chemicals of potential toxicity (benzene, arsenic, mercury and boric acid). At the Geysers, in California, the H_2S level in steam is about 220 ppm. Nearby residents have complained of headaches, nausea and sinus congestion from H_2S exposure.[19] However, assessment of the air quality standard for the area indicates that it would be exceeded only 1% of the time over a 580-square-mile area for a 3,000-MW_e geothermal plant. Over the last 5 years, the mean H_2S levels in ambient air at the Geysers have been 0.022 ppm, although emissions

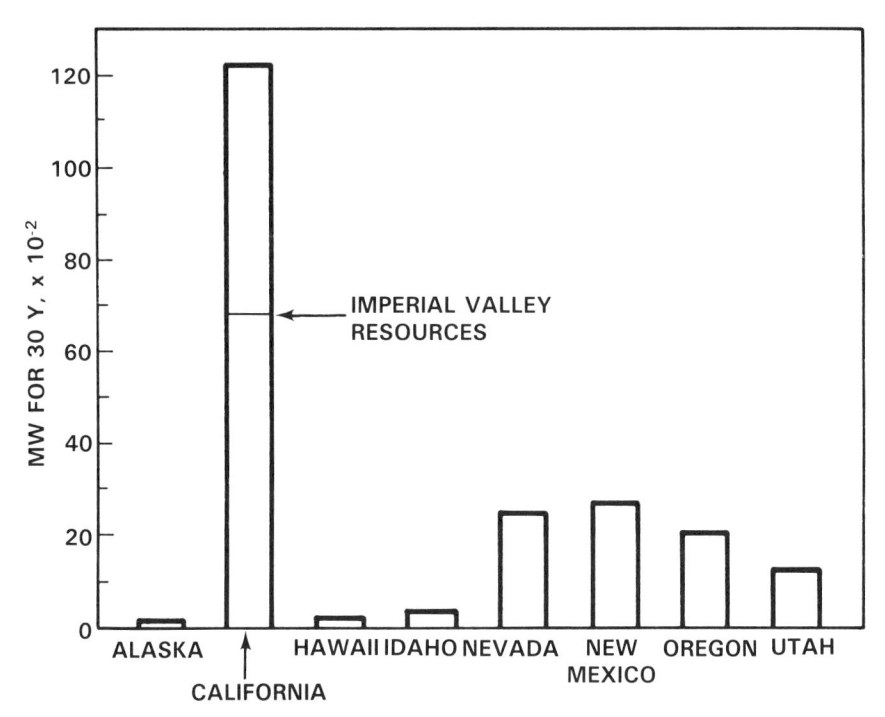

FIGURE 9.2 — Electrical energy potential for identified hot Mater geothermal resources in the western states (D. W. Layton, L. R. Anspaugh and K. D. O'Banion. 1981. Health and Environmental Effects Document of Geothermal Energy. UCRL-53232).

9 – Biomass, Solar and Geothermal

of H_2S from other geothermal resources in the world are considerably higher (Table 9.9).

Geothermal waters in California contain 1–100 ppm ammonia, levels which are not considered sufficiently high to be harmful in the air. The ammonia in geothermal fluids will react with hydrogen sulfide to yield ammonium sulfide. Benzene emissions from geothermal resources are thought to evolve from the thermal decomposition of organic matter.[17] Benzene levels in water from The Geysers varies from 1 to 45 ppm.

Because of its volatility, about 90% of the mercury in geothermal sources stays with the steam phase, passing into the atmosphere. In contrast, the much less volatile arsenic remains mostly in the hot brine. Arsenic emissions from geothermal plants can be several hundred times greater than from coal-fired power plants, on a comparable MW basis. Mercury concentrations are roughly 100–1000 times higher in steam condensates than in most natural waters and 10^5 to 10^6 times higher in noncondensable gases than in ambient air.[18] However, the atmospheric release of mercury from geothermal power plants is comparable to mercury releases from coal-fired plants, ranging from 0.3–18 µg/kg geothermal steam.[18,20] The Matsukawa geothermal power plant in Japan emits much higher levels of mercury than the California Geysers, even though the California Geysers

TABLE 9.9. Hydrogen sulfide emissions for geothermal sites of the world (D. W. Layton, L. R. Anspaugh and K. D. O'Banion. 1981. Health and environmental effects document of geothermal energy. UCRL–53232).

Resource Area	Estimated Emissions ($g/MW_e \times hour$)
In Liquids	
Brawley, California	2,424
Heber, California	20
Baca, New Mexico	2,125
Roosevelt Hot Springs, Utah	304
Wairakei, New Zealand	570
Matsukawa, Japan	5,050–20,800
Cerro Prieto, Mexico	32,000
In Steam	
Larderello, Italy	14,300
The Geysers, California	1,850

are located in one of the most productive mercury-mining districts of the U.S., due to the acidity of the waters in Japan which increases the extraction of mercury from rocks.

A hypothetical and theoretical estimate has been made of risk associated with exposure to various effluents from geothermal sources.[17] Individual annual risk of premature death from inhalation of H_2S from a 100-MW_e geothermal plant in California is estimated at 7.2×10^{-8}. For a 3,000-MW_e power complex, this would result in 4.8 premature deaths over 30 years without control of H_2S emissions. The lifetime risk of incurring leukemia from benzene exposure for a 100-MW_e plant is 2×10^{-8} at 10 kilometers distance; for a 3000-MW_e complex, this would result in an excess 0.01 deaths from leukemia in 30 years. A 3,000-MW_e complex is estimated to cause 7.6 cases of clinical severe neurologic disorders in 30 years resulting from mercury exposures to surrounding populations.[17]

References

1. S. E. Plotkin, 1980. Energy from biomass. Environment 22:6-3, 37-40.
2. Energy Research and Development Administration. 1977. Solar Program assessment: Environmental factors from biomass fuels. ERDA 77-477/7.
3. United States Environmental Protection Agency. 1975. Scientific and Technical Assessment Report on Particulate Polycyclic Organic Matter (PPOM). EPS-600/6-74-001.
4. T. B. Reed, T. Milne, J. Diebold and R. Desrosiers. 1981. Biomass gasification for production of gaseous and liquid fuels. Biotechnol. Bioeng. Symp. 11:151-169.
5. T. Milne. 1979. Pyrolysis–The thermal behavior of biomass below 600°C. In, "A Survey of Biomass Gasification". II. Principles of Gasification", SERI/TR-33-239, SERI, Golden, CO., pp. 11-95.
6. H. G. Davis, D. J. Kloden and L. L. Schaleger. 1981. Chemistry and stoichiometry of wood liquefaction. Biotechnol. Bioeng. Symp. 11:151-169.
7. L. J. Nuttall. 1981. Proceedings, 16th Intersociety Energy Conversion Engineering Conference, American Institute of Aeronautics and Astronautics, New York, p. 1425.
8. K. R. Williams. 1966. An introduction to fuel cells. Elsevier, Amsterdam.
9. S. Novick. 1976. The Electric War. Sierra Club Books, San Francisco, CA, pp. 139-144.
10. F. Dickens and W. Malherbe. 1942. Investigation into the possible carcinogenic activity of wood smoke. Cancer Res. 2:680.
11. R. B. Fick, E. S. Paul, W. W. Merrill, H. V. Reynolds and J. S. O. Loke.

1984. Alterations in the antibacterial properties of rabbit pulmonary macrophages exposed to wood smoke. Am. Rev. Respir. Dis. 129:76–81.
12. C. P. McCord. 1926. Occupational dermatitis from wood turpentine. J. Am. Med. Assoc. 86:1979.
13. P. N. Cheremisinoff and T. C. Regino. 1979. Principles and applications of Solar Energy, Ann Arbor Science Publishing, Inc., Ann Arbor, MI.
14. H. Rau. 1964. Solar Energy. Macmillan Publishing Co., New York.
15. W. J. Bradley. 1977. Designing and siting power plants. Consult. Eng. 48:80–84.
16. R. E. Gosselin et al. 1976. Clinical Toxicology of Commercial Products. Williams & Wilkins Co., Baltimore, MD.
17. D. W. Layton, L. R. Anspaugh and K. D. O'Banion. 1981. Health and environmental effects document of geothermal energy. UCRL-53232.
18. A. Y. Ellis. 1978. Geothermal fluid chemistry and human health. Geothermics 6:175–182.
19. L. R. Anspaugh and J. L. Hahn. 1979. Human health implications of geothermal energy. In, "Health Implications of New Energy Technologies", UCRL-83382.
20. D. E. Robertson, E. A. Crecelius, J. S. Fruchter and J. D. Ludwick. 1977. Mercury emissions from geothermal power plants. Science 196:1094.
21. A. Heller. 1984. Hydrogen-evolving solar cells. Science 223:1141–1148.
22. W. Summer. 1980. Photo-carcinogenicity: The physical basis of its exogenous causes. Br. J. Dermatol. 102:611–619.
23. F. Urbach et al. 1980. Ultraviolet radiation and skin cancer in man. Prevent. Med. 9:227–230.
24. R. B. Mellins and S. Park. 1975. Respiratory complications of smoke inhalation in victims of fires. J. Pediat. 87:1–7.
25. C. L. Hilado and D. P. Brauer. 1979. Effect of air flow on toxicity of pyrolysis gases from wood in USF toxicity test. J. Combust. Toxicol. 6:37–43.
26. D. R. Thorning, M. L. Howard, L. D. Hudson and R. L. Schumacher. 1982. Pulmonary responses to smoke inhalation: Morphologic changes in rabbits exposed to pine wood smoke. Human Pathol. 13:355–364.
27. K. L. Wong, M. F. Stock, D. E. Malek and Y. Alarie. 1984. Evaluation of the pulmonary effects of wood smoke in guinea pigs by repeated CO_2 challenges. Toxicol. Appl. Pharmacol. 75:69–80.
28. I. Alfheim et al. 1984. Short-term bioassays of fractionated emission samples from wood combustion. Teratogenesis, Carcinogenesis and Mutagenesis 4:459–475.
29. C. P. McCord. 1926. Occupational dermatitis from wood turpentine. J. Am. Med. Assoc. 86:1979.
30. A. Heller. 1984. Hydrogen-evolving solar cells. Science 223:1141–1148.

31. F. N. Dukes-Dobos. 1981. Hazards of heat exposure. A review. Scand. J. Work Environ. Health 7:73–83.
32. National Toxicology Program . 1985. Toxicology and Carcinogenesis Studies of Propylene Oxide in F344/N Rats and B6C3F1 Mice (Inhalation Studies). Technical Report Series No. 267, U.S. Department of Health and Human Services, Washington, D.C., 53 p.

CHAPTER 10

RADIOLOGICAL HEALTH

Health Physics and Ionizing Radiations

Ionizing radiations are a group of corpuscular and electromagnetic radiations which have sufficient energy to produce ionizations in matter. Because ionizing radiations exhibit both the properties of waves and particles, their forms are quite different. Corpuscular radiations include alpha and beta particles emitted from the nuclei of radioactive atoms. Electromagnetic radiations behave more like waves of energy, and include X-rays and gamma rays (Figure 10.1). The most energetic radiations are the X- and gamma rays, which are capable of producing ionizations in tissues. Other electromagnetic radiations are non-ionizing and include in decreasing order of energy: ultraviolet, visible light, infrared, microwave, radio waves and electric waves.

Electromagnetic radiations are emitted in discrete packets of energy called photons, and travel with the speed of light. The wavelength and frequency of electromagnetic radiations are inversely related. In radiological physics and radiation biology, electromagnetic radiations are characterized by their energy, which is usually expressed in units of millions of electron volts (MeV) or thousands of electron volts (keV). X- and gamma rays are identical, differing only in their mode of production. X-rays are produced outside the atomic nucleus and have energies up to millions of volts. Gamma rays are produced within the nucleus and are associated with the decay of a radioactive atom. Their energy usually ranges from a few hundred to a few million electron volts.

Beta particles are emitted from the nucleus of a wide variety of radioactive elements. A beta particle has the same mass as an electron. Alpha particles have energies ranging from about 4 to 8 MeV reducing the mass of the original nucleus by 4 mass units, and the atomic number by 2. Protons are hydrogen nuclei with a mass and positive charge of 1. Neutrons,

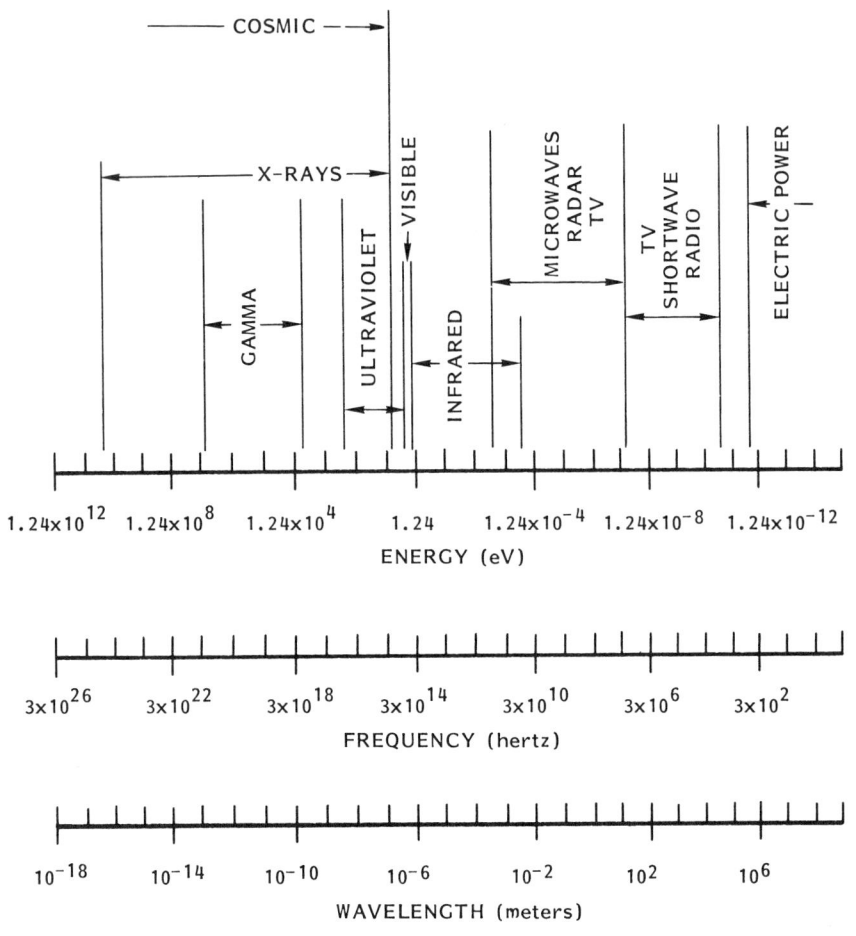

FIGURE 10.1—The electromagnetic spectrum, showing the relationship between energy, frequency, and wavelength (taken from Radiological Health Handbook, U.S. Department of Health, Education, and Welfare, Revised Edition, January 1970, p. 50).

on the other hand, are uncharged particles with a mass similar to the proton (Table 10.1).

Ionizing radiations interact with matter by giving up all or a portion of their energy as they traverse matter. About half of the energy deposited in tissue produces ionization, while the other half excites atoms and molecules, producing luminescence and phosphorescence as well as other effects. Alpha particles will travel about 50 microns in soft tissue, even though they have energies > 4 MeV. The radiation dose in the vicinity of the alpha track is very large. Beta particles have a much longer range in tissue; a 2 MeV beta particle will travel about 1 cm in soft tissue and about

TABLE 10.1. Properties of ionizing radiations.

Radiation	Atomic Mass	Charge	Comment
Electromagnetic	0	0	X- and gamma rays
Alpha	4	+2	Helium nucleus
Beta	1/1830	−1	Electron
Proton	1	+1	Hydrogen nucleus
Neutron	1	0	

8 meters in air. The ionization density and radiation dose along the beta particle track are much less than for alpha particles. Proton interactions are essentially identical to alpha interactions; the range of protons is about 4 times greater. Neutrons interact in a variety of ways, depending on their energy and the specific element they are traversing. Capture of a neutron by a atomic nucleus is a common reaction with low energy, or slow or thermal neutrons, producing radioactive elements. Other neutron reactions include recoil reactions in which a proton may be knocked loose from an atom, and fission, which is the breakup of a heavy nucleus into two or more smaller and usually radioactive nuclei along with the release of energy in the form of gamma rays and neutrons

Ionizing electromagnetic radiations most commonly interact with tissue in three ways: pair production, photoelectric effect and Compton effect. Unlike charged particles, electromagnetic radiations do not have a finite range in matter, but rather are absorbed exponentially, following the equation:

$$I = I_o e^{-\mu t}$$

in which I is the intensity of a photon beam after passing through an absorber of thickness t, I_o is the original intensity of the photon beam, and μ is a specific attenuation coefficient for a particular photon energy and absorbing material.

Radioactive elements continually emit radiations. Emissions occur at an ever decreasing rate, as the radioactive nuclei are transmuted or consumed, reducing their total number. The rate of radioactive decay, and hence emission of radiation, is predictable and unique for each radioactive element (radionuclide), according to the equation:

$$N = N_o e^{-\lambda t}$$

in which N is the number of radioactive atoms remaining after passage of a time interval t, N_0 is the initial number of radioactive atoms, and λ (lambda) is a constant specific for the radioactive element, called the decay constant. The physical or radioactive half-life (T_r) is the time required for 50% of the radioactive atoms to undergo decay.

Within the body, any given radionuclide will not only disappear as a result of radioactive decay, but also from biological turnover, translocation, metabolism and excretion. Thus, there is a biological half-life in the body as well as a radioactive half-life. The rate of removal or disappearance of any radioisotope is governed by both its radioactive and biological half-lives, which work in combination to produce an effective half-life. The relationship of the physical half-life (T_r), biological half-life (T_b) and effective half-life (T_{eff}) is:

$$T_{eff} = \frac{T_r \times T_b}{T_r + T_b}$$

In practice, there is considerable variation in the half-lives of radionuclides of biological interest (Table 10.2). For individual organs, the biological half-life may differ from that in the total body. Iodine, for example, has a biological half-life of 138 days in the thyroid gland but only a half-life of 7 days in the kidney and 14 days in bone. The retention and clearance of a radionuclide in an organ may be best measured by a single exponential expression or by a series of two or three exponential expressions.

TABLE 10.2. Half-lives of some biologically important radionuclides.

	Half-Life		
Nuclide	Physical	Biological	Effective
3-H	12.3 y	12 d	12 d
14-C	5600 y	10 d	10 d
24-Na	15 hr	11 d	14 hr
32-P	14 d	260 d	14 d
55-Fe	3 y	1,700 d	820 d
65-Zn	245 d	2,000 d	220 d
90-Sr	28 y	36 y	16 y
131-I	8 d	138 d	8 d
137-Cs	30 y	70 d	70 d
239-Pu	24,000 y	180 y	180 y

The unit of radioactivity has for many years been the curie (Ci), which is defined as being that quantity of activivty producing 3.7×10^{10} nuclear transformations or disintegrations per second (dps); a millicurie is 3.7×10^7, a microcurie is 3.7×10^4, a nanocurie is 3.7×10^1 and a picocurie is 3.7×10^{-2}. A newer SI unit (International System of Units), the becquerel (Bq) is now being substituted for the curie. One Bq = 1 disintegration per second.

Radiation dose is a measure of the energy deposited in a unit mass of tissue or material exposed. Dose is measured in units of rads. The rad is simply defined as that quantity of radiation that produces an energy deposition of 100 erg/g. The rad is applicable to any type of radiation and any absorbing media. It has also been replaced in the new SI system of units by the gray (Gy); 1 Gy = 100 rad.

Not all radiations produce an equivalent biological effect, even though the radiation dose per gram may be the same. The term, relative biological effectiveness (RBE), is used to define the ratio of the dose from a given radiation required to produce a given effect relative to the dose of some reference radiation (usually 250 kvP X-rays) required to produce the same effect. Linear energy transfer (LET) is an important factor in determining RBE. LET is a measure of energy deposition along the track of radiation through a material. In general, the higher the LET, the higher the RBE. X- and gamma rays and beta particles have low LETs and RBEs, while neutrons, protons and alpha particles have high LETs and RBEs. The rem dose is a method of normalizing biological effects from different radiations with different RBEs; thus rem is equivalent to rad × RBE. In the new SI system of units, the rem becomes the seivert (Sv); 1 Sv = 100 rem (Table 10.3).

Most of the naturally occurring radionuclides in our atmosphere are derived from radioactive decay of ^{238}U and ^{232}Th: ^{222}Rn and radon daughters from ^{238}U, and ^{220}Rn and thoron daughters from ^{232}Th. Typical soils, rocks and coals contain 1–5 ppm ^{238}U and ^{232}Th. Radon daughters in the air are mostly adsorbed onto small condensation nuclei or dust particles. Of the total natural background radiation dose to the human lung of about 2000 mrem/yr, about half is from inhaled radon and radon daughters.

Worldwide, natural ^{222}Rn levels are highly variable, ranging from about 0.1 to 10 pCi/liter. Levels of ^{220}Rn are considerably less than ^{222}Rn. Radon air levels vary inversely with barometric pressure as well as with humidity, time of day, season of the year, geographic location, amount of ^{238}U in the soil, rocks and building materials, type of building construction, altitude and ventilation rate in buildings.[1] Man-made sources of ^{222}Rn include uranium mines and uranium mine tailings, natural gas used in heating and cooking and cigarette smoke. Heavy smokers receive about 10 rem/yr to bronchial bifurcations as a result of inhaling radon daughters ^{210}Pb and ^{210}Po in smoke.

TABLE 10.3. Radiation quantities and units used in modern radiobiology.

Unit or Quantity	Symbol	Application
Curie	Ci	Quantity of radioactivity; 3.7×10^{10} dps
Becquerel	Bq	SI quantity of radioactivity; 1 dps
Rad	rad	Dose absorbed by matter
Gray	Gy	SI unit of dose; 100 rad
Linear Energy Transfer	LET	Energy deposition per unit of path length; usually in eV/micron
Relative Biological Effectiveness	RBE	Same effect from same dose of reference radiation; used in radiobiology
Quality Factor	Q	Biological effectiveness of radiations; used in radiation protection
Rem	rem	Unit of dose equivalent; rad × Q (or RBE) × other modifying factor
Sievert	Sv	SI unit of dose equivalent; Gy × Q (or RBE) × other modifying factor; 100 rem.

Radon and radon daughters are emitted from a variety of building materials as well as from the soil surrounding buildings, which may result in indoor radon levels that greatly exceed those outdoors. Diffusion of radon and radon daughters into enclosed spaces follows Pick's law of diffusion. Radon air levels are largely a function of ^{238}U and ^{226}Ra levels in building materials and soil, the porosity of the walls and soil, and radon accumulation and depletion by diffusion and ventilation. Indoor radon levels in well-insulated homes are higher than outside levels; for example, air concentrations in Swedish homes range from 1–12 nCi/m³, compared to outdoor levels of 0.1 nCi/m³. Air exchange rates of at least 0.5 air changes per hour are required to maintain indoor radon levels below 4 nCi/m³ in these circumstances.

In the U.S., the typical radiation whole body dose equivalent from natural radiation sources is about 120 mrem per year. To this is added dose from manmade sources, the largest of which, by far, is from medical exposure (Figure 10.2). In total, the typical American incurs a radiation dose equivalent of about 220 mrem per year – about 15 rem over a 70-year

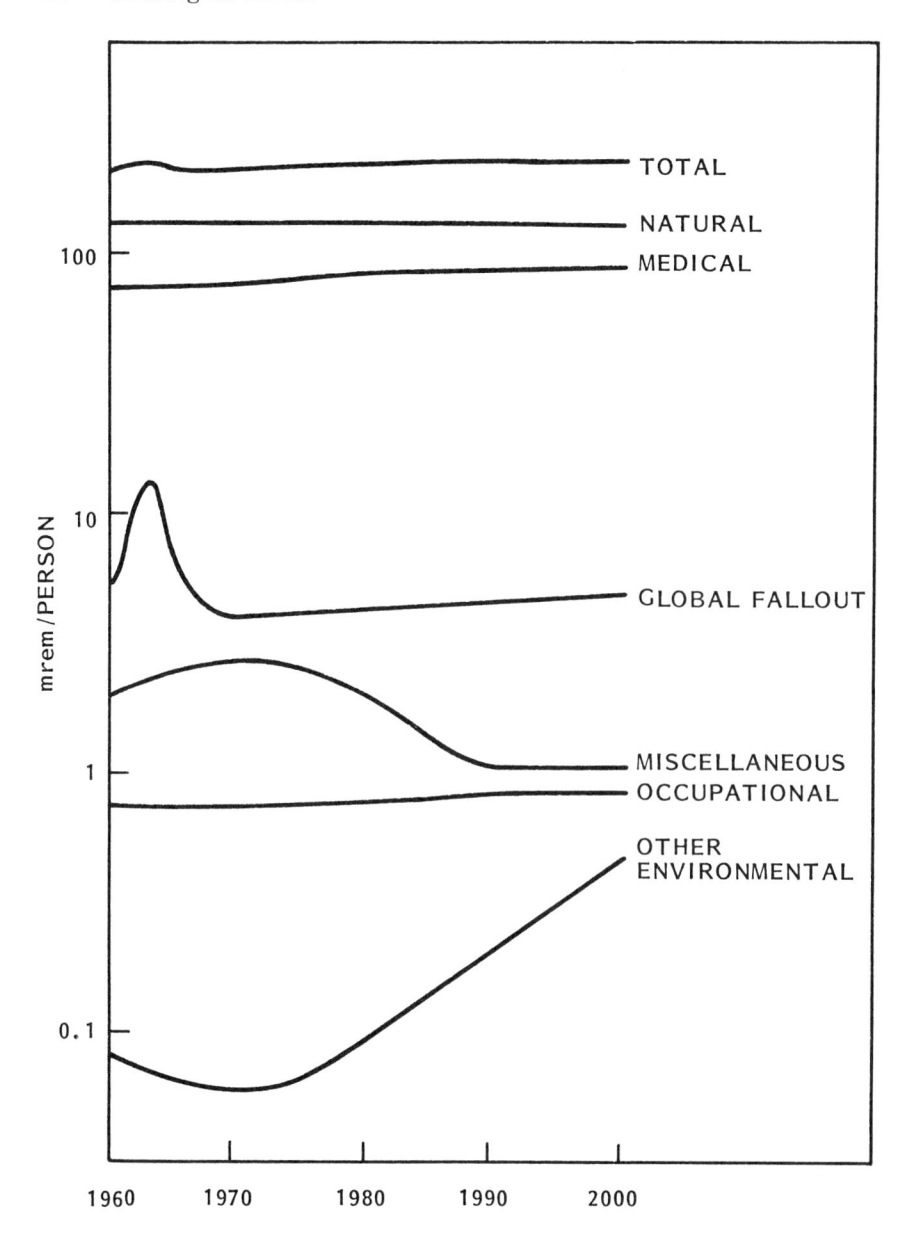

FIGURE 10.2 — Estimated average whole body radiation exposures in the United States, 1960–2000 (U.S. Environmental Protection Agency Publication ORP/CSD 72-1, August 1972, p. 167).

lifespan. Slightly less than 60% of this exposure is from natural sources.

Federal regulations limit the whole-body radiation dose at the boundary of nuclear power plants to a maximum of 5 mrem/yr, to 0.6 mrem/yr within 10 km of a plant. Actual levels at all 71 nuclear power plants in the U.S. are far below these maximum limits except in the event of nuclear accidents.

Basic Concepts in Radiobiology

Two French radiotherapists, Bergonie and Tribondeau in 1906 examined the available data in radiation biology and concluded that the sensitivity of mammalian cells to ionizing radiation was in direct proportion to their rate of cell division and inversely proportional to their degree of cell differentiation. With the exception of the hematocytoblast and lymphocyte, their observations hold true today.[50]

The nucleus of the cell is the most radiosensitive subcellular structure. Within the nucleus, the chromosomes are the most radiosensitive with DNA the most radiosensitive molecule. Factors that play an important role in determining the radiosensitivity of mammalian cells include radiation dose, dose rate and fractionation, LET and the presence or absence of oxygen, radioprotectors and radiosensitizers. Survival is a function of the portion of the cell cycle in which the cell is when exposed. Irradiation during mitosis is the most damaging, while exposure during G1 and the middle of S are the most radioresistant. Damage to cells is manifested by cell death after an abnormal mitosis, prior to mitosis (interphase death) or after several generations of normal mitosis, by delayed DNA synthesis and prolongation in cell cycle time (both for karyokinesis and cytokinesis), and by the production of chromosome aberrations. High LET radiations are more effective in producing chromosome aberrations (and of more severe types) than are low LET radiations (Figure 10.3).

In 1927, Muller irradiated male fruit flies and observed hundreds of different mutations following their fertilization with unirradiated female flies. Using the ClB marker gene, Muller was able to demonstrate a linear increase in the frequency of recessive lethal mutations with radiation dose. Muller concluded that any radiation dose would be genetically harmful; that is, that there was no apparent threshold dose below which no genetic damage would occur. The doubling dose (e.g., the dose required to double the spontaneous mutation rate) was 350 rad following exposure of the ovary and only 75 rad following exposure of the testis of fruit flies. Russel, using the specific locus technique, found that the genetic doubling dose was 50–100 rem in mice.[51] The National Academy of Sciences BEIR committee concluded that the doubling dose for humans was between 50 and 250 rem.[52]

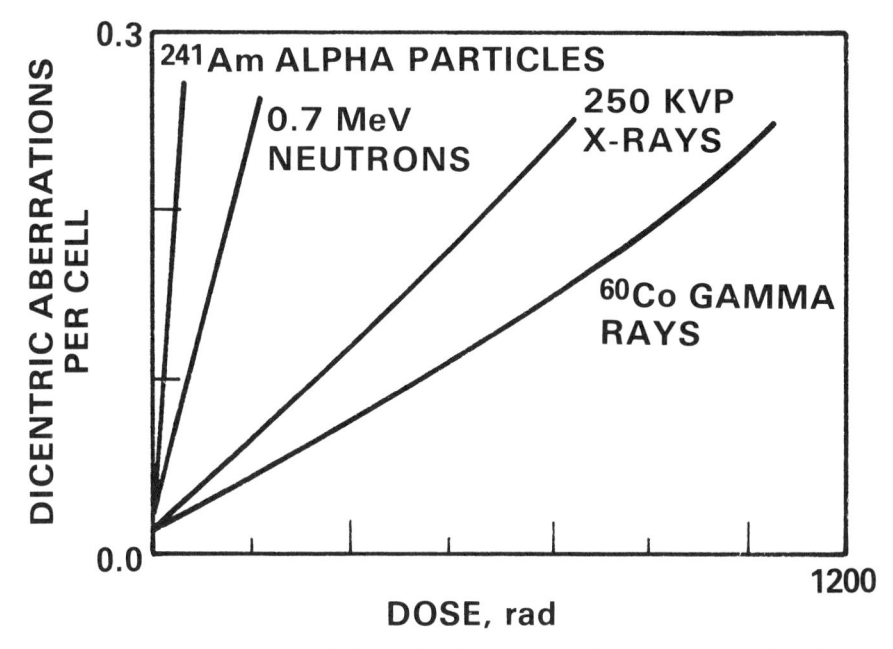

FIGURE 10.3 — Frequency of dicentric chromosome aberrations as a function of radiation dose and type (redrawn from: UNSCEAR, "Sources and Effects of Ionizing Radiation," United Nations, New York, 1977).

Radiation may interact with biological systems by directly damaging critical and sensitive targets (direct effect) or by producing radicals in solutions with the radicals migrating to sensitive targets and causing damage (indirect effect). The presence of oxygen during radiolysis increases the production of the peroxyl radical and hydrogen peroxide, enhancing damage from radiation; this is known as the oxygen effect.

Target theory is a quantitative method of describing the direct action of ionizing radiations on mammalian cells. According to target theory, a deleterious change in the cell is brought about by deposition of energy (e.g., ionizations) within a sensitive volume or target, assumed to be the nucleus. Ionizations in the target are called hits and follow a Poisson distribution pattern. Cell sterility (e.g., inability to produce viable daughter cells) results when the target(s) has received sufficient hit(s) to inactivate the target (e.g., the nucleus). If only one hit is required to sterilize the cell, a typical exponential curve (straight line on a semilog plot) will be seen (Figure 10.4). When more than one hit is required in the target, then a cell survival curve with a shoulder will be generated; the broader the shoulder, the more hits required. The slope of the straight line portion of

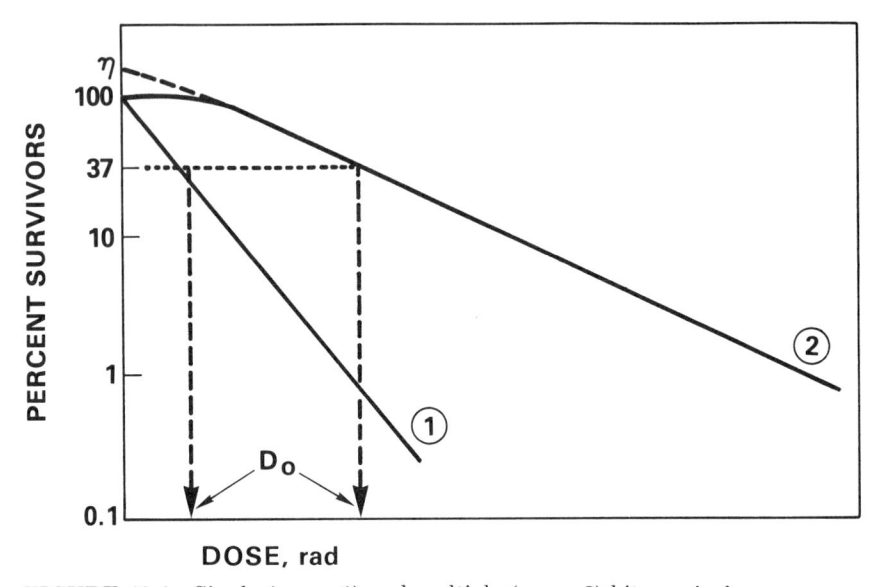

FIGURE 10.4 — Single (curve 1) and multiple (curve 2) hit survival curves seen following irradiation of a proliferating mammalian cell population (C. L. Sanders and R. L. Kathren. 1983. Ionizing Radiation. Tumorigenic and Tumoricidal Effects. Battelle Press, Columbus, OH., p.39).

the survival curve, k, is a measure of relative radiosensitivity. The D_o value is the dose that theoretically will kill all cells if the radiation is evenly distributed in the cell population; in practice this dose will kill only 63% of the cell population, because of statistical probabilities of killed cells being hit again by radiation. High LET radiations (e.g., alpha particles, neutrons, protons) will give a curve without a shoulder while low LET radiations (beta particles, X- and gamma rays) will give a survival curve with a shoulder, suggesting that more than one hit is required to kill the cell. Survival of cells given high LET radiations is not significantly altered by the presence of oxygen, radioprotectors or radiosensitizers; these factors do, however, significantly alter the survival of mammalians cells irradiated with low LET radiations. D_o doses for normal and tumor cells typically ranges between 100 and 200 rad (Table 10.4); D_o doses as high as 1000 rad have been observed for some brain tumor cells.

A measure of the ability to survive acute whole body exposure to external gamma irradiation is the LD_{50} dose, or the amount of radiation that will kill 50% of an exposed population. The $LD_{50(30)}$ dose is the amount of radiation that will kill 50% of the population in 30 days. $LD_{50(60)}$ is a better measure of acute radiation lethality in humans; this is estimated at 300–450 rad (Table 10.5). Acute radiation lethality is also a function

TABLE 10.4. D_o dose for normal and malignant mammalian cells of mice following irradiation with X- or gamma rays (G. Briganti and F. Mauro. 1979. Differences in radiation sensitivity in subpopulations of mammalian multicellular systems. Int. J. Rad. Oncol. Biol. Phys. 5:1095–1101).

Cell Type	D_o, rad
Skin	135
Jejenum crypt	131
Colon crypt	142–328
Testis	179
Cartilage	165
Fetal liver	146–164
Spleen CFUs	65–110
Bone marrow CFUs	87–100
Blood CFUs	140
Lymphoma	80–115
Colon adenocarcinoma	170
Melanoma	122–139

TABLE 10.5. Survival of vertebrate species following acute, whole-body exposure to X- or gamma irradiation as young adults.

Species	$LD_{50(30)}$, rad	Species	$LD_{50(30)}$, rad
Frog	700	Cricetid rodents	800–1,500
Tortoise	1,500	Burro	800
Lizard	1,000	Cow	500
Alligator	1,000	Sheep	400
Salmon	1,000	Goat	400
Goldfish	2,300	Swine	350
Bluebird	1,000	Dog	350
Chicken	600	Rabbit	800
Laboratory rat	700	Monkey	600
Wild rat	900	Human [$LD_{50(60)}$]	350

of dose rate or dose fractionation. For humans, acute radiation lethality is not seen at low dose rates <0.1–0.2 rad/minute. Shielding a portion of the body during whole-body, acute exposure can markedly improve survival. For example, shielding only one leg of the rat increases the $LD_{50(30)}$ two fold.

Acute radiation lethality in most mammals is divided into four categories based upon the radiation dose (Figure 10.5):

1. Molecular death within a few hours after >50,000 rad
2. Central nervous system death within a day after >5000 rad
3. Gastrointestinal death within 7–14 days after >1000 rad
4. Hematopoietic death within 60 days after 250–1000 rad.

Hematopoietic death is due to destruction of bone marrow cells that produce leukocytes, platelets and erythrocytes. Survival is dependent upon the ability of the bone marrow to repopulate the marrow before death from infection, bleeding or anemia occurs.

There is a surprising large body of data in humans exposed to acute radiation levels; these include therapeutic exposures for cancer, survivors of atomic bombings in Hiroshima and Nagasaki, radiation accident victims and fallout victims in the Marshallese Islands.

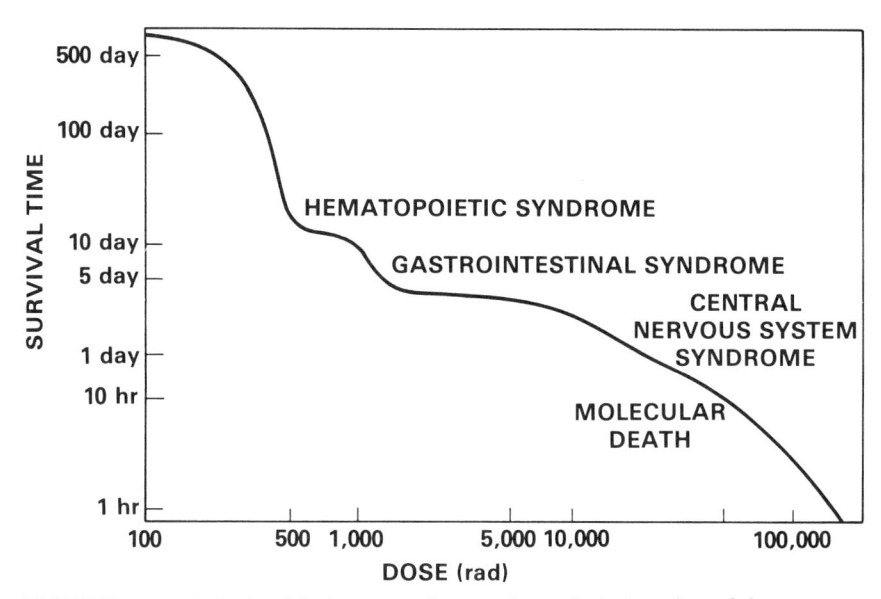

FIGURE 10.5—Relationship between dose and survival time for adult rats following a single whole-body exposure to X-rays. The mortality syndromes are indicated at approximate positions (A. P. Casarett. 1968. Radiation Biology. Prentice-Hall, Inc., Englewood Cliffs, NJ, p. 222).

Each radiation exposure may be divided into that portion which is rapidly repaired and that portion that causes irreparable damage to tissues. Radiation repair following low LET irradiation is mostly completed within an hour after exposure. Overall repair of high LET irradiation is much less. The effects of acute radiation exposure among Japanese A-bomb survivors did not indicate accelerated aging except for limited excess mortality from cancer. However, three significant effects of irradiation were observed on the Japanese embryo and fetus irradiated in utero: impaired growth, mental retardation and microencephaly.[52]

Radiation Carcinogenesis

Ionizing radiations are complete carcinogens, being both initiators and promoters of carcinogenesis. Ionizing radiation can induce benign and malignant tumors in any living mammalian cell type. There is, however, a large variation in the sensitivity of various cell types to tumor induction. For example, the radiation dose required to double the spontaneous incidence of leukemia in humans is about 25 rad, while about 800 rad is required to double the spontaneous incidence of stomach carcinomas.[16,50,52]

In general, there is a high sensitivity to radiation induced leukemia and other bone marrow tumors and to thyroid cancer. Sensitivity of the lung, liver, pancreas and lymphatic tissue to tumor formation is moderate. There is nothing unique about induction of tumors by ionizing radiation, since every tumor type induced by irradiation appears spontaneously in unirradiated individuals. Irradiation increases the tumor incidence and rate of appearance and in cases of high dose exposure, also decreases the latent period from time of exposure to first clincial appearance of tumor. The latency period following irradiation for cancer development is much shorter for radiation leukemogenesis than for most other tumor types (Table 10.6). Latency periods for solid tumors are typically 20 years but are only 2-4 years for leukemia.

The BEIR Committee[53] estimated the lifetime cancer mortality risk from a single whole-body dose of 10 rad low LET radiation, as increasing the naturally occuring cancer mortality by 0.5% to 1.4%. Continuous exposure for life to 1 rad/year would raise this estimate to 3-8%. This means that the current spontaneous cancer incidence of 16% would be increased only to 16.08% to 16.22% by a 1 rad/year continuous exposure. Most risk estimates of radiation induced cancer are based upon the assumption that the dose response is linear, without a threshold. Natural background exposure is estimated to cause 3,500 cancer cases a year in the U.S., or about 1% of all cancers seen in one year.

The association of in utero irradiation and subsequent development of cancer in children has been documented. One study in the United Kingdom

TABLE 10.6. Estimates of cancer mortality per million person-rem (BEIR-National Academy of Sciences. 1972. Effect on Populations of Exposures to Low Levels of Ionizing Radiation. BEIR Report, Washington, D.C.; UNSCEAR-United Nations Scientific Committee on the Effects of Atomic Radiation. 1977. Sources and Effects of Ionizing Radiation. United Nations, NY; ICRP -International Commission on Radiological Protection, Publication 26, Recommendations of the International Commission on Radiological Protection, 1977, Pergamon Press, Oxford; EPA-U.S. Environmental Protection Agency. 1976. Environmental radiation protection for nuclear power operations. Proposed Standards [40CFR19D]. Supplementary information. Washington, D.C.; NRC-U.S. Nuclear Regulatory Commission. 1975. Reactor Safety Study. Appendix VI. Calculation of Reactor Accident Consequences. WASH-1400 [NUREG 75/014]).

	Reporting Agency Fatal Cancer Cases per Million Person-Rem				
Type of Cancer	BEIR	UNSCEAR	ICRP	EPA	NRC
Leukemia	26–37	15–25	20	54	6–28
Lung	16–19	25–50	20	50	4–22
Bone	2–3	2–5	5	16	1–7
Thyroid	–	5–15	5	15	3–13
Female Breast	90	60	25	–	–
GI Tract	30	25	–	–	–

indicated that a fetal exposure of only 1 rad would increase the leukemia rate by >300 cases per million births by age 10.[54] Other studies indicate a much less leukemogenic response in children irradiated in utero.[16,52] The leukemia risk is 5 times greater if the exposure is given during the first trimester of pregnancy. Because of the overall high radiosensitivity of the fetus, the National Council on Radiation Protection and Measurements (NCRP) has recommended that radiation exposure of pregnant women and all women of child-bearing capacity be carefully considered with respect to medical diagnostic tests because of potential danger to the developing embryo and fetus from radiation.

Radionuclide Toxicology

Internally deposited radionuclides cause unique problems in the analysis of toxicity, particularly in evaluation of carcinogenic risk. One important problem is evaluating spatial-temporal radiation dose distribution patterns from radionuclides deposited in tissues. The following factors are important in evaluating radionuclide carcinogenesis:

- Physicochemical form of the radionuclide
- Type and energy of emitted radiations
- Amount and distribution of radionuclide in tissues
- Route of intake
- Translocation and excretion patterns of radionuclide
- Physical and biological half-lives (effective half-life) of radionuclide
- Radiological characteristics of daughter radionuclides
- Sensitivity to cancer formation of tissues receiving highest doses.

Tumor development is usually enhanced by dispersal of the radionuclide throughout an organ rather than by highly concentrating the same acitivity in a small volume of the organ. Irradiation of tissues by deposited radionuclides is often of a chronic and continuous nature. Thus maximum histopathological effects are produced more slowly and recovery is less evident than with acute external exposures. Potential carcinogenic risks for various radionuclides are listed in Table 10.7.

Uranium Mine Exposure

Uranium served in a variety of minor industrial uses prior to the discovery of nuclear fission. Since then most uranium is used as a fuel in nuclear power and weapons reactors. U.S. reserves of commercial uranium ore (>700 ppm U) amount to 3.5 million tons equivalent U^3O^8; an additional 13 million tons is present in shales that contain 25–700 ppm U.

The history of nuclear energy discovery and development is very recent. Neutrons were first identified in 1932. The ability of the neutron to cause certain heavy elements such as uranium-235 to split into two parts (nuclear fission) was discovered in 1939. Uranium-235 constitutes about 0.7% of all natural uranium in geological formations; >99% is the much longer-lived and hard to fission uranium-238. A large amount of energy is released when an atom of ^{235}U is fissioned by a neutron, in addition to the release of 2–3 more neutrons from the split uranium atom. It is these additional neutrons that produce a chain reaction, providing sufficient numbers of ^{235}U atoms are present in a geometrically confined space to sustain such a reaction. A few uranium-238 atoms will absorb a neutron and eventually become ^{239}Pu, which is even easier to fission than ^{235}U. Most atomic

TABLE 10.7. Relative carcinogenic risk of selected radionuclides based upon tritium as unity (J. N. Stannard. 1975. HASL Environmental Quarterly, HASL-291, ERDA, NY).

Radionuclide	MPC^w	Relative Risk	MPC^a	Relative Risk
^{3}H	0.03	1	2×10^{-6}	1
^{137}Cs	2×10^{-4}	150	2×10^{-8}	100
^{144}CE	1×10^{-4}	300	2×10^{-9}	1,000
^{131}I	2×10^{-8}	1,500	3×10	667
^{90}Sr	4×10^{-6}	7,500	4×10^{-10}	5,000
^{87}Kr	–	–	2×10^{-7}	10
^{233}U	4×10^{-5}	750	4×10^{-11}	50,000
^{239}Pu	5×10^{-5}	600	6×10^{13}	3.3×10^6 (bone)
			1×10^{-11}	2.0×10^5 (lung)
^{241}AM	5×10^5	600	6×10^{-12}	3.3×10^5 (bone)
			1×10^{-10}	20,000

MPC^w – maximum permissible concentration in water, μCi/ml
MPC^a – maximum permissible concentration in air, μCi/ml

warheads and hydrogen bombs use ^{239}Pu, while most nuclear reactors use uranium enriched in ^{235}U as fuel. The H-bomb is triggered by a ^{239}Pu fission bomb, causing fusion of hydrogen isotopes. The high energy neutrons released by the fusion process cause fission of a ^{238}U blanket surrounding the device, greatly enhancing its yield.

The nuclear fuel cycle starts with the mining of uranium. From here the uranium ore is processed and purified. The content of ^{235}U is enriched less for nuclear reactor fuel than for weapons use. Nuclear reactor fuel is made into oxide pellets and inserted in long rods into the reactor core. Reactors are run at high power levels for the production of electricity and at lower power levels to maximize the production of plutonium. Nuclear fuel is then removed from the reactor and separated into fission products, plutonium and unused uranium fuel. Most high level nuclear waste is stored above ground in tanks or unprocessed fuel elements are stored underwater. A typical large, light-water nuclear reactor requires about 100 tons of uranium oxide fuel, enriched to >3% ^{235}U. A standard 1000 MW nuclear power reactor produces about 500 pounds Pu^{239} per year, sufficient to produce 10 or more nuclear warheads. At present there are 700 nuclear reactors in the world on land and 330 reactors based in ships.

The unit of dose for exposure to radon and radon daughters is the working level (WL); 1 WL = 1.3×10^5 MeV/liter air of alpha decay energy or 100 pCi/liter for radon in equilibrium with its daughters. One WL exposure for 170 hours gives 1 working level month (WLM). Total exposure is measured in cumulative working level months (CWLM). The current International Commission on Radiological Protection (ICRP) maximum permissible concentration of radon and radon daughters in the air of mines is 0.3 WL. Background exposure levels are about 0.001 WL.

A variety of air pollutants are present in uranium mine atmospheres, including diesel exhaust, silica-containing dusts and radon and radon daughters. Deposition of radon and radon daughters in the lung depends on particle size, tidal volume, breathing frequency and the fraction of radon daughters adsorbed onto dust particle surfaces. In most mine atmospheres, nearly all radon daughters in the air are adsorbed onto particles in the respirable range, delivering a highest mean estimated dose of 1.5 rad/WLM to the subsegmental bronchi.[4] The alveoli of humans receive about one-fifth of that dose.

Since 1410, metal mines have been operated in Schneeberg and Joachimsthal, Czechoslovakia. In 1926, it was noted that over half of miners in these two mines were dying of lung cancer, with a latent period of only 10 years. Mean radon levels in the mines were about 2,000 pCi/liter, with poorly ventilated areas of the mines exhibiting levels as high as 50,000 pCi/liter, or 500 WL. Exposures in the U.S. uranium mines in 1957 ranged from 2 to 50 WL, with a mean of 7 WL; because of improved ventilation, this mean level decreased to 2.3 WL by 1964. In 1968, an exposure standard of 4 WLM/yr for a 30-year career was established for underground mines. Non-uranium mine miners such as Newfoundland fluorspar miners, British underground miners and U.S. hardrock miners have also experienced radon levels in mine air of 1 to 10 WL in 1968. By comparison, U.S. coal miners averaged only 0.01 WL exposures.

Beginning in the 1950s there was a growing awareness in the U.S. of the high incidence of lung cancer in European uranium miners. By 1959 it was noted that exposure in the mines for more than 3 years resulted in significantly higher mortality from lung cancer than for non-mining populations. By 1964, a 10-fold higher lung cancer mortality was noted in miners who had worked for more than 5 years in uranium mines; a 29-fold higher rate of lung cancer was also noted in the Newfoundland fluorspar miners.[7] A significant higher rate of lung-cancer mortality, irrespective of cigarette smoking history, is seen in U.S. uranium miners with >120 CWLM, in U.S. hardrock miners after doses of 90–360 CWLM, and in Swedish zinc miners at 900 CWLM.

An unusually high incidence of small cell undifferentiated carcinoma (oat cell or small cell carcinoma) was noted in uranium miners. In one study, nearly 60% of all lung tumors in uranium miners were of the small cell variety, compared to 18% in nonmining smokers and 6% frequency in nonmining nonsmokers.[5] The symptoms of small cell carcinoma developed rapidly in miners and were fatal. The primary site of first tumor appearance was in the primary to subsegmental bronchi, corresponding to the region of highest radiation dose. Sputum exfoliative cytology has been used to detect the early development of precancerous and cancerous lesions in the lung which are too small or limited to detect by radiographic methods.[6] Most of the uranium miners that developed lung tumors were cigarette smokers who had worked, early in their careers, in small, poorly ventilated mines.

The synergistic interaction of radon-daughter and cigarette-smoke exposures on lung tumor development has been experimentally demonstrated in rats[8] and, epidemiologically, in humans.[5]

In Swedish zinc and lead miners, the lifetime risk of lung cancer for nonsmokers appears to be higher than for smokers. The average time of appearance of lung tumors in smokers, however, was nine years earlier than in nonsmokers. A possible protective effect of smoking, due to a thickened mucus barrier in airways of smokers, is supported by data obtained in experimental animals. Some investigators feel that insufficient data are presently available to accurately model U.S. smokers versus nonsmokers in uranium mines accurately.[46]

The lung cancer rate for miners who worked during the period 1951–1971 is greater than that in nonminers by an excess 0.6 cases per million person years per rem delivered to the bronchial epithelium.

Two models have been developed for the purpose of computing lung cancer mortality due to radon exposure; the NIOSH model uses a 10 year median latent period, whereas the Biological Effects of Ionizing Radiation (BEIR) model incorporates a 5 year delay period. According to the BEIR model, an annual 100 WLM dose will result in a 27% incidence of lung cancer for miners who worked for 30 years. The NIOSH model estimates a 23% lung cancer incidence over this period of occupational exposure. The "virtually safe" dose for uranium miners was calculated as 0.2 CWLM if only one lung cancer death due to radon exposure was expected for every 100 million miners; 20 CWLM corresponds to 1 death in 10,000 miners and 120 CWLM to 1 death in 500 miners. Doubling of the lung cancer mortality rate in nonminers occurs in uranium miners at a dose of about 100 CWLM.

Comparison of Coal and Nuclear Power Plants

Several studies have dealt with comparisons of radionuclide emissions from coal- and nuclear-fueled power plants.[9-11] Coal contains 1–5 ppm ^{238}U and 1–5 ppm ^{232}Th; both radionuclides are enriched in coal fly ash by a factor of about 10. Annual releases from the stack of a 1000-MW$_e$ coal power plant are estimated at 23 kg of uranium and 46 kg of thorium.

Exposure from the discharge of radionuclides from coal-fired power plants can occur externally from the radioactivity in the air or ground, and internally from inhalation or ingestion of these radionuclides and their daughter products. A hypothetical, modern, 2000-MW$_e$ coal-fired power plant would deliver nearly all its radiation dose to those living downwind from the stack, by ingestion of ^{210}Pb, ^{210}Po and ^{231}Pa.[44] During coal combustion, U, Th and their radioactive decay products, other than Rn gas, are concentrated in ash particles with the greatest concentration occurring on the smaller sized particles.[58]

Some analyses[10] estimate that coal-fired power plants release more radioactivity into the air than comparably powered nuclear plants under normal operating conditions. Others indicate that the radiological risks from coal-fired plants are small in comparison to the risks from nuclear plants.[57] Carbon-14 is thought to be the major contributor to whole-body radiation dose from nuclear plant emissions. Other investigators[11] believe that nuclear power plants have greater radiological impacts than coal power plants by a factor of 100 if krypton-85 and tritium emissions are uncontrolled, and by a factor of 40 if only krypton-85 emissions are controlled. Most comparisons, however, ignore the radiological impact of short-lived xenon radionuclides emitted from nuclear power plants. Taking this into account, coal may be expected to produce 10-fold fewer health effects from radioactive emissions than nuclear plants.[11] In spite of apparent differences and even inconsistencies in evaluation of the radiological hazards of coal and nuclear power plants, it is apparent that the potential for radiological risk to human health is quite small for a coal power plant compared to that for a nuclear power plant.[45] Opinion, however, is varied.

Nuclear Fusion

Thermonuclear fusion has the potential to supply the world's energy needs for thousands of years. Early fusion reactors will likely use fuel comprising a mixture of deuterium and tritium which is heated to thermonuclear temperatures to form a plasma and confined to a vacuum chamber by strong magnetic fields. In the fusion reaction, deterium and tritium fuse to form 3.5 MeV He atoms and 14.1 MeV neutrons. Neutrons produced in the plasma are absorbed by a surface blanket which releases heat to a coolant, which is then passed through a heat exchanger to produce electricity by conventional methods. Neutrons react with lithium in the blanket to produce more tritium, which is later separated and reused as a fuel for the fusion reaction (Figure 10.6).

Tritium inventories in a fusion plant will range from 1–10 kg/gigawatt electric. One kg of tritium is equivalent to 9.6×10^6 curies, or more than the entire world's annual production of tritium by cosmic irradiation

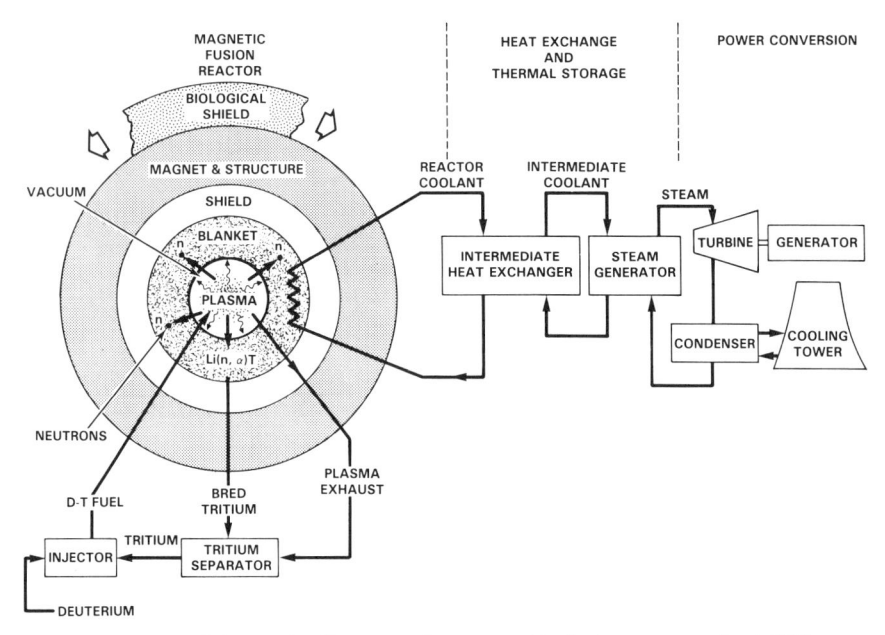

FIGURE 10.6. — Schematic diagram of a magnetic fusion power plant (U.S. Department of Energy. 1979. Environmental Development Plan Magnetic Fusion. DOE/EDP-0052, Washington, D.C.).

($4-8 \times 10^6$ curies).[12] Tritium is a pure beta-emitter ($E_{max} = .018$ MeV) with a physical half-life of 12.3 years. Tritium would be released into the environment mostly through leaks in the heat-exchange system, particularly into the cooling water. Afterheat, or the amount of heat remaining after termination of a nuclear reaction, is much less in a fusion reactor than in a fission reactor, so that loss of coolant has much less severe implications for a fusion reactor than for a fission reactor.

The 14.1 MeV neutrons produced by the fusion reaction will cause activation or induction of radioactivity in exposed elements in the fusion plant, thereby creating a radiological health hazard. Some radionuclides will be mobilized from the metal walls of containment vessels; radioactive gases, such as $^{14}CO_2$, ^{13}N and ^{37}Ar, will build up in the air of the containment vessel. After a few years of reactor operation, the stainless steel first containment wall will give a contact dose of 10^5 to 10^7 rad/hr, immediately after shutdown.[13] Replacement of the first wall will be required every 2 to 10 years of plant operation. Estimated annual amounts of neutron-activated solid radwastes are in the range of 250 metric tons/GW_e.[12]

Nonradioactive toxic materials associated with fusion reactors include large amounts of lithium, used for tritium breeding; lead or beryllium, used for neutron multiplication and shielding; mercury, in vacuum pumps; and

10 – Radiological Health   273

a number of alloy metals (beryllium, nickel, bismuth, manganese, molybdenum and chromium), used in the steel-wall containment vessel. The estimated toxicity of inhaled nonradioactive metals is several orders of magnitude less than for inhaled radionuclides in fusion plants.[14] Beryllium is the most favored neutron multiplier for tritium breeding; however, the amount of beryllium needed for a 1000-GW$_e$ fusion industry would consume the entire U.S. resource of beryllium. Because of its scarcity and also because of the high toxicity of beryllium, it is likely that lead will be used for this purpose.

Lithium will be present in fusion reactors (40–1000 metric tons Li/GW$_e$) as a tritium breeder.[15] Lithium is used in neuropsychiatric conditions because of its action on the CNS. The kidney appears the next most sensitive organ to the toxic actions of lithium. Toxicity of lithium in humans is seen after an oral dose of only 7 mg/kg body weight.

Large magnetic fields will be used in fusion reactors. Tokamak and mirror reactor magnets produce magnetic fields extending beyond the plasma containment vessel which decrease as the inverse cube of the distance from the magnet. Occupational exposures of up to 0.05 tesla (1 tesla = 10,000 gauss) may be anticipated in the immediate fusion reactor environment. However, the magnetic fields are expected to decrease to background levels at several hundred meters from the reactor core. Large electric fields will also be found in the vicinity of fusion reactors as well as a spectrum of microwave radiations, expected to deliver doses of < 10 mW/cm^2.

The current world inventory of natural tritium is estimated at 70 megacuries (MCi). In addition, a total of 4500 MCi of tritium has been released, mostly in the northern hemisphere, from the testing of thermonuclear weapons. Emissions of tritium from the nuclear fuel cycle will eventually become the most common source of anthropogenic tritium, surpassing tritium production from weapons testing by the mid-80s. The current mean whole-body dose rate from tritium in humans from naturally produced tritium is about 10^{-3} mrad per year. Anthropogenic tritium will produce dose rates of 10^{-2} mrad per year by the year 2000.[16,17]

Tritium is rapidly distributed in the tissues of the body by the blood after oral or inhalation uptake, with about 99% of inhaled tritiated water (HTO) being absorbed from the lung. HTO in humans exhibits a three exponential clearance from the body. The first clearance phase occurs with a biological half-time of 10 hours, representing tritium in body water. The slower next two clearance phases, with biological half-lives of 30 and 300 days, represent exchange with labile hydrogen in molecules and incorporation into cellular constituents, respectively.[18]

The LD$_{50}$ dose for tritium in young adult mice following injection is 1 mCig body weight, which is equivalent to a whole-body dose of 700 rad acute ^{60}Co gamma irradiation. Two human cases of fatal exposure to tritium have been documented. In one case, the whole-body dose was estimated at 300 rem, delivered over a 6-year period; the other case received about 1000 rem over a 3-year period. In both cases, severe damage was noted in the bone marrow.[19]

Doses of HTO that deliver 30 rad to pregnant rats resulted in increased frequencies of CNS birth defects, growth stunting and reduced litter sizes. Teratogenicity occurred at relatively high doses but at levels of exposure that were considerably less than that required to produce acute lethality in the dams. A dose of 20 µCi HTO/g body weight caused a 27% reduction in the number of sperm in mice; this corresponds to a gonadal dose of only 12 rad. The LD_{50} dose for primary oocytes in mice exposed continuously from conception to 14 days of gestation was 2 uCi/ml in drinking water.[19,20]

Genetic damage to DNA from tritium occurs by incorporation of tritiated precursors, such as tritiated thymidine, into DNA; by tritiated water surrounding DNA; and by hydrogen exchange with tritium in the DNA molecule. The biological effects of tritium decay are due mostly to ionization events in DNA from emitted beta particles. A small amount of DNA damage is also due to transmutational events in which incorporated tritium converts to unstable ^{3}He.

Tritium increases tumor production in animals at doses >1 µCi/g body weight.[21] Tritium does not appear to represent a special or unique carcinogenic hazard compared to other radionuclides and has been assigned a quality factor (QF) of 1. Estimates of tumor risk in humans due to tritium exposure from nuclear power generation are 8×10^{-5} fatal cancers per reference reactor year and 1.8×10^{-2} fatal cancers per year from nuclear fuel reprocessing.[22]

Nonionizing Radiations

Electromagnetic radiation is divided into nonionizing and ionizing radiations according to the energy required to eject electrons from molecules. The ionization potentials of carbon, hydrogen, nitrogen and oxygen range from 10 to 20 electron volts (eV). Radiowave and microwave radiations have very low energy levels, on the order of 10^{-7} to 10^{-3} eV, and are therefore not capable of producing ionizations. Microwaves include that portion of the electromagnetic spectrum with a frequency range of 0.1–300 GHz (1 GHz = 10^9 Hz), corresponding to a wavelength of 1 mm to 3 m. Radiowaves range in frequency from 1–100 Mhz (1 MHz = 10^6 Hz); ultrasound occurs at 0.1–10 MHz. All three types of radiation produce hyperthermia in tissues. Like light, they travel in a straight line, may be blocked by objects and transmitted through objects by focused beams as well as be reflected by objects.

The hyperthermic response to nonionizing radiations is the primary problem associated with exposure to these radiations. The thermoregulatory responses to nonionizing radiations may or may not be accompanied by endocrine, neurochemical and behavioral changes.[24] The hyperthermic effect of nonionizing radiation is a function of frequency, time relationships (pulsed or continuous exposure) and power density or intensity. Power density is measured in watts/cm^2 at the skin surface. Biologically damaging power densities are in the 100–100,000 W/cm^2 region for ultrasound with pulsed exposures ranging up to 10 seconds and >10 mW/cm^2 for radiowaves and microwaves. Power density for commercial diagnostic units range from 1–100 W/cm^2 for ultrasound with pulsed microsecond exposures.

Ultrasound is used for a variety of diagnostic procedures, including placental localization and estimation of fetal placement and size. Continuous-wave ultrasound, employing the Doppler-shift phenomenon, is used to monitor fetal function and viability and to examine the heart in echocardiology.[23] The embryo is considered the most sensitive stage to ultrasound exposure. Continuous-wave ultrasound produces malformations in Drosophila and in amphibian embryos, but is not mutagenic. The threshold dose for prenatal mortality in mice is 3–6 W/cm^2 at 3.2-MHz continuous-wave ultrasound exposure (Figure 10.7). Daily exposure to 6000-MHz microwave radiation at 35 mW/cm^2 in pregnant rats results in subtle neurophysiological alterations (fetal weight, eye-opening time, behavioral changes) but is not overtly teratogenic.[43]

The CNS is the most sensitive tissue to ultrasound and other nonionizing radiations. Power densities of >100 W/cm^2 produce pathological changes in the CNS in animals; however, the output of ultrasound from most commercial diagnostic instruments is far below the level of CNS sensitivity (Figure 10.8). Threshold exposures for biological damage range from 2×10^4 W/cm^2 for a 10^{-4} sec duration of exposure, to 100 W/cm^2 for 10 sec exposure over a frequency range of 1–6 MHz. Chorioretinal lesions and cataracts form in the eye following exposure to ultrasound, causing the eye to be quite sensitive to exposure. Cataracts were produced in the rabbit lens with an exposure time of 10 sec at an intensity of 3 W/cm^2 and a frequency of 5 MHz.[25]

The current U.S. occupational exposure maximum for microwaves is 10 mW/cm^2. In comparison, the U.S.S.R. exposure limit is 0.01 mW/cm^2, based on their belief that reversible, nonthermal, neurological and behavioral changes occur at very low doses of microwaves in humans. Western countries have not, however, been able to reproduce the Soviet data.

Most environmental exposures to microwaves occur from broadcast stations at intensities of up to 120 µW/cm^2 near transmitters. Less than 1% of the population in an urban area is exposed to microwaves in the 50–900

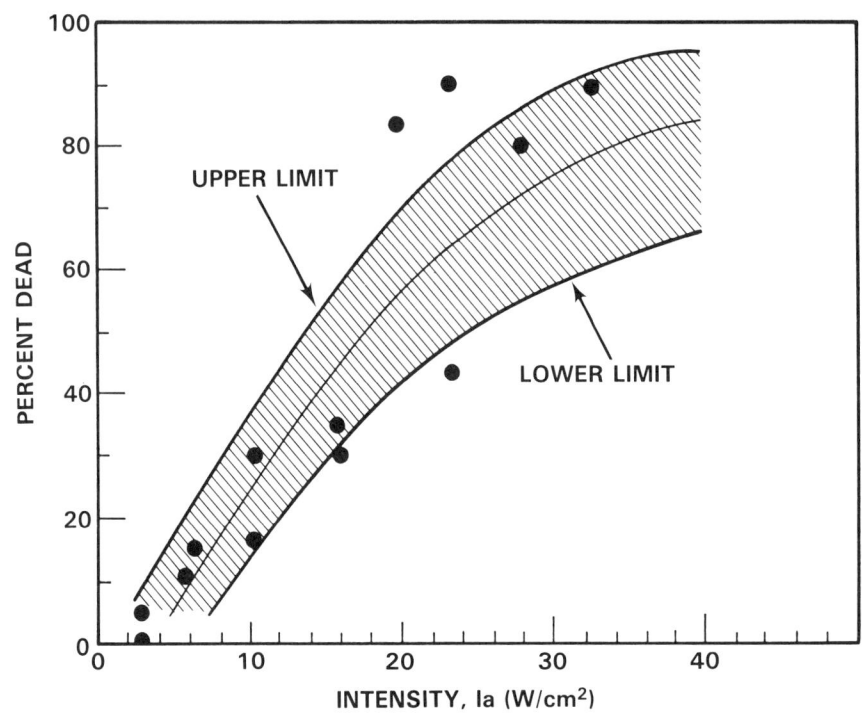

FIGURE 10.7 — Prenatal mortality of rats exposed to 3.2-MHz continuous-wave ultrasound as a function of power intensity. Upper and lower limits are 95% confidence limits for combined exposure times of 5 minutes and 15 minutes (M. R. Sikov and B. P. Hildebrand. 1976. Effects of ultrasound on the prenatal development of the rat. Part 1. 3.2-MHz continuous wave at nine days of gestation. J. Clin. Ultrasound 4:357–363).

MHz range at intensities of >1 µW/cm².[26] Microwave ovens, which operate at 2450 MHz, must conform to a performance standard of maximum leakage of 5 mW/cm² at any point 5 cm from the oven surface. This means that the maximum exposure at an arm's length from the oven would be 1 µW/cm² or less.

The bioeffects of microwave exposure at intensities of >10 mW/cm² are due to thermal processes such that normal thermoregulation cannot handle the excess heat generated in tissues of the body. The CNS appears to be the critical tissue for microwave exposure since nerve fiber transmission is highly temperature-dependent. A 5°C rise in temperature causes doubling of the heart rate. In humans, the CNS ceases to function at a brain temperature of 44–45°C, and the heart stops at 48°C. Fatal hyperthermia occurs in rodents, rabbits and dogs after 40–80 minute exposure to 2800 MHz at 165 mW/cm².[27]

Microwaves cause cataracts in animals; a retrolental temperature of

10 − Radiological Health

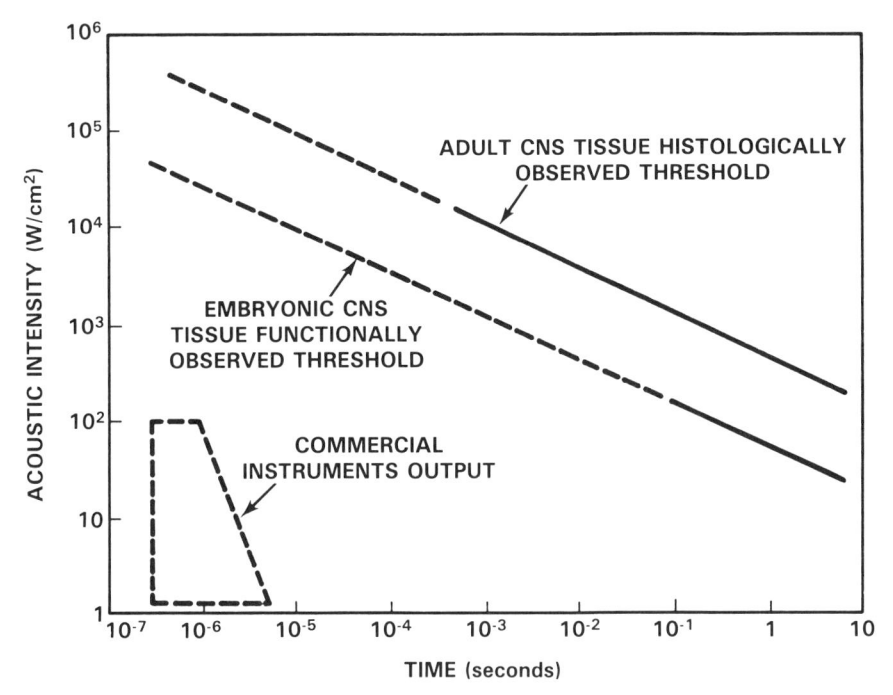

FIGURE 10.8—Estimates of threshold dosage curves for central nervous system tissues in mammals following exposure to 1–6 MHz ultrasound for varying periods of time (F. Dunn and F. J. Fry. 1971. Ultrasound threshold dosages for mammalian nervous system. IEEE Trans. Biomed. 18:253–256).

at least 42°C is required to induce opacities at a threshold intensity of 100 mW/cm^2.[28] The transformation of lymphocytes in culture by mitogens is depressed by 30-MHz microwave exposure. A 15-minute exposure to 2450 MHz continuous-wave irradiation reduced the number of colony-forming units in human and murine bone marrow.[29] However, a 8 hour exposure to 2450 MHz at an average power density of 30 mW/cm^2 to pregnant mice had no effect on hematopoiesis.[56]

In animals, alterations in electrophysiology, calcium-binding capacity of membranes and in the permeability of the blood-brain barrier have been observed following microwave exposure. Blood-brain permeability increased at intensities of 0.03–2.5 mW/cm^2 for a continuous, 30-minute exposure; the effect lasted for up to 4 hours after termination of microwave exposure (Figure 10.9).

Exposure of rats to low dose levels of microwaves for 10 hr/night caused a reduction in total food intake relative to sham-exposed controls. However, no other behavioral or physiological effects of subchronic exposure to microwaves were noted.[55] No "toxic" effects of such low continuous exposures have been noted in animals.

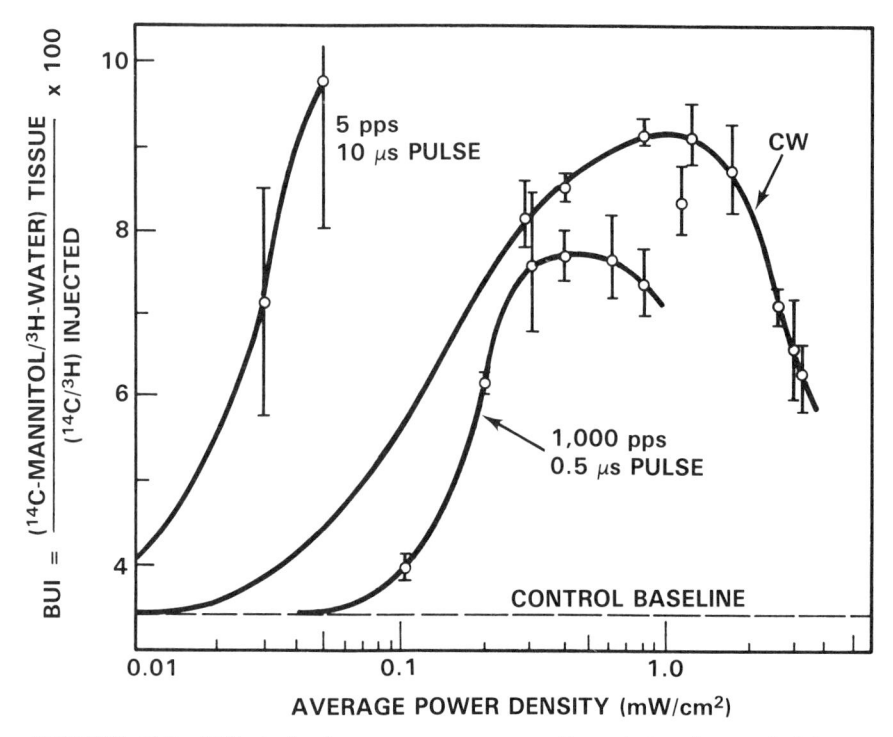

FIGURE 10.9 — Effect of microwave exposure on the uptake of mannitol in the brain of rats as a function of power density (K. J. Oscar and T. D. Hawkins. 1977. Microwave alteration on the blood-brain barrier system of rats. Brain Res. 126:281-293).

Hyperthermia has been used for centuries as a method of cancer therapy. Temperatures above 41°C kill some mammalian cells as well as sensitizing them to ionizing radiation and to chemotherapeutic anti-cancer drugs. Tumor cells appear more sensitive to hyperthermia and hyperthermia-enhanced damage from ionizing radiation and drugs than do normal cells. Tumor temperature is elevated by exposure of microwaves and ultrasound. The thermal enhancement ratio typically ranges from 1.5 to 2.5 for combined hyperthermia and ionizing radiation of tumors. Overall, hyperthermia appears to offer an excellent method for tumor control in humans.[30]

Electromagnetic Fields

In North America there are about 200,000 miles of overhead electric transmission lines rated from 138 kV to 765 kV. Experimental studies are being carried out with 2.2 mV lines. About 14,000 miles of transmission lines are rated at 500-kV and a few thousand miles are rated at 765 kV. A single 765-kV line can transmit as much power as 30–138 kV lines, requiring only 1/30th as much land to transmit the same amount of power as the 138-kV lines. Electric fields are, of course, higher under these high-voltage lines, ranging from 5–10 kV/m under 765 kV lines.

A field strength of >2500 kV/m will ionize air and cause a coronal discharge. The threshold perception of electric fields due to hair stimulation occurs at field strengths >50 kV/m. Electric field strengths of 70–400 kV/m have failed to produce organic lesions or significant biomedical effects in experimental animals.[31] No evidence of stress reactions, electrolyte shifts or metabolic disorders have been noted at exposures of 20 kV/m. Studies in the U.S.S.R., however, describe functional CNS and cardiovascular changes in humans (listlessness, excitability, headache and fatigue) at exposures <10 kV/m. Exposures of rodents to 60Hz electric fields at 40–100 kV/m for up to 150 days showed few significant, reproducible effects. Apparent bioeffects included increased synaptic excitability, increased recovery from muscle fatigue and reduced testosterone levels.[32]

Ozone production due to coronal discharge around transmission lines may account for levels of 8 ppb (parts per billion) ozone under static air conditions; in contrast, the natural background levels of ozone are about 50 ppb.[33] Other variables that may influence biological responses to electric fields under transmission lines are microshocks, acoustical noise and vibration. A few patients with heart pacemakers may experience interference under transmission lines.

Magnetic fields are associated with cyclotrons, fusion reactors, high voltage transmission lines, electrolytic industries, magnetohydrodynamics, energy storage systems and isotope separation facilities; occupational exposures have exceeded 10^4 gauss in some cases. In contrast, the magnetic field around a transmission line is about 0.6 gauss for a field strength of 10 kV/m. Magnetic fields around some household appliances ranges from 10^{-3} to 25 gauss.

Irreversible bioeffects have not been observed in animals exposed to magnetic fields of up to 10,000 gauss.[34-36] A decrease in heart rate, increase in sinus arrhythmia and T-wave amplitude were seen in monkeys exposed to 20,000–100,000 gauss. Genetic effects have been noted in microorganisms exposed to electric and magnetic fields. However, the results are highly variable and their significance in mammals have not been demonstrated. Other effects include improvement in thrombophlebitis at 200–400 gauss, increased resistance to infection at 200 gauss and faster healing of bone fractures at 500 gauss.[13] Consistent and powerful effects of pulsed magnetic fields have been observed on the growth and development of chick embryos at 1.2 to 120 milligauss.[47] In one study of 320 workers who spend the major portion of their workday in magnetic fields produced by direct current through electrolytic cells, a decrease in leukocyte counts was observed with a shift in white blood cell type to an abundance of lymphocytes. Recent studies have suggested a possible link between magnetic field exposure and leukemia formation in experimental animals. There is considerable controversy concerning the significance of these studies in humans.[48,49]

Magnetohydrodynamics

Magnetohydrodynamics (MHD) is a potentially useful method for producing electricity directly from thermal energy, thus eliminating the conversion step from thermal to mechanical energy inherent in conventional steam-electric-turbine generators. The efficiency of MHD generators is estimated at 50% or greater, compared to <40% for conventional fossil-fuel power plants. MHD is being developed as a method of more efficient utilization of coal for power production, while reducing by >95% the pollution of a coal-fired, conventional power plant.[37]

MHD is the branch of fluid mechanics dealing with the flow of an electrically conducting fluid in the presence of a magnetic field. In accordance with Faraday's law of electromagnetic induction, the interaction of the electrically conducting fluid with a magnetic field produces a voltage proportional to the product of the flow velocity and the magnetic field strength. An ionized plasma is produced by burning fine particles of coal dust at high temperatures. The ionized gaseous fluid combustion products are seeded with potassium to facilitate the ionization of the combustion plasma as they pass through the MHD system.[38] Electricity is drawn off directly from the plasma and also from turbines driven by steam from excess heat in the MHD system. MHD requires a large superconducting magnet around the generator to extract energy from the plasma. The intensity of the magnetic field in an MHD plant will probably range from 30,000–60,000 gauss.[42] Total emissions of SO_2 and particulates are considerably less in MHD than from conventional coal-fired power plants. The high temperatures of MHD will produce about 10 times more NO_x than conventional coal power plants.[39] MHD plants emit more mercury, arsenic, selenium, zinc, antimony, lead and cadmium than conventional coal-fired power plants.[40,41]

References

1. W. M. Cox, R. L. Blanchard and B. Kahn. 1970. Relation of radon concentration in the atmosphere to total moisture detention in soil and atmospheric thermal stability. Adv. Chem. Ser. 93:436.

2. R. J. W. Melia, C. du V. Florey, S. C. Darby, E. D. Palms and B. D. Goldstein. 1978. Differences in NO_2 levels in kitchens and gas or electric cookers. Atmos. Environ. 12:1379-1381.
3. R. J. W. Melia, C. du V. Florey, D. S. Altman and A. V. Swan. 1977. Association between gas cooking and respiratory disease in children. Br. Med. J. 2:149-152.
4. F. T. Cross. 1979. Exposure standards for uranium mining. Health Phys. 37:765-772.
5. V. E. Archer and J. K. Wagoner. 1973. Lung cancer among uranium miners in the United States. Health Phys. 25:351-371.
6. G. Saccomanno, R. P. Sanders, V. E. Archer, O. Auerbach and L. Brennan. 1970. Metaplasia to neoplasia. In, "Morphology of Experimental Respiratory Carcinogenesis," CONF-70050, NTIS, Springfield, VA, pp. 63-80.
7. A. J. de Villiers and J. P. Windish. 1964. Lung cancer in a fluospar mining community. Br. J. Ind. Med. 21:94-109.
8. J. Chameaud, R. Perraud, J. Chretien, R. Masse and J. LaFuma. 1980. Combined effects of inhalation of radon daughter products and tobacco smoke. In, "Pulmonary Toxicology of Respirable Particles," CONF-91002, NTIS, Springfield, VA, pp. 551-557.
9. J. P. McBride, R. E. Moore, J. P. Witherspoon and R. E. Blanco. 1978. Radiological impact of airborne effluents of coal and nuclear plants. Science 202:1045-1050.
10. M. Eisenbud and H. G. Petrow. 1964. Radioactivity in the atmopsheric effluents of power plants that use fossil fuels. Science 144:288-289.
11. J. E. Martin. 1974. Comparative population dose commitments of nuclear and fossil fuel electric power plants. In, "Symposium on Population Exposures," CONF-741018, NTIS, Springfield, VA, 317-326.
12. C. E. Easterly, K. E. Shank and R. L. Shoup. 1977. Radiological and environmental aspects of fusion power. Nucl. Safety 18:203-215.
13. H. J. Willenberg and W. E. Bickford. 1978. Safety aspects of activation products in a compact Tokamak fusion power plant. PNL-2823, NTIS, Springfield, VA.
14. W. Hafele et al. 1976. Fusion and fast breeder reactors draft document. Vol. 2, International Institute for Applied Systems Analysis, Luxemburg, Austria, pp. 271-506.
15. J. P. Holdren. 1978. Fusion energy in context: Its fitness for the long term. Science 200:168-180.
16. UNSCEAR. 1977. United Nations Scientific Committee on the Effects of Atomic Radiations. Ionizing Radiation: Levels and Effects, Vol. II, United Nations, New York, 447 p.
17. NCRP. 1979. National Council on Radiation Protection and Measurements. Tritium in the environment. NCRP Report No. 62, Washington, D.C.

18. H. Q. Woodard. 1970. The biological effects of tritium. Report No. HASL-229, USAEC, Health and Safety Laboratory, New York.
19. W. Seelentag. 1973. Two cases of death by panmyelophthisia after incorporation of tritium. In, "Tritium Report," CONF-710809, NTIS, Springfield, VA, p. 267.
20. B. E. Lambert. 1969. Cytological damage produced in the mouse testis by tritiated thymidine, tritiated water and X-rays. Health Phys. 17:547.
21. E. P. Cronkite, J. S. Robertson and L. E. Feindendegen. 1973. Somatic and teratogenic effects of tritium. In, "Tritium Report," CONF-710809, NTIS, Springfield, VA, p. 198.
22. W. H. M. Ellett and A. C. B. Richardson. 1976. Estimate of the cancer risk due to nuclear electric power generation. EPA Technical Note ORP/OSD-76-2, Washington, D.C.
23. M. R. Sikov and B. P. Hildebrand. 1976. Effects of ultrasound on the prenatal development of the rat. Part l. 3.2 MHz continuous wave at nine days of gestation. J. Clin. Ultrasound 4:357-363.
24. S. M. Michaelson. 1982. The influence of radiofrequency/microwave energy absorption on physiological regulation. Br. J. Cancer (Suppl. V) 45:101-107.
25. R. T. Torchia, E. W. Purnell and A. Sokollu. 1967. Cataract production by ultrasound. Am. J. Ophthal. 64:305-309.
26. R. A. Tell and D. E. Janes. 1976. Broadcast radiation—A second look. In, "Biological Effects of Electromagnetic Waves," FDA 77-8010, pp. 363-388.
27. S. M. Michaelson. 1974. Thermal effects of single and repeated exposure to microwaves—A review. In, "Biologic Effects and Health Hazards of Microwave Radiation," Polish Medical Publishers, Warsaw, Poland, pp. 1-14.
28. P. O. Kramar, A. F. Emery, A. W. Guy and Y. C. Lin. 1975. The ocular effects of microwaves on hypothermic rabbits: A study of microwave cataractogenic mechanisms. Ann. N.Y. Acad. Sci. 247:155-165.
29. Y. C. Lin. 1979. Health aspects of radio and microwave radiation. J. Environ. Pathol. Toxicol. 2:1413-1432.
30. N. M. Bleehen. 1982. Hyperthermia in the treatment of cancer. Br. J. Cancer (Suppl. IV) 45:96-100.
31. S. M. Michaelson. 1979. Human responses to power-frequency exposures. In, "Biological Effects of Extremely Low Frequency Electromagnetic Fields," CONF-781016, NTIS, Springfield, VA, pp. 1-20.
32. R. P. Phillips, L. E. Anderson and W. T. Kaune. 1981. Biological effects of high-strength electric fields on small laboratory animals. DEO/RL/o1830-T7, NTIS, Springfield, VA.
33 D. D. Mahlum. 1977. Biomagnetic effects: A consideration in fusion reactor development. Environ. Health Perspect. 20:131-140.
35. R. P. Phillips, M. F. Gillis, W. T. Kaune and D. D. Mahlum. Editors.

1979. Biological Effects of Extremely Low Frequency Electromagnetic Fields. CONF-781016, NTIS, Springfield, VA, 577 p.
36. J. L. Marsh, T. J. Armstrong, A. P. Jacobson and R. G. Smith. 1982. Health effects of occupational exposure to steady magnetic fields. Am. Ind. Hyg. Assoc. 43:87–394.
37. R. S. Detra. 1978. MHD Electric power generation. Energy 3:
38. M. Petrick. 1979. Advances in MHD technology. In, "Symposium on Advances in Coal Utilization Technology," Louisville, KE.
39. S. S. Strom et al. 1978. Controlling NO_x from a coal fired MHD process. 13th Intersociety Energy Conversion Engineering Conference, San Diego, CA.
40. E. W. Schmidt et al. 1976. Size distribution of fine particulate emissions from a coal fired power plant. Atmos. Environ. 10:1065–1069.
41. Montana State University. 1978. Review and Assessments of Potential Environmental Health and Safety Aspects of MHD Technology. DOE.
42. D. B. Montgomery and J. E. C. Williams. 1976. The technology base for large MHD superconducting magnets. Third U.S.–U.S.S.R. Colloquim on MHD Electric Power Generation, Moscow, U.S.S.R., pp. 241–258.
43. R. P. Jensh. 1984. Studies of the teratogenic potential of exposure of rats to 6000-MHz microwave radiation. I. Morphologic analysis at term. II. Postnatal psycho-physiological evaluation. Radiat. Res. 97:272–301.
44. J. L. Smith-Briggs and M. J. Crick. 1982. Natural radionuclides near a coal-fired power station. Radiological Protection Bulletin No. 49, ISSN 0308-4272. National Radiological Protection Board, Chilton, Didcot, England.
45. V. De Santis and I. Longo. 1984. Coal energy vs. nuclear energy. A comparison of the radiological risks. Health Phys. 46:73–84.
46. National Council on Radiation Protection and Measurements. 1984. Evaluation of occupational and environmental exposures to radon and radon daughters in the United States. NCRP Report No. 78, NCRP, Washington, D.C., pp. 159–160.
47. J. M. R. Delgado, J. Leal, J. L. Monteagudo and M. G. Gracia. 1982. Embryological changes induced by weak, extremely low frequency electromagnetic fields. J. Anat. 134:533–551.
48. N. Wertheimer and E. Leeper. 1979. Electrical wiring configurations and childhood cancer. Am. J. Epidemiol. 109:273–284.
49. J. P. Fulton et al. 1980. Electrical wiring configurations and childhood leukemia in Rhode Island. Am. J. Epidemiol. 111:292–296.
50. C. L. Sanders and R. L. Kathren. 1983. Ionizing Radiation. Tumorigenic and Tumoricidal Effects. Battelle Press, Columbus, OH, 335 p.
51. W. L. Russell. 1973. Proc. Natl. Acad. Sci. 74:3523.
52. National Academy of Sciences. 1980. The effects on populations of

exposure to low levels of ionizing radiation. Washington, D.C., p. 118.
53. National Academy of Sciences. 1972. Effect on populations of exposures to low levels of ionizing radiation. Washignton, D.C.
54. A. Stewart and G. W. Kneale. 1970. Lancet 1:1185.
55. R. H. Lovely, S. J. Y. Mizumori, R. B. Johnson and A. W. Guy. 1983. Subtle consequences of exposure to weak microwave fields: Are there nonthermal effects? In, "Microwaves and Thermoregulation," Academic Press, Inc., NY, pp. 401–429.
56. M. J. Galvin, G. L. MacNichols and D. I. McRee. 1984. Effect of 2450 MHz microwave radiation on hematopoiesis of pregnant mice. Rad. Res. 100:412–417.
57. V. De Santis and I. Longo. 1984. Coal energy vs nuclear energy: A comparison of the radiological risks. Health Physics 46:73–84.
58. D. G. Coles, R. C. Ragaini and J. M. Ondov. 1978. Behavior of natural radionuclides in western coal-fired power plants. Environ. Sci. Technol. 12:442–446.

GLOSSARY OF ENERGY-RELATED TERMS

Angstrom (Å) – a unit of length; 1 Å = 10^{-9} meter
Absorption – the penetration of one substance into another
Acclimation – the physiological and behavioral adaptation to changes in environment
Acid gas – hydrogen sulfide and carbon dioxide
Acute exposure – an exposure of such intensity that acute clinical effects appear within a short period of time, usually <4 days
Acute toxicity – an exposure(s) to toxic agent(s) that results in pronounced toxicity within 1–4 days
Adaptation – a change in the morphology, habit or behavior of an organism that improves its environmental compatibility
Adenoma – a benign tumor derived from glandular epithelium
Adsorption – adherence of elements or compounds to the surface of another substance
Aerobic – biological life or processes that can exist only in the presence of oxygen
Aerosol – a suspension of liquid or solid particles in air
AHH – see aryl hydrocarbon hydroxylase
Air pollution – air that has become contaminated with toxic substances from anthropogenic or from natural sources
Air sampling – the collection and analysis of air samples for the detection and quantification of toxic substances
Alicyclic – a group of chemicals characterized by formation of carbon atoms in closed-ring formations
Aliphatic – organic compounds characterized by straight chains of carbon atoms
Alkene – an aliphatic hydrocarbon containing one double bond (e.g., ethylene)
Alkylation – the addition or substitution of one or more alky radicals into an organic compound

Alloy – a mixture of metals

Amalgamation – the process of alloying metals with mercury; particularly used in extraction of silver and gold from their ores

Ambient air quality – outdoor atmospheric levels of specific air pollutants expressed as concentration and duration of exposure in immediate proximity to pollutant emission sources

Anaerobic – biological life and processes that cannot exist in the presence of oxygen

Anoxic – oxygen deficiency

Antagonism – counteraction of a process by an agent

Anthracite – hard coal containing a high percentage of carbon and low amount of volatile matter and sulfur

Anticarcinogen – a substance that counteracts a carcinogen

Aromatic – group of unsaturated cyclic hydrocarbons containing one or more benzene rings

Aromatization – the use of hydrogen in the presence of pressure, heat and a catalyst (e.g., platinum) to convert petroleum to fractions such as gasoline

Aryl hydrocarbon hydroxylase – AHH, microsomal oxidase involved in biotransformation of polycyclic aromatic hydrocarbons and other lipophilic organic compounds

Ash – inorganic residue remaining after combustion of a material

Asphaltene – a component of bitumen, soluble in carbon disulfide, consisting of high-molecular-weight hydrocarbons

Background concentration – level of substances due to natural sources

Bag house – A building or room containing bag filters for the removal of arsenic, lead and sulfur fumes from smelter flues

BaP (benzo(a)pyrene) – common polycyclic aromatic hydrocarbon

Barrel (bbl) – 42 American gallons or 306 pounds

Barrels per day – bbl/day

Benzene – liquid hydrocarbon derived from coal tar and naphthenes used for a variety of industrial purposes (C_6H_6); old term for gasoline and naphtha

Beryl ore – a silicate ore containing beryllium and aluminum

Bi-gas – a process of coal gasification

Bioaccumulation – the concentration of chemicals in living organisms (e.g., cadmium in marine invertebrates)

Bioassay – a method for determining a biological response to toxic substances

Bioconversion – the change of one form of energy into another by plants or microorganisms (e.g., synthesis of starch from carbon dioxide and water by plants in presence of sunlight, or digestion of sewage sludge by microorganisms to yield methane)

Biological half-life – the time required for half of a deposited substance to be eliminated from an organ or the whole body

Biomass – total weight of a given population or species in an environment
Biosphere – the portion of the earth and its atmosphere capable of supporting life
Biosynthesis – the formation of complex molecules by living organisms from simpler molecules
Biota – the sum total of all living organisms in any area
Bitumen – a natural dark colored substance composed mostly of hydrocarbons; specifically a tar-like substance surrounding sand particles in tar sands.
Bituminous coal – the most plentiful type of coal with intermediate BTU content and higher volatile and sulfur contents than anthracite coal
Blast furnace – a furnace in which combustion is forced by air under pressure
Blood-brain barrier – the barrier to chemical transport of hydrophilic molecules created by cell membranes of the central nervous system
Body burden – the total amount of a substance in the body at any time
British Thermal Unit – Btu, the quantity of heat necessary to raise the temperature of 1 pound of water 1°F; 1 Btu = 252 cal/g
Brown coal – lignite
Butane – C_4H_{10}, a colorless gas used in petrochemical synthesis and in bottled gas
By-product – a residual or secondary product obtain in the processing of a raw material
°C – degrees centigrade (Celsius)
Calorie – the amount of heat required to raise the temperature of 1 kg water by 1°C (large calorie)
Cancer – a malignant tumor originating from any cell type
Carbon black – pure carbon produced by the incomplete combustion of hydrocarbons
Carbon oxides – compounds of carbon and oxygen, notably carbon dioxide and carbon monoxide, produced by the combustion of organic material
Carcinogen – an agent that induces cancer
Carcinogenesis – the initiation, promotion and development of cancer
Carcinoma – a cancer derived from epithelial cells
Catalyst – a chemical used to speed up the rate of a chemical reaction, which is itself not consumed in the reaction
Catalytic cracking – a process of cracking in which the catalyst provides for a high reaction rate but at lower pressure and temperature
Ceramic – pottery, brick and tile products manufactured from clay and fired at high temperatures
Cubic feet per hour – cfh
Cubic feet per minute – cfm
Cubic feet per second – cfs
Char – the solid, carbonaceous residue remaining after incomplete combustion of organic material

Char-Oil-Energy Development process (COED) – a process for low-temperature distillation of coal products

Chelate – association of a metal by covalent binding to one or more other molecules or ions so that heterocyclic rings are formed with the central metal atom as part of each ring.

Chromosome – subcellular structures in nuclei of cells, comprised of DNA and histone protein, that distributed genetic material during cell division

Chronic exposure – long-term, continuous or protracted exposures

Clone – a cell population derived from a single cell

Centimeter – cm

Cubic centimeter – cm^3

Coal – a solid, combustible organic material comprised of condensed, aromatic, high-molecular-weight chemicals

Coal conversion – transformation of coal to a liquid or gas

Coal gas – gas that comes from retorts or ovens during the distillation of coal

Coal gasification – conversion of coal to a gas to be used as fuel

Coal liquefaction – conversion of coal into a liquid by hydrogenation to be used principally as a fuel

Coal oil – oil obtained by distillation of bituminous coal (also used in the past for kerosene made from oil)

Coal seam – a bed or layer of coal between layers of rock

Coal tar – a gummy, black by-product of bituminous coal distillation

Cocarcinogen – a substance or agent that enhances or potentiates carcinogen

COED – see Char-Oil-Energy Development process

Coke – a porous, solid residue of carbon remaining after the incomplete combustion of coal

Coke-oven gas – the gas obtained from coke ovens during the production of coke

Coking – high temperature carbonization of coal, coal tar, pitch or crude oil residues to produce coke

Connective tissue – tissues that support parenchymal cells and organs, binding them together; also serve in hematopoiesis, food storage and defense against parasitism and cancer invasion. examples are collagen, elastin, reticulin, adipose tissue, bone, cartilage, blood vessels, lymphatics, fibroblasts

Convection – the transfer of heat by circulation in a gas or a liquid

Conversion factors – Anthracite = 25.4 million Btu/ton
Bituminous = 26.2 million Btu/ton
Subbituminous = 19.0 million Btu/ton
Lignite = 13.4 million Btu/ton
Crude petroleum = 5.5 million Btu/bbl

Residual fuel oil = 6.3 million Btu/bbl
Gasoline = 5.3 million Btu/bbl
Jet Fuel = 5.4 million Btu/bbl
Kerosene = 5.7 million Btu/bbl
Liquid propane = 2500 Btu/ft^3
Natural gas = 1030 Btu/ft^3 at STP^{235}
Uranium = 74 million Btu/g

Cooling tower – a structure or tower used for cooling water before it passes into the air, much as a radiator cools a motor vehicle engine

Cracking – refining process which changes the nature of oil by heating and pressure, with or without a catalyst, to provide a variety of breakdown products, such as gasoline

Creosote – oily liquid containing a mixture of phenols; obtained from the incomplete combustion of wood

Crude oil – petroleum as it comes from the ground

Cryogenics – procedures, generally at temperatures < 100°C

Cycloolefin (cycloalkene) – an alicyclic hydrocarbon having two or more double bonds

Cycloparaffin (naphthene) – an alicyclic hydrocarbon in which > 2 of the carbon atoms are combined in a ring structure and each ring carbon atom is joined to 2 hydrogen atoms or alkyl groups

Decomposition – the degradation of a substance into its basic parts

Dehydrogenation – the process of hydrogen removal from compounds by chemical methods

Deoxyribonucleic acid (DNA) – a nucleic acid containing deoxyribose sugar, phosphate and purine (adenine, guanine) and pyrimidine (thymidine, cytosine) bases as a polymerized structure; present in the chromosomes of cells, and chemical basis of genes and heredity

Deposition – the depositing of mineral or particulate matter

Desulfurization – the removal of sulfur and sulfur compounds from fossil fuels

Detection limit – the threshold concentration for detection of a substance

Detergent – a chemical that reduces the surface tension of water, exerting emulsifying action at oil-water interfaces

Diesel oil – the oil fraction after petroleum and kerosene have been distilled from crude oil

Diffusion – movement of a chemical from a zone of high concentration to a zone of lower concentration due to molecular motion

Diol – a chemical compound containing two hydroxyl groups

Dispersant – a chemical used to break up concentrations of organic matter, such as oil spills

DMBA – dimethylbenzo(a)anthracene

Dust – micron-sized and smaller diameter solid particles in the air

Ecology – the study of living organisms and their environment

Ecosystem – the complex associations of living organisms and their environment forming a functional unit (e.g., a marine ecosystem of a tidal pool)

Efficiency, thermal – available heat that is converted to useful purposes (Efficiency, E = Btu output/Btu input)

Effluent – discharge of raw (untreated), partly treated or completely treated emission streams containing pollutants from industrial, urban, etc. sources

Electrolysis – chemical change produced by passing an electric current through a chemical solution

Electrostatic precipitator – air pollution control device that removes particulate matter by imparting an electrical charge to the particles and collecting the particles on an electrode

Endothermic – a slow chemical reaction that absorbs heat

Environment – the sum of external factors that influences biological life

Epidemiology – the study of disease in human populations

Epithelium – cell layers forming the epidermis of the skin and surface of numerous mucous and serous membranes; epithelium may be flatten (squamous), cube-shaped (cuboidal) and cylindrical (columnar). Tissues lined with epithelial cells include the respiratory tract, gastrointestinal tract, bile ducts, uterus, salivary glands, skin and pancreas.

Epoxide – a cyclic ether; a metabolic intermediate in PAH biotransformation reactions

Ethylene – a colorless, flammable, olefinic gas having a characteristic sweet odor and taste, derived from cracking of petroleum

Exothermic – a reaction in which heat is liberated, usually rapidly and sometimes explosively

°F – degrees Fahrenheit

Firedamp – a mixture of methane and air in coal mines (5–15% methane in air may result in explosion)

Fischer-Tropsch process – production of complex hydrocarbons like gasoline by combining carbon monoxide and hydrogen over a catalyst at moderate temperature and pressure

Fluidization – suspension of particles of a solid in a gaseous or liquid stream of sufficient velocity to cause separation

Fluidized bed combustion – combustion of coal particles on a moving bed of air or gas of sufficient velocity to keep the particles suspended in air

Flue gas – stack gas, gases resulting from the combustion of a fossil fuel

Fly ash – fine, solid particles or residue resulting from combustion of fossil fuels

Food chain – the transfer of elements or nutrients from one group of organisms to another (e.g., plants →herbivores →carnivores)

Fossil fuel – naturally occurring, organic substances derived from biological organisms; includes coal, oil, shale, tar sands and natural gas

Fractional distillation – a fundamental process of separation of crude fossil fuels by refining distillation; separation occurs primarily because of differences in volatility of components and volatility depends on the boiling points of the constituents. Separation of crude oil into fractions by progressive distillation gives the following in order of increasing boiling points:
    propane
    butane
    naphtha
    gasoline
    kerosene
    No. 2 heating oil
    Lubricating oil stocks
    Residuals including tars, pitches, chars

Fuel – a material or chemical used principally to produce heat, either by combustion of by nuclear reactions

Fuel gas – synthetic, low-BTU gas used for heating; obtained from coal

Fuel oil – liquid petroleum product burned for production of heat or use in engines; may be divided as follows:
1. residual fuel oil – oil from the bottom of the crude oil barrel that cannot be further refined; remains as solid at $<120°F$
2. distillate fuels – distillates derived directly or indirectly from crude oil
3. blended fuels – Nos. 3, 4, or 5 fuels produced by blending residual and distillate fuel oils

Fume – fine, solid particles resulting from the condensation of gases or vapors

G – gram; microgram (μg); milligram (mg); kilogram (kg)

Gallon – a unit of liquid measure which, in the U.S., contains 3.785 liters and weighs 8.3 pounds; one U.S. gallon = 0.83 Imperial gallon

Gas, natural – a naturally occurring mixture of hydrocarbon gases found in porous geological formations

Gas synthesis – a method of obtaining liquid fuels from coal by oxidation, converting pulverized coal into a gas mixture, which is then passed over a catalyst

Gas, synthetic – a gas obtained by the destructive distillation of coal, by thermal decomposition of oil, or by the reaction of steam passing through a bed of heated coal or coke

Gasification – conversion of coal into a high-BTU synthetic gas

Gasoline – petroleum fraction consisting of small-branched-chain, cyclic and aromatic hydrocarbons

Geothermal energy – heat energy available in hot water, steam and rocks in geological formations

Germ cell – egg or sperm cells

Glomerulus – a clump of capillaries covered by epithelial cells as point of origin of the nephron in the mammalian kidney cortex; filters the blood and passes filtrate to proximal convoluted tubule; thousands of glomeruli are present in each human kidney

Gallons per minute – gpm

Grab sample – a sample obtained over a very short period of time and used to analyze its contents

Granulocyte – a white-blood cell; polymorphonuclear leukocyte or neutrophil

Gross heating value – the total heat released when a fuel is burned

Groundwater – subsurface water which feeds wells and springs

H-Coal process – a method to convert high-sulfur coal to a low-sulfur, low-ash fuel by catalytic hydrogenation

Heavy metals – metallic elements such as mercury, lead and cadmium that are toxic to biological organisms

Heavy hydrocarbons – high boiling-point, high density hydrocarbons having >6 carbon atoms

Heavy oils – fuel oil, heavy distillate oil, heavy furnace oil and other oils derived from crude oil; boiling points >650°F

Hepatocellular injury – damage to the cells of the liver

Heterocyclic – a cyclic or ring structure in which one or more of the atoms in the ring is an element other than carbon (e.g., pyridine, pyrrole, furan, thiophene, purine)

High-Btu gas – fuel gas with a heating value of >1,000 Btu/scf

High-sulfur coal – coal that contains >1 wt% sulfur

Hydrocarbon – organic compounds consisting only of carbon and hydrogen:
  A. Aliphatic (straight-chain)
   1. paraffins (alkanes)
   2. olefins
      a. alkenes
      b. alkadienes
   3. acetylenes
   4. acyclic terpenes
  B. Cyclic (closed-ring)
   1. alicyclic
      a. cycloparaffins
      b. cyclo-olefins
      c. cycloacetylenes
   2. aromatic
      a. benzene group (1-ring)
      b. naphthalene group (2-rings)
      c. polycyclics

Hydrogenation – a process of treating coal with oil and hydrogen under heat and pressure and separating the resultant liquid mixture into fractions

Hydrophilic – having a strong affinity for binding or absorbing water
Hydrophobic – antagonistic to water or incapable of dissolving in water
Hydrosphere – water on the surface of the earth in contrast to the lithosphere, the land
HYGAS – a process to produce pipeline quality gas by hydrogasification of coal
Hyperplasia – increased proliferation of normal cells in normal locations of a tissue or organ
Inducibility – increased de novo biosynthesis or activation of an enzyme, such that the activity of the enzyme in tissues is increased
Infrared radiation – electromagnetic radiation with wavelengths of 0.78 to 1,000 microns
Ion exchange – reversible chemical reaction between a solid (the ion exchanger) and a water solution, by means of which ions may be interchanged from one substance to another
Kilocalorie – kcal
Kerogen – resinous hydrocarbon material that is the chief organic component of oil shale; when heated to 450–600 °C, releases vapors that are converted to raw shale oil
Kerosene – petroleum fraction containing hydrocarbons slightly heavier than those in gasoline or naphtha
Laws of Thermodynamics – the First Law of Thermodynamics states that energy can neither be created nor destroyed. The Second Law of Thermodynamics states that when a free exchange of heat takes place between two bodies, the heat is always transferred from the warmer to the cooler body.
LC_{50} – median lethal concentration
LD_{50} – median lethal dose
$LD_{50}(30)$ – dose that kills 50% of a population within 30 days
$LD_{50}(60)$ – dose that kills 50% of a population within 60 days
Leaching – extraction of a soluble component from a mixture, resulting in eventual separation
Light distillates – naphthas or fractions used directly in the production of gasoline
Light hydrocarbons – low-boiling, low-density hydrocarbons having one to six carbon atoms
Lignite – brown coal of low BTU content intermediate between subluminous coal and peat
Liquefaction – conversion of coal to a liquid
Liquified Natural Gas (LNG) – natural gas that has been changed into a liquid by cooling to about -260°F, at which point it occupies about 1/600 of its gaseous volume at STP
Liquid Petroleum Gas (LPG) – bottled gas, consisting mostly of propanes and butanes extracted from refinery and natural gases
Liter – l

Low Btu gas – fuel gas have a heating value of <450 Btu/scf
Low-sulfur coal and oil – coal or oil with a sulfur content of <1 wt%
Lurgi process – main commercial process for coal gasification
Meter – m; micrometer, μm; millimeter, mm; kilometer, km
Mercaptans – organic compounds resembling alcohol, but with sulfur replacing the oxygen of hydroxyl groups (thiols); strong, skunk-like odor
Mesentery – double folded membrane lined with mesothelial cells that encloses the peretoneum (peritoneal mesenteries) and the thoracic cavity (thoracic mesenteries)
Metabolite – a product of metabolism
Metallurgical coal – coal with strong coking properties, containing no more than 8.0 wt% ash and 1.25 wt% sulfur
Metaplasia – change in shape and form to a cell that is not normal for that anatomical location in a tissue (e.g., squamous cells in place of normal type I and II epithelium in alveoli of the lung, termed squamous cell metaplasia)
Methanation – the catalytic combination of carbon monoxide and hydrogen to produce methane and water
Methane – CH_4, major fraction of marsh gas and natural gas
Methylation – replacement of one or more hydrogens in a compound with a methyl group
Middle distillate – hydrocarbons derived from crude oil (C_{10} to C_{15}) that boil at 350–650°F (e.g., kerosene, diesel oil, jet fuel)
Mist – gaseous dispersion of liquid particles of less than 50 microns diameter
Mitochondria – subcellular organelles responsible for cellular oxidative metabolism
Mutagen – an agent that increases the incidence of mutations in an organism
Mutagenesis – the process of producing mutations which are inherited and passed on to offspring (genetic mutation) or cause disease such as cancer in the exposed individual (somatic mutation)
Mutation – a transmissible change in DNA passed on to daughter cells following cell division
Naphtha – petroleum fraction containing mostly aliphatic hydrocarbons
Naphthalene – $C_{10}H_8$, aromatic hydrocarbon that is the most abundant component of coal tar
Natural gas – natural hydrocarbon gas comprised of methane, ethane, butane, propane and other organic gases
Necrosis – pathological processes leading to death of cells or tissues
Neoplasm – a benign or malignant tumor
Nitrification – oxidative process that converts ammonium salts to nitrites and nitrates
Nutrient – a chemical substance needed for life and health of plants and animals

Oil shale – a sedimentary rock containing kerogen
Olefin (alkene) – a class of unsaturated aliphatic hydrocarbons having the general formula CnH_2n that contain one or more double bonds
Oncogenesis – the process of tumor formation
Osmoregulation – the regulation of osmotic pressure within cells and fluids of the body
Oxic – having sufficient supplies of oxygen
Oxidant – any oxygen-containing substance that reacts in the air to produce new substances
Oxides of nitrogen – primarily nitric oxide (NO), which oxidizes slowly in air but in the presence of sunlight and hydrocarbons, oxidizes rapidly to nitrogen dioxide (NO_2)
Oxides of sulfur – sulfur dioxide (SO_2) and sulfur trioxide (SO_3) react with water to form sulfurous and sulfuric acids, respectively
Ozone – O_3, a toxic, oxidizing gas produced in photochemical smog and found naturally in the air
PAH – polycyclic aromatic hydrocarbons
PAN – peroxyacetylnitrate, one of a series of similar compounds irritating to the eyes and highly toxic to some plants; product of photochemical smog
Papilloma – a branching or lobulated benign tumor arising from epithelial cells
Paraffin – a class of aliphatic hydrocarbons occurring mostly in Pennsylvania and mid-continent petroleum particulates
Particulate matter – solid particles, such as ash or soot, or wind-borne soil particles, suspended in air
Peat – one of the earliest stages in coal formation; a low BTU fuel that contains large amounts of water and the visible remains of plant material
Peritoneum – the serous membrane comprised of connective tissues, mesothelial cells and "milky spots" or aggregates of mononuclear cells that lines the abdominal cavity and abdomenal organs
Petrochemicals – chemicals obtained from processing crude oil or natural gas
Petroleum – an oily, flammable bituminous liquid found in underground geological formations, and comprised of a complex mixture of hydrocarbons
Petroleum refiner – an industrial plant the converts crude oil into numerous fractions, usually by fractional distillation
pH – a measure of acidity or alkalinity, on a scale of 0 to 14 with 0 representing absolute acidity, 14 absolute alkalinity and 7 the neutral state; $pH = 1/\log[H^+]$
Phenol – aromatic organic compounds in which one or more hydroxyl groups are attached to a benzen ring

Photochemical oxidants – primarily gaseous pollutants formed by the action of sunlight on nitrogen oxides and hydrocarbons in air
Photochemical smog – a complex mixture of oxidant gases and hydrocarbons produced by the action of sunlight in polluted air
Photosynthesis – the biosynthesis of complex organic compounds (e.g., starches, sugars) by plants using carbon dioxide and water via chlorophyll exposed to sunlight
Phytotoxic – agents toxic to plants
Pitch – a black or dark brown semi-solid or solid residue obtained by evaporation or distillation of tars
Plume – visible emission from a flue or chimney
Polar solvent – a solvent having a high dielectric constant
Pollution – the accumulation of waste by-products from human or natural activity that are normally considered harmful
Polymerization – a chemical reaction that produces large molecules by a repetitive addition of smaller molecules
Polyploidy – increased number of sets of chromosomes in the cell nucleus greater than the 2n number present in G_1 cells
POM – polycyclic organic matter
PPOM – particulate polycyclic organic matter
Probable reserves – a realistic assessment of the fossil fuel reserves that can be recovered using current technology
Process stream – any material stream within the coal-conversion processing area
Propane – C_3H_8, an easily liquefiable hydrocarbon gas
Pounds per square inch – psi
Pulverized coal – fine coal dust, >5% of which will pass a 200-mesh screen
Pyrolysis – the transformation of a substance by heating into one or more other substances, Quad – 1×10^{15} BTU
Quality factor – QF, unit of relative biological effectiveness of ionizing radiations, used in radiological protection
Quenching – sudden cooling
Rad – unit of ionizing radiation dose; 1 rad = 100 erg/g
Radiomimetic – characterizes biological effects from chemicals similar to those produced by ionizing radiations
Rainout – the removal or scavenging of a pollutant with clouds by precipitation
Rank – the separation of coal types according to carbonization and Btu content (peat → sublignite → lignite → subbituminous coal → bituminous coal → anthracite coal; the degree of coal metamorphism
Rem – unit of ionizing radiation dose equivalent; 1 rem = 1 rad × RBE
Renal – pertaining to the kidney
Residence time – the period of time during which a agent resides in a specific area

Glossary of Energy Related Terms 297

Residual fuel oil – viscous residue left from processing of crude oil
Residue – the remaining material left after fractionation, distillation or evaporation of crude oil
Respirable particles – particles of <5 microns diameter that are most likely to be deposited in the lung because of their small size
RNA – ribonucleic acid, present in cells and participating in the synthesis of proteins; comprise of ribose sugar, phosphate, and purine (adenine and guanine) and pyrimidine (uracil and cytosine) bases
Roasting – a high temperature, ore-refining operation used to remove impurities such as sulfur or promote oxidation
Roentgen (R) – a unit of radiation exposure in air; $1 R = 2.58 \times 10^{-4}$ coulomb per kg air
Sarcoma – a malignant tumor that arises from connective tissues, muscle, bone, blood vessels, adipose tissue and lymphatics
Scavenging – removal of a pollutant from the air to the ground by rain or other natural meteorological processes; removal from stack by process such as by electrostatic precipitation
Scrubber – an air-pollution control device that uses a liquid spray to remove pollutants from a gas stream by absorption or chemical reaction
Shale oil – a viscous oil produced by the distillation of oil shale
Smelting – the melting or fusing of ore to separate metals
Smog – polluted air, as contraction of smoke and fog
Smoke – solid or liquid particles formed by the incomplete combustion of materials
SNG – synthetic natural gas – manufactured gas that is 97% methane with a heating value of about 1,000 Btu/scf
Solubility – the ability of one substance to blend with another
Solvent – a liquid that dissolves other substances
Solvent extraction – method to convert coal and hydrogen under pressure to a low-sulfur liquid fuel, by first solubilizing the coal with a liquid solvent
Solvent-Refined-Coal – a process to remove ash, sulfur and other potential pollutants from coal; the end fuel product has a heating value of 16,000 Btu/lb, an ash content of 0.1% and a sulfur content of 0.5%
Soot – aggregation of tar-impregnated carbon particles that form when carbonaceous material undergoes incomplete combustion
Sour gas – a gas containing hydrogen sulfide
Steroid – a class of chemicals synthesized from cholesterol or similar polycyclic structures; includes hormones from adrenal cortex, ovary and testis, sterols, vitamin D, bile acids and glycosides
Strip mining – the mining of coal by removing surface covering material and stripping the entire underlying coal seam
Surfactant – a compound that reduces surface tension when dissolved in water or that reduces interfacial tension between two liquids or between a liquid and a solid

Synergism – interaction of two agents producing an effect that is greater than the sum of the effect produced by either agent alone

Synfuels – synthetic fuels produced from the gasification or liquefaction of coal

Syngas – synthetic gas (see SNG)

Synthane – a coal gasification process for production of pipeline quality gas

Tailings – the rock residue remaining from mining operations after most of the metal has been recovered

Tar – very viscous distillate fractions from wood and fossil fuels

Tar sand – sand particles derived from sedimentary rock, surrounded by bitumen

Teratogenesis – the production of malformations (birth defects) during embryogenesis due to the action of an agent

Terpene – unsaturated hydrocarbon found in oils and resins of plants

Thermal efficiency – the percentage of the total heating value input to a plant that is recovered as product and by-product heating value or equivalent energy

Thermal pollution – emission of heat into the air or water, resulting in the elevation of temperature above normal ambient values

Thiophene – CHCHCHCHS; a highly reactive cyclic organosulfur

Threshold dose – minimum amount of a substance required to produce a measurable biological effect

Threshold limit value (TLV) – a time-weighted concentration of an airborne material believed to be safe for nearly all workers exposed continuously or repeatedly during an 8-hour day, 40-hour workweek

TNT – trinitrotoluene; a powerful explosive

Tolerance – capability of an organism to endure unfavorable environmental conditions

Ton – a unit of weight equivalent, in the U.S., to 2,000 pounds. A metric ton is equivalent to 1,000 kg or 2,204 pounds. The U.S. ton is also called the short ton

Toxicant – an agent that causes biological damage

Toxicity – the degree of biological damage caused by a toxicant

Toxic wastes – waste products that contain toxic substances in sufficient concentration to cause injury to surrounding biota

Trace metals – small amounts of metals found in materials

Translocation – movement of substances or chemicals within a plant or animal

Trophic level – levels of feeding (e.g., producers or plants, primary consumers or herbivores)

Ultraviolet radiation – region of the electromagnetic spectrum from 10–380 nm; 200–380 nm is the region of greatest biological effect in sunlight

Vapor – gaseous substance present in air as a mixture of gas and liquid

Venting – release of gases or vapors into the atmosphere under pressure
Volatility – property of a liquid that allows it to vaporize
Weathering – chemical, biological and meteorological processes
Welding – binding together pieces of metals, using electric arc, oxyacetylene, spot, inert or shielded gas welding techniques. Health hazards result from fumes of the weld metal, gases produced during welding, fumes from flux and exposure of the eyes to intense ultraviolet radiation
White damp – in mining, carbon monoxide characteristic
X-ray diffraction – a technique utilizing the characteristic diffraction of X-rays by different crystals; used to determine the presence or absence of quartz and other types of crystalline silica in dusts
Xenobiotics – anthropogenic substances or chemicals that enter the environment
Xylene – a hydrocarbon liquid derived from the distillation of petroleum, coal tar or coal gas
Zooplankton – the microscopic animal life found in bodies of water

INDEX

A

Acetaldehyde, 87, 97
Acetone, 87
Acetylcysteine, 72
Acetylene, 87
Acid rain, 187–189
Acids in air pollution, 111
Acinar functional units, 50
Acrolein, 59, 87, 97
Acute upper respiratory disease, 70
Aerosols,
    particle size, log-normal distribution, 66
    statistical description of, 65
Aflatoxin, liver and, 50
Agricola, 79, 105
Air pollution, 105–148
    acute episodes of, 106
    airborne contaminants standards and, 113
    ammonia and, 142–143
    aromatic amines and, 120–121
    carbon dioxide and, 140–142
    carbon monoxide and, 139–140
    carcinogens in urban and rural, 109–110
    fossil fuel plant emitted, 111–112, 114–115
    general aspects of, 107–112
    history of, 105–107
    hydrogen cyanide and, 143–144
    hydrogen sulfide and, 137–139
    nitrogen dioxide and, 125–129
    ozone and, 129–132
    particles and, 121–125
    polycyclic aromatic hydrocarbons and, 116–119
    stationary sources of, 108
Airborne particles, sizes of common 65
Alcohol,
    aliphatic, 88–89
    effects or in reproduction, 60
    in air pollution, 111
Aldehydes in air pollution, 111
Aliphatic alcohols, 88–89
Aliphatic hydrocarbons, 196
Alkylbenzenes, 92
Allergic pneumonitis, 74
Alpha-chloroacetophenone (MACE), 58
Aluminum,
    in air pollution, 111
    toxicology of, 154
Alveolitis, 73
American Conference of Governmental Industrial Hygienists, 113
Amines in air pollution, 111
Ammonia,
    air pollution and, 111, 142–143
    fossil fuel discharges after combustion, 143
The Amount of Smoke Produced by Tobacco and Its Absorption in Smoking as Determined by Electric Precipitation, 63
Antibodies, classes of, 29
Anticoagulants, 72
Antimony,
    in air pollution, 111
    pulmonary lesions and, 155
    toxicology of, 154–155
AQS standards, 116
Apresoline, 72

Aromatic amines, air pollution and, 120–121
Arsenic,
　absorption of, 150
　biotransformation of, 156
　eczematoid eruptive lesions and, 156
　hyperkeratosis and, 156
　in air pollution, 111
　metal toxicology and, 155–157
　nervous system and, 56
　portal hypertension and, 156
　skin cancer and, 156
　therapy for syphilis, 155
Arthus's reaction, 29–30
Aryl mercury, 171
Asbestos in air pollution, 111
Asbestosis, 75, 77, 79
Aspiration chemical pneumonitis, 88
Aspirin, 72
Asthma, 71
Atelectasis, 70
Atmospheric suspended particles, 121–125
Auramine, liver and, 50
Axothiopurine, 72

B

Barium in air pollution, 111
BCNG, 72
BCNU, 72
Benz(a)anthracene, 97
Benzene, 87, 90–91
　effects of in reproduction, 60
　in air pollution, 111
　nervous system and, 57
Benzidine, liver and, 50
Benzo(a)pyrene (BaP), 23, 24, 96, 97
　annual atmospheric emission of in U.S., 197
　carcinogenicity of on mouse skin, 55
　content of energy-related substances, 195
　in air pollution, 111
　sources of emissions, 118
Benzo(b)fluoranthrene, 97
Berenblum, 24–25
Berylliosis, chronic, beryllium dust produced, 157

Beryllium,
　chronic berylliosis caused by, 157
　in air pollution, 111
　toxicology of, 157–158
Beryllium Case Registry, 157
Beryllium disease, 157
Beta-naphthylamine, 97
Biofuel cells, 240–241
Biological Effects of Ionizing, Radiation, 42
Biomass conversion, 235–237, 240
　environmental effects of, 237
Biotransformation, 19–20,
　arsenic and, 156
Bis (chloromethyl) ether, 96
Blackfoot disease, 156
Bleomycin, 72
Blood, 72
Bloom's syndrome, 33
B-lymphocytes, 29
Boeck's sarcoidosis, 157
Bone, toxicology and, 55–56
Botulinum, nervous system and, 56
Bromine, 111
Bronchial constriction, 70
Bronchiectasis, 70
Bronchiolitis, 70
Bronchiolo-alveolar carcinomas, 94
Bronchitis, 70
Bronchogenic epidermoid carcinoma, 94
Bronchopneumonia, 70
Brownian motion, 67
Busulfan, 72
Butane, 87, 88
Butene, 87

C

Cadmium, 97
　absorption of, 150
　arsenic therapy, 155
　effects in reproduction, 60
　in air pollution, 111
　kidneys and, 49
　prostate gland cancer and, 159
　toxicology, 158–162
Cancer,
　arsenic therapy, 155
　biology of, 31–34
　chemical agents indicated as causes of, 21

cigarette smoking and, 92–93, 95, 97–98
epidemiology, 42–45
lung, 92–94
  aluminum reduction workers and, 154
  arsenic trioxide exposed copper workers and, 157
  cadmium and, 159
  hematite miners and, 164
  metals and, 153–154
  organic compoundings causing, 96
  rural and urban mortality caused, 120
mortality per million person-rem, 266
prostate gland, 159
scrotal, 20
skin, arsenic and, 156
Carbolic acid, 91–92
Carbomazepine, 72
Carbon bisulfide in air pollution, 111
Carbon disulfide, 87
Carbonyl sulfide, 87
Carbon dioxide,
  air pollution and, 114, 140–142
  inhalation effects in humans, 141
  production consequences, 184–186
Carbon disulfide, nervous system and, 56
Carbon monoxide, 97
  air pollution and, 114, 139–140
  effects of in reproduction, 60
  nervous system and, 56
Carbon tetrachloride, 87, 88
  kidneys and, 49
Carbon tetrachloride, 87
Carcinogenesis, radiation, 265–266
Carcinogenicity, 20–25
  of benzo(a)pyrene on mouse skin, 55
  predictability criteria, 23
  tests, 34–42
Carcinogens
  fossil-fuel sources, 194–196
  urban and rural atmospheric concentrations of, 109–110
Carcinoma,
  hepatocellular, 50
  hepatoma, 50
  liver, 50
Cataracts, caused by thallium salts in rats, 59

Cell injury, 25–26
Cell-mediated immunity (CMI), 31
Chemical agents,
  cancer causing, 21
  immunotoxic, 30
Chemicals, absorption, distribution and excretion of, 12–14
Chloral hydrate in air pollution, 114
Chlorambucil, 72
Chlordiazepoxide, 72
Chlorinated hydrocarbons in air pollution, 114,
  kidneys and, 49
Chlorine in air pollution, 114
Chloroform, 87
  kidneys and, 49
Chlorpromazine, 72
Chromium,
  hexavalent, 153
  in air pollution, 114
  toxicology, 162
Chromosomal abberrations, arsenic and, 156
Chronic bronchitis, 73
Chrysene, 97
Chrysotile asbestos, 77
Cigarette smoke, toxic agents in, 97
Cigarette smoking,
  cadmium and, 158
  lung cancer and, 92–93, 95, 97–98
Clean Air Act of 1977, 113
Coal,
  conversion technologies, 211–212
    processing routes for, 214
  fly ash, 189–194
  gasification,
    carcinogens in, 219
    diagram of, 215
  liquefaction and gasification, 211–219
  metals concentration in, 191
  tar, 96
  trace element emissions in, 216
Coal worker's pneumoconiosis, 181–184
Cobalt,
  in air pollution, 114
  toxicology, 162–163
Colistin, 72
Copper,
  in air pollution, 114
  toxicology, 163–164
Corticosteroids, 72

Count Median Diameter (CMD), particle size and, 65
Creosols in air pollution, 114
Cromolyn sodium, 72
Crude oil, 196, 198
    mutagenicity of, 222
Curie, defined, 257
Cyanogen, 87

D

Dialkylnitrosamines, 97
Dibenzanthracene, 96
Dibenz (a, h) anthracene, 97
Diesel fuel, 200
Diethylnitrosamine, 96
Diethylstilbesterol, effects of in reproduction, 60
Difluorodichloromethane, 87
Dimethyl sulfide, 87, 114
Dimethylacetamide, 114
Dimethylbenzanthracene, 96, 97
Dimethylformamide, 114
Dimethylmercury, 170
Dimethylsulfate, 114
Dioxins, effects of in reproduction, 60
Diphenyl sulfide, 114
Diphenylhydrantoin, 72
Diplococcus, 75
D-penicillamine, 72
Drug contaminants, 72

E

Eczematoid eruptive lesions, arsenic and, 156
Electromagnetic fields, 278–280
Electromagnetic radiations, 253
Electromagnetic spectrum, 254
Electrostatic precipitator, design for, 124
Emergency Exposure Limits (EEL), 113
Emphysema, 71, 73
    antimony and, 155
    coal miners and, 184
Energy resources, 1–6
Environmental pollutants, effects of on reproduction, 60
Environmental Protection Agency (EPA), 113

Epigenetic agents, 36
Essays on the Floating Matter of the Air, 63, 121
Ethane, 87, 88
Ethanol, 89–90
    blood concentration and toxicity, 89
Ethyl acrylate, 114
Ethyl alcohol, nervous system and, 56
Ethylene, 87
Eye, toxicology and, 58–59

F

Federal Coal Mine Health and Safety Act, 182
Fibrosis, 74
Fibrotic lung disease, aluminum and, 154
Fick's law, 52
5-Fluorouracil, 72
Fluorine and fluorides, 114
Fluorotrichloromethane, 87
Fly ash,
    coal, 189–194
    oil, 199–200
Food Additive Amendment, 8
Food, Drug and Cosmetic Act, 8
Formaldehyde, 86, 87
    inhalation effects in humans, 88
Formalin, 86
Formic acid, 87
Fossil fuel conversion, 211–234
Fossil fuel plants, air pollutants emitted by, 111–112, 114–115
Fossil fuels,
    carcinogens from, 194–196
    toxic effects of combustion, 201–205
Furodantin, 72

G

Gasoline, 200
    tetraethyl lead (TEL) and, 165–166
    vapors, 88
Genetic toxicology tests, 42
Genotoxic agents, 36
Gentamicin, 72

Geometric Standard Deviation (GSD), 65
Geothermal energy, 247–250
    hydrogen sulfide emissions and, 249
Glossary of energy related terms, 285–299
Gold, 72, 114
Granuloma, 71, 73
    pouch assay, 42

H

Hafnium, 114
Hematopoiesis, oxygen utilization and, 54
Hepatitis, toxic, 50
Hepatocellular adenoma, 50
Hepatocellular carcinoma, 50
Hepatoma carcinoma, 50
Hepatotoxins, 50
Heroin, 72
Hexamethonium, 72
Hexavalent chromium, 153
Hippocrates, 164
Host-parasite interactions, 28
Humoral-mediated immunity (HMI), 31
Hydrazine, 97
Hydrochloric acid, 114
Hydrogen cyanide, 97
    air pollution and, 143–144
Hydrogen sulfide,
    air pollution and, 114, 137–139
    geothermal energy emissions and, 249
    inhalation effects in humans, 138
Hydrogeneration, 217
Hyperkeratosis, arsenic and, 156
Hyperplastic nodules, 50

I

Immunity,
    cell-mediated (CMI), 31
    humoral-mediated (HMI), 31
Immunocellular stem cell numbers, 31
Immunoglobulins, classes of, 29
Immunopathologic responses to tissue injury, 32
Immunotoxic chemical agents, 30
Immunotoxicology, 28–31
In vitro transformation tests, 42
Indomethacin, 72
Inflammation, 26
Infrared radiation, 246–247
Inhalation toxicology, 63–103
    alkylbenzenes, 92
    asbestosis and, 75, 77, 79
    benzene, 90–91
    carcinogenesis, 92–99
    ethanol and methanol, 89–90
    formaldehyde, 86
    inhalation effects in humans, 88
    gases and vapors, 85–86
    inhaled particulates, deposition and fate of 64–70
    mixed-silicate pneumoconiosis, 83–84
    organic vapors and solvents, 87–89
    phenol, 91–92
    pulmonary pathology and, 70–74
        terms and definitions, 70–71, 73
    pulmonary structure and function, 63–64
    silicosis, 79–80, 83
    welder's pneumoconiosis, 83
Inhaled particulates, deposition and fate of, 64–70
    removed to respiratory tract by, 66–67
Internal Commission on Radiological Protection (ICRP), 68
International Agency for Research on Cancer (IARC), 91
International Commission for Protection Against Mutagens and Carcinogens, 42
Interstitial lung disease, 73
Ionizing radiations 253–260
    effects of on reproduction, 60
    properties of, 255
Iron,
    dust, 115
    toxicology, 164
Isoproterenol, 72
Itai-Itai, 159

K

Kanamycin, 72
Kepone, nervous system and, 57
Kerosene vapors, 88
Ketones, 115
Kidneys,
   antimony produced degenerative lesions, 155
   toxicology and, 49
Klebsiella, 75

L

Lacrimators, 58
Lead,
   effects of in reproduction, 60
   EPA and OSHA exposure guidelines, 168
   in air pollution, 115
   nervous system and, 56
   toxicology, 164–169
Leukemia, benzene and, 91
Leydig cell tumor formation, 159
Lithium, 115, 273
Liver,
   antimony produced degenerative lesions, 155
   carcinogenicity of chemicals, 51
   toxicology, 50
Lobar pneumonia, 70
Lung cancer, 92–94
   aluminum reduction workers and, 154
   arsenic trioxide exposure, copper workers and, 157
   cadmium and, 159
   hematite miners and, 164
   metals and, 153–154
   organic compoundings causing, 96
   rural and urban mortality compared, 120
   structure and function of, 63–64
   uranium miners and, 267–270
Lymphangiograms-contrast media, 72

M

Magnethohydrodynamics (MDH), 280
Malignant mesothelioma, 77, 93

Mammalian intrauterine development, 59
Manganese,
   in air pollution, 115
   nervous system and, 56
   toxicology, 169
Mecamylamine, 72
Melphalan, 72
Mephenesin, 72
Mercaptans, 115
Mercury, 115, 150
   effects of in reproduction, 60
   kidneys and, 49
   nervous system and, 56
   toxicology, 169–172
   common chemical forms of, 172
Mesothelioma, malignant, 77, 93
Metal fume fever, 164
Metal oxides,
   lung clearance of intratracheally installed, 150
   whole-body clearance of intratracheally installed, 151
Metals,
   absorption of, 149–154
   detoxification mechanisms 151
   toxicology of, 149–179
      aluminum, 154
      antimony, 154–155
      arsenic, 155–157
      beryllium, 157–158
      cadmium, 158–162
      characteristics, 153
      chromium, 162
      cobalt, 162–163
      copper, 163–164
      iron, 164
      lead, 164–169
      manganese, 169
      mercury, 169–172
      nickel, 171–172
      palladium, 172–173
      platinum, 172–173
      selenium, 173
      uranium, 173–174
      vanadium, 174–175
Methacrylates, 115
Methadone, 72
Methane, 87, 88
Methanol, 87, 89–90
Methotrexate, 72

Index 307

Methyl CCNU, 72
Methyl chloride, 87
Methyl mercaptan, 87
Methylcholanthrene, 96
Methylene chloride, 87
Methylmercury, 150, 170
Methysergide, 72
Microwaves, 274–278
  occupational exposure to, 275
Mineral oil, 72
Minimata disease, 170
Mitomycin, 72
Mixed-silicate pneumoconiosis, 83–84
Molar collector, 242
Molybdenum, 115
Monomethylmercury, 170
Mouse specific-locus test, 40
Mutagenicity,
  crude oil, 222
  short term, 34–42
Mycobacterium tuberculosis, 30

N

National Academy of Science, 116
National Primary and Secondary
  Ambient Air Quality Standards
  (AQS), 113
National Toxicology Program
  (NTP) 9
Neocarzinostatin, 72
Neomycin, 72
Nephrons, 49
Nephrotoxins, 49
  symptoms of, 57
Nervous system,
  chemical-induced structural
    damage, types of, 58
  toxicology and, 56–58
Neurotoxicology, 56–58
  chemical-induced structural
    damage, types of, 58
N-hexane, 88
Nickel compounds, 97
  in air pollution, 115
  toxicology, 171
Nicotine, 97
Niobium, 115
Nitriles, 115
Nitrobenzene, 115
Nitrofurantoin, 72

Nitrogen dioxide,
  air pollution and, 125–129
  histopathological effects of
    inhaled, 129
Nitrogen oxides, 97, 115
Nitrosopiperdine, 97
Nitrosopyrrolidine, 97
N-methyl, 96
N-nitroso compounds, 95, 96
N-Nitrosonornicotine, 97
Nodules,
  A, 50
  B, 50
  hyperplastic, 50
Nonionizing radiations, 274–278
Nuclear fusion, 271–274

O

Occupational Safety and Health Act
  (OSHA), 113
Oil
  crude, 196, 198
    mutagenicity of, 222
  fly ash, 199–200
  shale, 219–221
1,1,1-Trichloroethane, 87
Optic neuritis, exposure to
  thallium, 59
Oral contraceptives, 72
Organic chemicals,
  demonstrated induction of
    malignant lung, tumors
    from, 96
  pulmonary carcinogenesis from,
    94–95
Organic vapors and solvents
  inhalation toxicology and, 87–89
  atmospheric concentration
    range table, 87
Oxides of Nitrogen, 115
Oxygen, 72
Ozone,
  air pollution and, 115, 129–132
  effects of in reproduction, 60
  inhalation effects in humans, 131
Ozonized gasoline, 96

P

PAH compounds, 112

Palladium toxicology, 172-173
Para-aminosalicyclic acid, 72
Paraquat, 72
Particles, atmospheric suspended, 121-125
Particulate organic carbon (POC), 123
Pathobiology, 25-45
  cell injury, 25-26
  host-parasite interactions, 28
  immunotoxicology, 28-31
  inflammation, 26
Pathogenicity, factors associated with, 28
Penicillin, 72
Pentolinium, 72
Peroxyacetyl nitrate, 87
Petroleum, formation of, 1
Phagocytic Kupfer cells, 50
Phagocytosis, 14
Phenols, 91-92, 112
Phosgene, 112
Phosphine, 112
Phosphorus, 112
Pinocytosis, 14
Platinum toxicology, 172-173
Pleural mesothelial hyperplasia, 77
Pneumoconiosis, 74
  aluminosilicate dusts and, 154
  antimony and, 155
  coal workers and, 181-184
  mixed-silicate, 83
  welder's, 83
Pneumonitis, 73
  antimony trioxide and, 155
  aspiration chemical, 88
  beryllium dust produced, 157
  cadmium and, 159
Pollutants, environmental, effects of on reproduction, 60
Polonium-210, 97
Polychlorinated biphenyls,
  effects of on reproduction, 60
  nervous system and, 57
Polycyclic aromatic hydrocarbons, air pollution and, 116-119
Polymyxin B, 72
Portal hypertension, arsenic and, 156
Power plants, fossil fuel, air pollutants emitted by, 111-112, 114-115
Procainamaide, 72
Procarbazine, 72

Proctolal, 72
Propane, 87, 88
Propene, 87
Propoxyphene, 72
Propranolol, 72
Prostrate gland cancer, cadium and, 159
Pulmonary alveolar lipoproteinosis, 70, 80
Pulmonary carcinogenesis, organic chemicals and, 94-95
Pulmonary edema, 70
Pulmonary fibrosis, 71
Pulmonary lesions,
  antimony produced, 155
  pathological scheme of, 85
Pulmonary pathology, 70-74
  terms and definitions, 70-71, 73
Pulmonary system, structure and function, 63-64
Pulmonary thrombosis, 70

R

Rad, defined, 257
Radiation
  carcinogenesis, 265-266
  coal and nuclear power plants compared, 270-271
  exposure in the U.S., 259
Radiations,
  electromagnetic, 253
  half life formula, 256
  infrared, 246-247
  ionizing, 253-260
    effects of in reproduction, 60
    properties of, 255
  nonionizing, 274-278
  ultraviolet, 246-247
  uranium mine exposure, 267-270
Radiobiology
  basic concepts in, 260-265
  radiation quantities and units used in, 258
Radiological health, 253-283
Radionuclides toxicology, 267
  carcinogenic risk of selected, 268
Relative biological effectiveness (RBE), defined, 257
Reproduction,
  environmental pollutants effects on, 60

Reproductive toxicology, 59–61
Resource Conservation and Recovery Act, 113
Respiratory disease
  drugs known to produce, 72
  factors associated with non-cancerous, 71
  frequency of in U.S., 74
Respiratory track,
  inhaled particles,
    deposition and fate of, 64–70
    removed by, 66–67
  structure and function of, 63–64
Rubidium, 112

S

Salmonella/Ames test, 37
Samarium, 112
Scandium, 112
Schneeberg miners disease, 20
Scrotal cancer, 20
Selenium, 112
  effects of on reproduction, 60
  toxicology, 173
Sex-linked recessive lethal test, 37, 40
Shale oil, 219–221
Short-Term Exposure Limits (STL), 113
Silicosis, 79–80, 83
Silver, 112
Sister-chromatid exchange test (SEC), 40–41
Skin,
  cancer, arsenic and, 156
  toxicology and, 50, 52–54
Smoking, lung cancer and, 92–93, 95, 97–98
Solar energy, 241–247
Solar heat,
  toxicity problems of, 242, 244–245
  transfer fluids, 243
  toxicity of, 245
Streptomycin, 72
Styrene-vinyl benzene, 112
Sulfonamides, 72
Sulfur dichloride, 112
Sulfur dioxide, 96
  inhalation effects in humans, 135
Sulfur oxides,
  air pollution and, 112, 132–137

Sulfuric acid, 134
  inhalation effects in humans, 136
Synfuels,
  biological effects, 223–224, 226
  exposure to and carcinogenicity of, 226
Syphilis, arsenic therapy and, 155

T

Tar sand, 221
Task Group on Lung Dynamics, 65, 66
Tetraethyl lead (TEL), 165–166, 202
Thalidomide, effects of in reproduction, 60
Thallium salts,
  cause of cataracts in rats, 59
  optic neuritis and, 59
3-Nitro-3-hexane, 96
Threshold Limit Values (TLV), 113
Tin, 112
Tissue injury, immunopathologic responses to, 32
TLV standards, 116
T-lymphocytes, 29, 30
Tokamak, 273
Toluene, effects of in reproduction, 60, 112
Toxic agents
  absorption, distribution and excretion of, 12–14
  dose categories, 16
  renal damage and, 49
Toxic hepatitis, 50
Toxic Substances Control Act (TOSCA), 8, 113
Toxicity,
  biotransformation in human tissues and, 18
  chemical ratings, 16
Toxicokinetics, 14–19
Toxicology, 7–11
  bone and, 55–56
  defined, 7
  environmental, 7
  inhalation, 63–103
    inhaled particulates,
      disposition and fate of, 64–70
    pulmonary pathology and, 70–74

pulmonary structure and
function, 63–64
see also Inhalation toxicology
liver and, 50
nervous system and, 56–58
non-respiratory tract, 49–62
kidney, 49
liver, 50
radionuclide, 267
carcinogenic risk of
selected, 268
reproductive system and, 59–61
skin and, 50, 52–54
Toxicology of metals *see*
Metals, toxicology of
Treatise on Mining, 79,
Tritium, 273–274
Tuberculosis, mycobacterium, 30
Tumors, 33–34
liver in rodents, 50
2, 4-D, effects of in reproduction, 60
2, 4, 5-T, effects of in
reproduction, 60
2-Methylfluoranthrene, 97
2-naphthylamine, liver and, 50
2-Nitropropane, 97

U

Ultraviolet radiation, 246–247
*Untersuchungen uber Staubinhalation
und Staubmetastase*, 63, 121

Uranium, 112
kidneys and, 49
toxicology, 173–174
Uranium mines, exposure to
radiation in, 267–270

V

Vanadium, 112
toxicology, 174–175
Vineyard sprayer's lung, 163
Vinyl chloride, 96, 97
liver and, 50
Volatile phenols, 97

W

Water pollution, stationary sources
of, 108
Welder's pneumoconiosis, 83
Wilson's disease, 163
With Every Breath You Take, 107
Wood combustion, 237–240
steps in, 238

Z

Zinc, 112

NO LONGER THE PROPERTY
OF THE
UNIVERSITY OF R.I. LIBRARY